Eleven Seconds into the Unknown
A History of the Hyper-X Program

Eleven Seconds into the Unknown
A History of the Hyper-X Program

Curtis Peebles
TYBRIN Corporation
NASA Dryden Flight Research Center

Ned Allen, Editor-in-Chief
Lockheed Martin Corporation
Bethesda, Maryland

Published by
American Institute of Aeronautics and Astronautics, Inc.
1801 Alexander Bell Drive, Reston, VA 20191-4344

Unattributed photographs are held in the public domain.

Quotations at the beginning of each chapter, unless otherwise noted, are taken from *Great Aviation Quotes*, collected by Dave English at http://www.skygod.com/quotes/index.html.

American Institute of Aeronautics and Astronautics, Inc., Reston, Virginia

1 2 3 4 5

Library of Congress Cataloging-in-Publication Data

Peebles, Curtis.
  Eleven seconds into the unknown : a history of the hyper-X program / Curtis Peebles ; Ned Allen, editor-in-chief.
     p. cm.
  ISBN 978-1-60086-776-7
  1. Hypersonic planes--Research--United States. 2. X-43A (Hypersonic plane) 3. Dryden Flight Research Facility--History. I. Allen, Ned. II. Title.
  TL567.R47P4348 2010
  629.133'349--dc22

2010044977

Cover design by Gayle Machey.

The land and sea are long since named and mapped and parcelled out. Only the air and all beyond, the greatest mystery of all, was still unmastered and unknown when I was young. Now we have learned to shuffle about the house and even plan to visit the neighbours. A million starry mansions wink at us as if they knew our hopes and beckon us abroad. All that I shall not see. But at the start, the little lost beginning, I can say of one small part of it: "Here is a witness from my heart and hand and eye of how it was!"

Cecil Lewis, *Sagittarius Rising*

A NOTE FROM THE EDITOR-IN-CHIEF

Conceived early in the 20th century, the first fully successful flight of the scramjet did not occur for 100 years. Using a deceptively simple cycle, a scramjet is an extremely complex type of aircraft, one that poses extraordinary design challenges and constant fine-grained control to start, stay lit, and produce thrust. Many technologies had to mature, and Curtis Peebles recounts the fascinating story of each in this saga of NASA's experimental hypersonic "X-Planes" designed to bring down the supersonic limit and reach for the thermodynamic one: the Hyper-X projects.

The technical obstacles to the thermodynamic limit were not the most daunting problems. The political will to drive and sustain the Hyper-X projects was painfully inconsistent, thus progress slowed considerably. Curtis Peebles documents and thoughtfully analyzes the political context of the Hyper-X projects.

The scramjet may yet play a role in replacing today's heavy and costly rocket boosters with cheaper, smaller, and reusable airbreathers. Perhaps the drive to produce space systems more cost effectively will ultimately free the resources to complete scramjet maturation and scale it up to truly useful dimensions.

This is Curtis Peebles's second contribution to the Library of Flight. *Road to Mach 10: Lessons Learned from the X-43A Flight Research Program* was his first contribution, although not his first book. Curtis Peebles is a distinguished professional historian at NASA's Dryden Flight Research Center in Southern California and a very welcome contributor to the series.

Ned Allen
January 2011
Bethesda, Maryland

CONTENTS

xii

PREFACE

Eleven Seconds into the Unknown: A History of the Hyper-X Program is an extension of the author's first book on the X-43A/Hyper-X project, *Road to Mach 10: Lessons Learned from the X-43A Flight Research Program.* As the title suggests, this earlier book focused on flight research and the activities at the NASA Dryden Flight Research Center. In the course of the writing and reviewing of that book, it became clear that a much broader story remained to be told. Whereas the intended audience for *Road to Mach 10* was undergraduate engineering students, *Eleven Seconds into the Unknown* is written as a project history, focused on a larger audience including those interested in high-speed flight, aerospace history, the organization and management of technological projects, and the future of spaceflight.

Although the Hyper-X was a relatively small project, in terms of both cost and personnel, like all modern technological projects, it involved the efforts of numerous groups and organizations, both governmental and commercial, including several NASA centers; the U.S. Air Force, Navy, and Army; the FAA; and contractors who built the components for the boosters and the X-43A vehicles. Each of these had its own particular cultures, experiences, and traditions. To be successful, these disparate elements had to be combined into a single team working together toward the common goal. One central theme in the Hyper-X story is how this was accomplished.

Making the team's task more difficult was the history of scramjet development. The scramjet was seen as the engine of the future. The problem was, it never became the engine of today. Projects started, individuals made bold promises, and ambitious goals were laid out, but these efforts ended in constant disappointment. These include the Hypersonic Research Engine (HRE), the National Hypersonic Flight Research Facility (NHFRF), and the National Aero Space Plane (NASP). Not until the failure of NASP did the idea originate of building a small, simple scramjet-powered vehicle. This could be done at low cost, in a short period of time, and produce valuable test results. This idea became the Hyper-X. Even with this conceptual breakthrough, a prolonged, systematic effort was still needed, spanning several NASA centers, to transform the scramjet's potential into a workable engine.

The task the team faced became more difficult when the first launch ended in failure. This led to a 3-year halt in the Hyper-X project as the reasons for the failure were identified and corrected, and the solutions tested. The importance of leadership in holding the team together was underlined in this period. Finally, the efforts and courage was rewarded with two successful flights, showing the scramjet had become a reality.

Although the Hyper-X project was a story of technology, at the same time it involved politics, organizational cultures, economics, the ability to overcome disappointments, and the will to succeed. In the end, Hyper-X was as much about humans as it was technology.

Curtis Peebles
January 2011

ACKNOWLEDGMENTS

Writing the story of the first successful hypersonic flights of an aircraft powered by a supersonic combustion ramjet required the help and advice of many individuals. They include NASA Langley Research Center managers and researchers, NASA Dryden Flight Research Center managers, engineers, ground controllers, aircrews, and pilots, as well as contractors from the companies involved. Together, they formed the team which made the Hyper-X a success in the face of technological unknowns and repeated setbacks. It is with thanks I acknowledge their help and efforts:

Vince Rausch, Lawrence Huebner, Dave Reubush, David McAllister, Chuck McClinton, Earl Andrews, H. Keith Henry, Anthony Castrogiovanni, Randy Voland, Lowell Keel, Joel Sitz, Walter C. Engelund, Patricia Jewell, Erik Conway, Gail Langevin, Paul Reukauf, Dave Lux, Jerry Reedy, Griff Corpening, Yohan Lin, Gordon Fullerton, Linda Soden, and Laurie Grindle.

Thanks also go to the individuals who edited, suggested changes, and reviewed the draft manuscript. These include:

Sarah Merlin, Dr. Christian Gelzer, Peter Merlin, Everlyn Cruciani, Elizabeth Kissling, and Charles Rogers. Finally, thanks to Tony Springer at NASA Headquarters who made this all possible. If I overlooked anyone here, it was by error only.

It was a memorable experience, and I am honored to have been allowed to be a witness.

Chapter 1

THE QUEST FOR HYPERSONIC FLIGHT

> There is no excuse for an airplane unless it will fly fast!
>
> *Roscoe Turner, 1930s racing pilot*

The immediate aeronautical challenge facing engineers and scientists at the end of World War II (WWII) was supersonic flight. Some two decades would pass, with changes in virtually every aspect of aviation engineering and technology, before supersonic flight was fully mastered. But as this effort was beginning, an even greater unknown awaited—flight at speeds above Mach 5. This was referred to as "hypersonic" and involved flight not only in the atmosphere, but also at its outer fringes.

During WWII Nazi Germany developed the V-2 rocket, the first large, long-range liquid fuel rocket (Fig. 1.1). Unlike conventional aircraft, there was no defense against the supersonic V-2. Also, unlike the subsequent intercontinental ballistic missiles (ICBMs), the V-2 was a road-mobile missile, which could be launched from an unprepared site (Fig. 1.2). This made destruction of missiles prior to launch nearly impossible. Although the V-2's influence on the outcome of the war was minimal, V-2 technology served as the basis for Cold War missile development in both the United States and the Soviet Union—the potential the V-2 represented was great. A much larger rocket, fitted with an atomic warhead and having a range of thousands of miles, could strike a city on the other side of the world within a matter of minutes, and be unstoppable by any existing air defense. Before this concept could become a reality, however, many technical difficulties had to be overcome.

The V-2's range was only about 200 miles, far less than the range of existing bombers like the B-17. Its payload was only 1 metric ton, again less than existing bombers, or even tactical fighter aircraft. Although a bomber could strike at strategic targets such as a factory, the V-2's crude guidance system could hit only an area target like London. It was deployed as a terror weapon, to demoralize the population. Finally, the V-2 represented a new and immature technology and was unreliable. Many rockets failed within sight of their launch pads; others broke up in midair as they descended.

For the United States to build a rocket capable of delivering a nuclear weapon to the Soviet Union from within the continental United States, the

Fig. 1.1   A V-2 is launched from the Peenemünde test site in the 1940s.

Fig. 1.2   A V-2 rocket on
its transport trailer.

V-2's range would have to be increased from 200 miles to more than 5000 miles, its payload increased by up to a factor of five, and its accuracy improved from an area the size of a city to only a third of a mile from the target.

Among the technological hurdles the long-range rocket would face were those of hypersonic flight. The metal skin of the V-2 was capable of withstanding the supersonic descent of its short-range flight. To reach a target on the other side of the world, however, the rocket would need to achieve a speed of about Mach 20 before its engines shut down. This would require the warhead to follow a ballistic trajectory taking it hundreds of miles into space. No existing material could survive the frictional heating of atmospheric reentry.

The difficulties in building a long-range rocket led influential voices in the scientific community to reject the concept. Among them was Vannevar Bush, president of the Carnegie Institute of Washington, D.C., and director of the Office of Scientific Research and Development during WWII. Bush told the Senate Special Committee on Atomic Energy on December 3, 1945 [1],

> There has been a great deal said about a 3000 mile high-angle rocket. In my opinion such a thing is impossible and will be impossible for many years. . . . I say, technically, I don't think anyone in the world knows how to do such a thing. And I feel confident it will not be done for a very long period to come. I think we can leave that out of our thinking. I wish the American public would leave that out of their thinking.

In 1949, Dr. Bush wrote, "No other nation will have the atomic bomb tomorrow . . .". In August of that year, the Soviets tested their first A-bomb.

U.S. ROCKET DEVELOPMENT AND SCIENTIFIC RESEARCH

The most significant rocket development work undertaken by the United States during the early postwar period was the Army V-2 launches from White Sands, New Mexico, between 1946 and 1952. As Nazi Germany was collapsing in the spring of 1945, Wernher von Braun and most of the engineers who had built the V-2 surrendered to the U.S. Army to avoid possible capture by the Soviets. U.S. technical intelligence units recovered V-2 hardware and documents connected with the V-2 program. The recovered V-2s were shipped to New Mexico. The von Braun team was also brought to the United States, and began reassembling the rockets and assisting with test launches. The V-2 flights were intended to provide the Army with large-rocket experience, but they also brought together Army, Navy, and academic scientists to undertake research activities in such fields as magnetism, cosmic rays, upper atmosphere research, and solar astronomy.

The V-2 was also combined with early U.S. rocket technology to enable the first hypersonic flight of a manmade object. This involved use of a V-2 with a WAC-Corporal fitted to its nose, producing the first two-stage rocket.

The single-stage WAC-Corporal was a small, test version of a liquid-fuel research rocket built for the Army by the California Institute of Technology Jet Propulsion Laboratory. When the V-2 ran out of fuel, the WAC-Corporal rocket motor fired, the stages separated, and the WAC-Corporal rocket motor boosted the second stage to a higher speed and altitude. On February 24, 1949, a Bumper WAC rocket, as the combination was called, lifted off from White Sands. The WAC-Corporal reached an altitude of 244 miles, and was tracked at a top speed of 5150 miles per hour [2]. In all, a total of six Bumper WAC rockets were fired from White Sands in 1948 and 1949; two more were launched from the Air Force Eastern Test Range at Cape Canaveral, Florida.

With the supply of V-2s limited, replacement sounding rockets were needed if high-altitude research was to continue. A scaled-up version of the two-stage WAC-Corporal B, which would carry scientific instruments to high altitudes for research purposes, served as the basis for the Aerobee sounding rocket. Built by Aerojet with Navy funding, it was slightly larger than the WAC-Corporal B and had a modified aerodynamic design. Development took only 18 months and three launches with dummy payloads; the first research flight was made November 24, 1947. The Aerobee became the primary sounding rocket used for upper atmospheric research in the 1950s.

The Navy also funded the larger Viking rocket. This liquid-fueled rocket could carry multiple payloads to higher altitudes. The Viking also had a gimbaled rocket engine for steering, and proved more flexible than the older V-2 rockets. However, the Viking rocket was also more expensive than the small Aerobees, and far fewer were launched [3].

Although scientific research drove much of the early U.S. rocket activity, it did little to make a long-range missile development program a reality. Only a few scientists and universities were involved with the early rocket flights, and many dropped out soon after work began due to institutional pressures, launch failures, and poor results. There was also a lack of support and respect by the larger scientific community.* Most scientists were satisfied with the long-established research techniques they were using, and viewed the rocket results as questionable. As a result, the rocket data were rarely cited in scientific papers. Those scientists who remained active in rocket-based research were too few in number and had too little academic influence to mount a more extensive effort [3, pp. 69, 94, 207, 213, 214, 230–232].

The U.S. military's post-WWII plan to develop intercontinental delivery systems for nuclear weapons was formulated by Theodore von Kármán, a

---

*The aeronautical scientists were mostly members of the Institute of Aeronautical Sciences, organized in 1932, and they exchanged ideas and viewpoints at meetings of that society. The rocket scientists could be found at meetings of the more radical American Rocket Society, which was organized in 1930. The two groups had little regard for each other, and had little in common until the very late 1950s when the field of hypersonics brought them together. The two organizations merged in 1963 to form the American Institute of Aeronautics and Astronautics (AIAA).

scientific advisor to the Army Air Forces. The first step was development of manned jet bombers. The development of "pilotless aircraft," known today as cruise missiles, would follow the manned bombers. Finally, long-range rocket-powered ballistic missiles would be developed. The advantage of this incremental approach was that it made maximum use of existing technology. Jet bombers like the B-52, which were able to carry atomic weapons, were already in the planning stage, and would be ready within a few years. The second generation of pilotless aircraft would be powered by turbojet or ramjet engines. They made use of existing aircraft technology, enhanced by new electronic guidance systems that directed them to a target. Only later would ballistic missiles be developed.

Pilotless aircraft were smaller and weighed less than jet bombers or ballistic missiles and needed only half the range of bombers, because they flew one-way missions. They were also more fuel-efficient than ballistic missiles because they were powered by air-breathing (turbojet or ramjet) engines rather than by rockets. Although a rocket flew much faster than a pilotless aircraft, it had to carry its own supply of oxidizer, whereas pilotless aircraft engines used oxygen from the atmosphere. In order to carry a significant payload, the rocket's structure had to be as light as possible, and therefore had to leave Earth's atmosphere as quickly as possible to avoid the aerodynamic and thermal loads associated with atmospheric flight.

With the emerging Cold War between the United States and the Soviet Union, delivery systems that would be available in a short time frame were the pressing requirement. Both jet bombers and pilotless aircraft met this short-term need. Ballistic missiles, with their numerous technological unknowns and resulting higher risks, did not [4].

The Air Force began development of the subsonic turbojet-powered Snark and the supersonic ramjet-powered Navaho. Both pilotless aircraft were able to reach targets in the Soviet Union from the U.S. mainland and carried nuclear warheads. The Navy developed the Regulus, a shorter-range pilotless aircraft designed for launch from submarines and large surface ships. It also had a nuclear warhead [5, 6].[†]

Although these piloted and pilotless aircraft were viewed as being within the state of the art, ballistic missiles were a different matter. Convair had been working since the late 1940s on the design of a long-range rocket called the Atlas, the most advanced rocket under consideration by the U.S. military. The problems it faced were not limited to postwar budget cuts and an uncertain political environment. The original missile's design showed the technological challenges of building such a vehicle.

[†]The Snark spent a decade in development before finally entering service in January 1961. It was phased out 6 months later in an economy move. The Navaho was cancelled while still in the test phase in 1957, after repeated failures. The Regulus was operational from 1956 to 1964. It was phased out as Polaris submarines came into service.

The initial Atlas design, in 1951, was for a missile 160 ft tall and 12 ft in diameter powered by seven rocket engines. It was to have a range of 5000 nautical miles, carry an 8000-lb atomic bomb, and be accurate to within 1500 ft of a target. By the fall of 1953, the Atlas design was revised to that of a five-engine missile that was 110 ft tall and 12 ft in diameter, weighed 440,000 lb, and carried a 3000-lb atomic warhead. Its range was increased to 5500 nautical miles, but the accuracy was still to within 1500 ft. The Atlas was still too large, complicated, and expensive to be a practical weapon [7, pp. 70, 78].

FIRST STEPS TO HYPERSONIC FLIGHT

Despite the perceived impracticality of ballistic missiles, hypersonic research was begun using both ground facilities and flight tests in the late 1940s and early 1950s. The initial work on ground facilities that could simulate hypersonic conditions was done between 1945 and 1949 by National Advisory Committee for Aeronautics (NACA) researchers John Becker and Alfred Eggers.

Becker's design at the Langley Memorial Aeronautical Laboratory (now the NASA Langley Research Center) used the "blowdown" method. A tank of air was pressurized to 50 atm and heated, using electrical resistance, to 900°F to prevent the air from becoming a liquid at low temperatures after expansion caused by the nozzle producing the high test Mach numbers. The air was released through a nozzle into the wind tunnel's 11-in. test section. With high air pressure on one side and low pressure on the other, a huge difference of pressure could be created as air passed through the nozzle and expanded. The air was then recovered in a low-pressure tank and passed through a pump and dryer unit before returning to the high-pressure tank. The flow could be maintained for about 100 s. Its first successful run was made on November 26, 1947.

Eggers's hypersonic tunnel was at the NACA's Ames Aeronautical Laboratory. The flow path in the hypersonic tunnel was complex. Pressurizing pumps in the adjoining 12-ft pressure tunnel forced air through electrical heaters and through a nozzle into the 10-by-14-in. test section where the model was positioned. The air then passed through a variable-geometry nozzle and diffuser downstream of the test section, and finally through a battery of large vacuum pumps. This created a continuous airflow velocity of Mach 3 to 6 in the test section, though doing so took considerable effort. The tunnel was started at a low Mach number, with the normal shock wave located far back in the diffuser. This shock wave, at a right angle to the flow, dissipated a great deal of power, reducing the tunnel's efficiency. Once the tunnel was running, the nozzle and diffuser throats were narrowed, which increased the Mach number of the airflow but not the losses from the shock wave in the diffuser. The Ames tunnel went into operation during 1949 and, like the

Langley tunnel, provided researchers with experience in both hypersonic aerodynamics and the design of future hypersonic tunnels [8].

The other device used for hypersonic tests was not a wind tunnel in the traditional sense, but rather a "shock tube" (Fig. 1.3). This was a long, constant-diameter tube containing the high-pressure *driver gas* and low-pressure *working gas*, separated by a diaphragm. When the diaphragm was ruptured, a high-speed shock wave (the incident shock) was created in the working gas, which heated and compressed the gas through which it traveled and accelerated it to a speed close to the shock speed. This gas passed over a model in the tube. The test period lasted for a few milliseconds and ended when the contact surface that separated the driver gas from the working gas reached the model. The speed of the shock wave was proportional to the initial pressure ratio between the driver and working gases, and the speed of sound of the driver gas. For this reason, shock tubes were often operated with high-pressure helium gas as the driver gas. The working gas was usually air, but other gases could also be used. Much higher temperatures, which were needed to simulate hypersonic flight, could be achieved in shock tubes than in blow-down or continuous-flow tubes.

The device had a long history. French chemist Paul Vielle built the first shock tube before 1900, using it to demonstrate that shock waves traveled faster than the speed of sound. He suggested that shock tubes would have applications in understanding mine-tunnel explosions. The most significant development of the shock tube in the period before WWII was made by British researcher William Payman, who used diaphragm pressures as great as 1100 psi that were contained in a large compressed-air tank. Payman used high-speed photography to record the behavior of the shocked flows. This research focused on verification of gas-dynamic equations that predicted the behavior of shock waves.

Walter Bleakney at Princeton University greatly expanded the shock tube's value as a research tool during WWII. He found that the shock tube could produce repeatable and predictable results, based on the initial conditions of the "shot." He noted that the velocity of the incident shock could be controlled with an accuracy of 0.1 percent. The shock tube was thus able to meet the basic scientific requirements for research in high-temperature gas dynamics. He also pointed out that shock tubes could be used in a wide range of research experiments, including observations of the shocks themselves and detonations and related phenomena. More presciently, Bleakney also noted, "The tube may be used as a wind tunnel with a Mach number variable over an enormous range" [9, p. 32].

W.D. Crozier and William Hume of the New Mexico School of Mines began development of a *light gas gun* in 1946 under a study contract from the Navy Bureau of Ordnance. The school undertook research on meteorites, and the light gas gun was designed to study what occurred as they entered Earth's

a)

b)

Fig. 1.3 A shock tube and related instrumentation.

atmosphere at high velocities. A nitrocellulose propellant charge was placed in the back of a cylinder. When detonated it drove a piston forward, compressing hydrogen gas, which in turn propelled a free-flying model down the tube at speeds of about Mach 12.

The light gas gun had direct application for determining the temperatures and dynamic pressures a missile warhead would experience during reentry. This method also produced more realistic Reynolds numbers and real gas effects on models than did existing tunnel designs.

Crozier and Hume's design was successful, and other hypersonic researchers built similar light gas guns. Raymond Seegerat of the Naval Ordnance Laboratory at the Aberdeen Proving Grounds (Maryland) built one, as did R. N. Hollyer at the University of Michigan. Another was constructed at the Ames Aeronautical Laboratory. However, the data that could be obtained were exclusively optical because instrumentation of the model was out of the question.

A more advanced version of the shock tube is the *shock tunnel*. Although the shock tube could produce very high pressures, temperatures, and velocities in the working gas, the Mach number was limited to about 2. In the simplest shock tunnel, the working gas is expanded through a diverging nozzle at the end of the tube to increase the Mach number, at the cost of reducing the test interval. To overcome this limitation and further increase the range of test conditions, the *reflected shock tunnel* was developed. In this design, a converging–diverging nozzle was appended to the tube, with the nozzle throat diameter much smaller than the tube diameter. This arrangement was much the same as in a conventional wind tunnel. The useful test time was basically determined by the time required to drain the reservoir of shock-heated gas. The period of steady flow in a reflected shock tunnel typically lasted only a few milliseconds.

Beyond the difficulties of building and operating a hypersonic wind tunnel or shock tunnel, there were also problems with optical instruments used to collect flow visualization data from the shock waves. The three methods traditionally used were *shadowgraph*, which measured the rate of change of the density gradient; *schlieren*, which showed the density gradient; and *interferometry*, which measured the air density itself. Shadowgraph was the simplest to use, but provided the least data. Schlieren became the most common method used, because it balanced data, cost, and difficulty. Interferometry provided the best data, but was the most costly and difficult to use.

All three methods became ineffective at low atmospheric pressures, however. In the 11-in. tunnel at Langley, Mach 7 airflows had a static pressure of less than a thousandth of an atmosphere. This simulated the conditions at 150,000 ft, where hypersonic vehicles would fly. Becker indicated that this was the limit for the schlieren method. If hypersonic measurements were to be made at higher simulated altitudes, a new procedure would be required.

Joseph Kaplan, a consultant, suggested using nitrogen as the test gas, and exposing it to an electrical discharge. This caused the gas to glow momentarily in the dark tunnel. Photographing the flow field and shock waves was no longer a problem. Becker concluded that use of nitrogen gas and the discharge afterglow could be used at static pressures as low as a ten-thousandth of an atmosphere, equivalent to an altitude of about 230,000 ft—close to that of space itself.

The 11-in. hypersonic tunnel at Langley was initially used for very basic aerodynamic experiments. One set of early tests involved different shapes at Mach 6.86 including cones, cylinders with flat ends, and cone/cylinders, all of which were tested at different angles of attack. Another set of tests involved boundary-layer measurements of a triangular wing, a square wing with several airfoil sections, and a flat plate. The test data were compared with existing theoretical calculations of lift, drag, pressure distribution, and flow fields at hypersonic speed.

Most of the tests showed good matches between theoretical models and wind-tunnel measurements. There were exceptions, however. One theoretical model showed accurate values for drag, but its predictions overestimated the measured values for pressure and lift. Test results showed that hypersonic boundary layers are thick due to large temperature fluctuations. The boundary-layer effects could result in a greater than three-fold increase in the vehicle's drag coefficient at hypersonic speeds, compared to calculations reflecting only inviscid flow effects [2, pp. xxxi, xxxii, xliv].

An overall assessment of the hypersonic wind-tunnel results confirmed that, at Mach 7, the aerodynamic characteristics of wings and bodies "can be predicted by available theoretical methods with the same order of accuracy usually obtainable at lower speeds, at least for cases where the boundary layer is laminar" [9, p. 19].

While the hypersonic wind tunnels and shock tubes were providing ground-based data on hypersonic flight, rocket-boosted models were also undergoing tests. The initial NACA hypersonic flight tests were undertaken by the Pilotless Aircraft Research Division (PARD), based at Langley. Efforts there involved ground-launched test vehicles on multi-stage rocket boosters based on the Nike surface-to-air missile. The launch site for the tests was Wallops Island, Virginia, where the NACA had a research station. On November 20, 1953, a two-stage test vehicle was fired carrying a small parabolic nose cone made of Inconel alloy. The vehicle reached Mach 5, and provided a limited amount of hypersonic heat-transfer data during the flight.

The following year, the PARD created a four-stage rocket called the Nike-Nike-T40-T55. It had Nike first and second stages, with the smaller T40 and T55 rockets as the third and fourth stages. This rocket assembly reached Mach 10.4 at 86,000 ft on October 14, 1954. After the fourth-stage rocket burned out, it coasted to a peak altitude of 219 miles before impacting some

400 miles downrange in the Atlantic Ocean. The fourth stage also served as the test vehicle. It was built of steel and Inconel, and had a flared base for hypersonic stability. This structural design performed like a badminton birdie, with the drag on the flared base providing a symmetrical force that kept the test vehicle from tumbling. During the flight, the test vehicle transmitted heat-transfer data to a ground station, and detected the transition from turbulent flow to laminar flow on the vehicle [2, pp. lxiv, lxv].

The physics of hypersonic airflows posed challenges to researchers using either ground wind tunnels or rocket-boosted models. Hypersonic bow shock waves wrap closely around the surface of the vehicle. Near the nose of the vehicle it creates a zone of locally subsonic flow, which is further compressed as gas stagnates at the nose. As a consequence, the test model is subject to very high stagnation temperature. Literally all of the vehicle's kinetic energy is converted into heat at the nose. The heat generated is astonishing. At a speed of about Mach 7 at 200,000 ft, stagnation temperatures melt steel and titanium. Just below Mach 15 and at the same altitude, hafnium carbide melts. At speeds approaching Mach 25 at 200,000 ft stagnation, temperatures are above 6000 K, which is hotter than the surface of the sun.

The high stagnation temperatures created by hypersonic flow have several effects on the oxygen and nitrogen molecules that are the primary constituents of air. Below Mach 5, they act like simplified "ideal gases"—the molecules behave like billiard balls. Above Mach 5, the diatomic oxygen and nitrogen molecules begin to act more like dumbbells that vibrate, bend, and break, dissociating the molecules (first the oxygen, later the nitrogen) into their component atoms, and thermochemical effects begin to occur. As the vehicle approaches Mach 15 the gas begins to ionize, and free electrons are liberated into the airflow. A significant fraction of the heat generated by stagnating the flow goes into dissociating the gas molecules, reducing the temperature relative to an ideal gas value. However, the atoms are likely to recombine on the vehicle's surface, transferring the heat back into it. This tendency depends on the catalytic effect of the surface material, with metallic surfaces tending to be highly catalytic and glassy surfaces, noncatalytic. Thus the heat-transfer rate at hypersonic conditions can depend on the surface material used in the vehicle construction—a phenomenon unknown at subsonic and supersonic speeds.

In general, the high inertia of air at hypersonic conditions tends to stabilize the boundary layer, reducing the tendency for the flow to transition from smooth, laminar flow to turbulent flow. However, chemical effects caused by the heating can also mitigate this effect, causing the boundary-layer flow to become turbulent sooner than otherwise expected. A sharp increase in the heat-transfer rate accompanies the transition to turbulent flow, potentially becoming large enough to destroy the vehicle, or at least increase dramatically the weight of the required thermal protective materials.

Researchers of hypersonic flow face a wide range of high-energy phenomena that could affect air's behavior. These include free atoms, electrons, and molecules in excited states, which could create additional energy-transfer modes and which carry the energy of dissociation through the boundary layer. As the particles relax from their excited states, energy is released that could also contribute to radiative heat transfer. Another possibility is heat transfer from electrons and ions themselves. Additionally, evaluating hypersonic flow quantities is complicated because many of these parameters are interdependent.

Yet another factor in evaluating hypersonic flows is the speed at which the different processes, such as dissociation and recombining of the dissociated oxygen and nitrogen molecules and of the ionized molecules and free electrons, occur. Changes on the atomic level may occur very quickly, but there may also be a delay in the transfer of energy. The extreme cases are "equilibrium flow" and "frozen flow." In the first, the dissociation and recombination processes occur instantaneously in response to the changing temperatures and pressures in the flow. In the frozen flow (which may actually occur at thousands of degrees Kelvin), the changes in local conditions occur so fast, due to the high velocity, that the chemical and/or vibrational relaxation processes cannot keep up and remain essentially frozen at their initial conditions [10, pp. 30, 31, 34, 35]. Often, the time frames for the flow processes and the chemical processes are comparable, resulting in the more general case of *nonequilibrium flow*.

In October 1954, the state of the art in hypersonic simulation was summed up in a report by the Aircraft Panel of the Air Force Scientific Advisory Board to the Air Force Chief of Staff. Board members included Clark B. Millikan of the California Institute of Technology (Caltech) and Robert R. Gilruth of the NACA. The report highlighted the unknowns of hypersonic flight and the shortcomings of existing knowledge [2, pp. xxvi–xxvii]:

> In the aerodynamic field, it seems to us pretty clear that over the next ten years the most important and vital subject for research and development is the field of hypersonic flows; and in particular, hypersonic flows with stagnation temperatures which may run up to the order of thousands of degrees. This is one of the fields in which an ingenious and clever application of the existing laws of mechanics is probably not adequate. It is one in which much of the necessary physical knowledge still remains unknown at present and must still be developed before we arrive at a true understanding and competence. The reason for this is that the temperatures which are associated with these velocities are higher than temperatures which have been produced on the globe... and that there are problems of dissociation, relaxation time, etc., about which the basic physics is still unknown. The experimental techniques which we believe will be important in this field are several. First, the use of supersonic wind tunnels; these are intrinsically and basically limited in stagnation temperature. They cannot simulate the stagnation temperatures

that will occur in hypersonic flight within the atmosphere.... A second experimental technique involves the use of shock tubes and other devices for producing extremely strong shocks. The characteristic of this type of technique is that the time available for measurements is measured in the order of milliseconds and this requires very special experimental procedures. It is our belief that in the near future there should be intensive pushing of this type of facility but that the most useful ones will in general be relatively small scale and inexpensive....

Among the issues involved in hypersonic flight that the panel examined were structures and materials. These involved materials able to withstand both steady-state heating and brief, very intense heating that would create severe thermal stresses. The panel wrote [2, pp. xxvi–xxix]:

There is a specific recommendation which the Panel makes in connection with these two fields. There should be immediate emphasis on the development of non-isotropic or composite materials of the following types, among others: Laminated Materials made of several alloys or combinations of metals, or from plastics; Honeycomb Composites of several types of alloys; Fibrous Materials with metallic or nonmetallic wires, and Pre-Stressed Ceramics. All of these we believe should be investigated and should be made subjects of continuing research.

The second of the structural problems is involved with the non-steady, extreme temperatures associated with the reentry problems for long-range ballistic rockets. This will involve not only structures but aerodynamics, materials, cooling, and a large number of other fields. As far as we can see, no very elaborate ground facilities will be used in this kind of investigation although a number of small-scale facilities will be required. The difficulty is that nobody sees how to design ground facilities that will match the conditions. Accordingly, the major experimental technique will almost certainly be rocket test vehicles.

... we feel that a fairly specific step-wise program should now be undertaken. We visualize three steps in this program. (1) The use of existing solid-propellant high performance rockets which have been developed within the last few years. As a specific example, the "LOKI" rocket is one which might be employed. These could be used in multiple end stages. Some studies which have been made indicate that by the use of a two-staged cluster of LOKIs one could get a 30-pound warhead or test head to a Mach number of about 10 to 12. This we believe is a step in which both the rockets and the instrumentation required are now available. (2) The second step, we believe, should be a more ambitious program using larger scale liquid-propellant rockets, at least for the boost phase. This would permit larger payloads and, hence, more instrumentation and tests. (3) The third stage in this step-wise program of rocket test vehicles would be the Atlas test vehicle, which is already in essence proposed as one of the elements in the Atlas test program.

The panel's comments and recommendations came at a time when ballistic-missile development was undergoing rapid changes. When the Atlas was first proposed in 1951, it was designed to carry a heavy atomic bomb. This meant that the rocket would have to be very large, and the relatively low yield of an atomic weapon also required that the missile be very accurate. This was unattainable with existing guidance-system technology. In November 1952, the United States successfully tested the first hydrogen bomb. This meant a very high-yield weapon could be built that was much smaller than an atomic bomb. With a lighter warhead, the missile's size and weight could be much smaller. Because the weapon's yield was far greater, the accuracy require-ments were eased.

The October 1954 final design of the Atlas had three rocket engines (two boosters and a sustainer engine); at 240,000 lb it was just over half the launch weight of the earlier design, was 10 ft in diameter, and was accurate within 5 n miles. The Atomic Energy Commission predicted that the warhead would weigh 1500 to 1700 lb and have a yield of 1 megaton (a million tons of TNT). The new design was approved for development in December 1954 [4, pp. 66, 78, 94, 117].

Although a long-range ballistic missile was now a practical concept, the issue of hypersonic reentry remained a major unknown. The solution would be nonintuitive, and provide another indication of the complexities of hyper-sonic flight.

REFERENCES

[1] Bush, V., U.S. Senate, Special Committee on Atomic Energy, *Hearings Atomic Energy*, November 27–30 and December 3, 1945, pp. 179–180. (On file at the NASA Headquarters History Office.)

[2] Hallion, R. P. (ed.), *The Hypersonic Revolution: Case Studies in the History of Hypersonic Technology*, Vol. I, Air Force History and Museums Program, Washington, DC, 1998, p. lxiv.

[3] DeVorkin, D. H., *Science with a Vengeance: How the Military Created the US Space Sciences After World War II*, Springer-Verlag, New York, 1992, pp. 170–177.

[4] Neufeld, J., *Ballistic Missiles in the United States Air Force 1945–1960*, Office of Air Force History, Washington, DC, 1990, p. 24.

[5] Lonnquest, J. C., and Winkler, D. F., *To Defend and Deter*, USACERL Special Report 97/01, Washington, DC, November 1996, p. 25.

[6] Stumpf, D. K., *Regulus: The Forgotten Weapon*, Turner Publishing Company, Paducah, KY, 1996, pp. 84–151.

[7] Gibson, J. N. *The Navaho Missile Project*, Schiffer Military/Aviation History, Atglen, PA, 1996, pp. 50–79.

[8] Baals, D. D., and Corliss, W. R., *Wind Tunnels of NASA*, NASA SP-440, Washington, DC, 1981, pp. 56–58.

[9] Heppenheimer, T. A., *Facing the Heat Barrier: A History of Hypersonics*, NASA SP-2007-4332, Washington, DC, pp. 13–20, 30–32.

[10] Allen, J. E., *Aerodynamics: The Science of Air in Motion*, 2nd ed., Granada, London, 1982, pp. 42–44.

Chapter 2

HYPERSONIC REENTRY

That was a real fireball.

John Glenn, describing his reentry aboard a Mercury capsule

The first half-century of aircraft development was focused on reducing drag to improve speed, range, and efficiency. Wood and fabric biplanes with wire-braced wings gave way to all-metal monoplanes, and finally to jet aircraft with long fuselages and thin, swept-back wings. Their pointed shape reduced drag. Initially, concepts for hypersonic vehicles followed this design philosophy. The early Atlas missile design had a very slender, sharply pointed nose, which faired into a long cone that carried the atomic warhead. The idea entered popular culture with Chesley Bonestell's space paintings and such films as *Destination Moon*. These depicted needle-shaped rockets resembling an enlarged V-2 [1]. These V-2 like missiles were sometimes referred to as "pregnant needles."

BLUNT BODIES, SHOCK TUBES, AND REENTRY TESTING

In 1953, H. Julian Allen and Alfred Eggers at the Ames Aeronautical Laboratory made a mathematical study of reentry problems, published as NACA TN-1381. Their approach was to look at the total heat input that various nose-cone shapes would experience as they reentered the atmosphere. The goal was to discover what shapes would minimize this input. Allen and Eggers made the counterintuitive discovery that for a lightweight vehicle, which was slowed by drag during reentry, a long, thin shape actually maximized heat input. In contrast, minimum heat input occurred with a blunt-shaped nose cone.

Allen and Eggers also looked at the heat-transfer rate averaged over the entire missile, rather than at just the nose cone. Their calculations showed that either a very sharp- or a very blunt-shaped missile would minimize the peak heat-transfer rate. Missile shapes between the two extremes showed much higher heat-transfer rates. The most critical part of the missile regarding reentry heating was the nose, where heating rates were at their highest. A blunt shape with the maximum possible radius was preferable. The optimal shape for the forward surface of the nose cone would be like that of the underside of a shallow bowl.

The blunt shape maximized the strength of the shock waves, increasing the amount of kinetic energy imparted to the air in the form of heat. At the same time, this also minimized the amount of heat absorbed by the nose-cone structure. With a blunt shape, the "stand-off" distance between the bow shock and the vehicle surface was maximized. The gas passing through the blunt portion of the bow shock formed an entropy layer of hot gas that flowed over the vehicle above the boundary layer. As much as 99 percent of the heat was radiated away or left in the vehicle wake. The thick boundary layer formed on the vehicle surface and the entropy layer above it effectively insulated the nose cone; only the remaining 1 percent had to be absorbed by a heat shield. This could still result in the nose cone having to endure stagnation temperatures of 6000 K.

In contrast, a sharply pointed nose cone formed a shock wave very close to its tip. The shock lay close to the surface, and the entropy layer was very thin and was quickly swallowed by the boundary layer. As a result, the boundary layer was too thin to provide effective insulation for the nose cone. More of the total heat input was absorbed by the sharp nose cone [2, 3].

Neither the Langley nor the Ames hypersonic wind tunnels were suitable for ground tests of the blunt-body concept, because the Mach numbers and temperatures that could be achieved in them were far too low as compared with those experienced by a missile warhead. As a result, the physical, chemical, and aerodynamic phenomena that occurred in hypersonic reentry could not be replicated. Only shock tubes could reproduce reentry conditions.

In the years following World War II (WWII), Arthur Kantrowitz, at Cornell University, did extensive research on high-temperature gases using shock tubes. Rather than high-pressure helium as the driver gas, he used a combustible mixture of oxygen and hydrogen gas together with helium. Under the right conditions (e.g., 70 percent helium), a smooth burn of the mixture could be obtained without detonating the gas, and the procedure was easily replicated. Following combustion, the driver gas had a higher speed than unheated helium, and hence could drive stronger shocks into the test gas. Kantrowitz also changed the test gas from air to argon. At the high temperatures he was attempting to create, the diatomic oxygen molecules in air would dissociate into two oxygen atoms. In the process, they would absorb energy and reduce the temperature. Argon, which is a monatomic gas, would neither dissociate nor lower the temperature. By using argon, Kantrowitz was able to reach temperatures of 18,000 K.

Following in the steps of earlier researchers, Kantrowitz focused on gas dynamics. He studied the dissociation of air and nitrogen, confirming that the degree of dissociation increased with higher temperatures and lower pressures. Argon was used to study thermal ionization. In this process, hot atoms lose their electrons, causing them to become electrically charged. Kantrowitz, like Walter Bleakney before him, realized that the shock tube had applications in hypersonic flight. Kantrowitz wrote in 1952 that " ... it is possible to obtain shock Mach numbers* in the neighborhood of 25 with reasonable pressures and shock tube sizes,"† more than enough for studies of ballistic missile warheads [3, p. 33]. Indeed, Mach 25 was the speed of a vehicle reentering the atmosphere from Earth orbit.

Kantrowitz subsequently joined the Avco Research Laboratory, which opened in early 1955, to work on reentry simulations. This work required the highest possible Mach numbers, which Kantrowitz obtained by using a hydrogen-oxygen-helium mixture as the driver gas. The hydrogen and oxygen were burned, burst a diaphragm in the shock tube, and accelerated the test gas to create the shock wave, which subsequently flowed over the model in the shock tube. To ensure a controlled burn, the helium had to represent 70 percent by weight of the gas mixture. The burning oxygen/hydrogen mixture heated the helium gas, which prevented detonations that would render test results unreliable.

To collect the data, a standard shadowgraph, schlieren photographs, and an interferometer were used. Measurements of the shock wave's speed were made using pickups mounted along the length of the tube. These measured electrical resistance, which changed as the shock wave passed over them. This apparatus provided reliable data up to Mach 17.5.

Although hypersonic flow conditions were achieved in the shock tubes, the test duration experienced by a reentry vehicle model mounted in the tube—the interval between the passage of the incident shock and the arrival of the driver gas—was measured in milliseconds (thousandths of a second). Moreover, although the air stream velocity after passage of the incident shock was close to the shock velocity, the local Mach number was, at most, about 2 due to the high temperature the incident shock produced. Although the shock tube could produce the temperatures and pressures downstream of the reentry vehicle's bow shock, the aerodynamic simulation of the bow shock itself was imperfect.

In spite of the obvious disadvantages of very short test time in a shock tube, it had three very important advantages:

- Reentry-level temperatures could be produced and contained without melting the test facility.
- Heat-transfer rates could be measured accurately using transient-heat-conduction methods.
- Optical access to the test section was possible with uncooled windows.

---

*Shock Mach number is the ratio of the speed at which the shock wave travels in the tube to the speed of sound in the gas ahead of the shock.

†The boundary layer that grows on the tube wall downstream of the shock reduces the shock speed. Larger tubes have relatively thinner boundary layers and hence can produce nearly ideal shock speeds.

Kantrowitz assembled a team of aerophysicists to work on reentry problems. Shao-Chi Lin investigated a related heating issue, that of the ionization of atmospheric gasses. Because the ionized air was electrically conductive, novel shock-tube and instrumentation designs were required. The tube itself was made of nonconductive glass, and a strong magnetic field was used to measure the ionization of the gasses. As ionized air passed down the tube, it altered the direction and strength of the magnetic field. Lin tested the system by firing a piece of metal with a known conductivity down the tube at a known speed. When the magnetic measurements of the ionized air were compared with calculated values, there was good agreement in a speed range of Mach 10 to 17.5. At the maximum Mach number, the ionized air had conductivity of an order of magnitude greater than seawater [3, pp. 31–36].

Researchers at the Avco Lab also drew upon resources from neighboring universities. The formula developed by James Fay of the Massachusetts Institute of Technology and Frederick Riddell at Avco for predicting stagnation-point heat transfer (the Fay-Riddell stagnation heat transfer method) became the standard against which shock-tube data were compared.

These early shock tube tests provided an understanding of the physics of hypersonic reentry. Equations were derived to predict the point at which a reentry vehicle would experience, first, free-molecular flow as it first penetrated the atmosphere, then noncontinuum flow (a transitional state between free-molecular and continuum flows), and finally continuum flow (in which the molecules behaved in accordance with conventional theories). The continuum regimes encompassed the most severe heating conditions as well as drag on the vehicle, and extended downward to the point of impact. The state of the gas in the continuum regime could be chemically frozen in some portions of the flowfield (the time in which reactions could occur was far longer than the resident time in the flow), in a nonequilibrum state in others (where reaction times are comparable to residence times), and in still others could be in equilibrium (reaction times are fast enough that they maintain chemical equilibrium). The circumstances, it was learned, depended on both vehicle speed and shape. The circumstances could also involve reactions between the air and the gases that entered the flowfield.

The next step in the development of reentry vehicles was actual flight tests using rocket-boosted subscale models in high-speed reentries. To do this required a more powerful version of the multi-stage solid fuel rockets used by PARD in the early 1950s. Lockheed X-17 solid-fuel three-stage rockets launched from Cape Canaveral, Florida, were used for these tests. The X-17 consisted of a Thiokol XM-20 rocket as the first stage, a cluster of three Thiokol XM-19 rockets as the second stage, and a modified XM-19E1 rocket as the third stage. The X-17 could reach a speed of over Mach 14 and experience stagnation temperatures of 15,000 to 20,000°F as well as heat-transfer rates of several thousand BTUs per square foot per second. To achieve these results an unusual over-the-top profile was used in which the rockets were fired during both descent and ascent.

During the X-17's ascent, it was stabilized by two spin rockets on the first stage. After the first stage burned out, it remained attached to the upper stages. The X-17 continued upward to a peak altitude of 330,000 to 500,000 ft—in space, where there were no aerodynamic forces acting on the booster. As the X-17 underwent the long fall toward the Atlantic Ocean, air density increased and the four large fins on the first stage stabilized the vehicle in a nose-down attitude.

The complete rocket would descend until its altitude was less than 100,000 ft. The first stage then separated, and the second and third stages fired in rapid succession. At burnout, the test vehicle separated from the spent booster. This flight profile allowed control of both the peak Mach number and the altitude at which it occurred. By waiting to fire the second and third stages until the X-17 was at a relatively low altitude, the nose cone would not be decelerated at too great an altitude.

A wide range of reentry vehicle designs were tested in the course of the X-17 flights, including blunt, parabolic, conical, and hemispheric shapes. Six of the vehicles were designed and built by Avco's Research and Advanced Development Division (which had been spun off from Kantrowitz's research lab), and another six were designed and built by General Electric (at a newly established Research Division). The X-17 flights did not attempt to simulate the actual reentry profile of a warhead, but were designed to gain basic data. The first research launch, on July 17, 1956, carried a hemispherical reentry vehicle that reached a top speed of Mach 12.4 at only 40,000 ft. The August 23, 1956, flight, made with a blunt reentry shape following two failed launches, reached Mach 12.4 at a mere 38,500 ft. The highest speed reached was Mach 14.4, at 57,000 ft. This was accomplished on February 7, 1957, with a General Electric blunt-cone shape. Many of the tests were made at speeds around Mach 12, at altitudes of 40,000 ft and lower.

The test vehicles radioed telemetry on heat transfer, boundary-layer conditions, and pressure distribution to receiving stations at Cape Canaveral. The X-17-powered reentry vehicles were not treated with heat protection, but were intended for hypersonic aerodynamic measurements. Given that the reentry vehicles' speeds were about half those of an orbiting satellite, at altitudes achieved by existing jet fighters, they vanished in a fiery streak before impacting the ocean.

The flight rate of the X-17 reentry tests underlined the new priority of the U.S. ballistic-missile program. The Air Force launched 26 X-17s between April 1956 and August 1957. Seventeen of the launches were made in the first eight months; six were failures.

From the ground simulations, theoretical calculations, and X-17 flight data, the initial Air Force reentry vehicle design emerged. This was the Mark

Fig. 2.1 The preparation and launch of an Atlas D ICBM from Vandenberg Air Force Base. Like the Thor, it carried a Mark 2 reentry vehicle.

2 blunt body, carried by the Thor intermediate-range ballistic missile (IRBM) and the early Atlas D intercontinental ballistic missile (ICBM; see Fig. 2.1). The heat shield was a heavy slab of copper, which had both high heat capacity and high conductivity and absorbed sufficient heat to keep the reentry vehicle's overall structure below the melting point. Although the high drag of its blunt shape was effective in carrying away much of the heat, it also slowed the warhead to subsonic speeds at low altitudes, which was undesirable. A copper heat shield was also very heavy, reducing the payload, and was limited in the amount of heat it could absorb before disintegrating. It was clearly only a stepping-stone in development of a practical thermal protection system. Figure 2.2 shows the evolution from the highly streamlined initial concept to the initial blunt-body design. The bottom row shows the heat sink reentry vehicle and the initial manned capsule concept, which featured an ablative heat shield on a blunt body shape.

ABLATION

While the Air Force was developing blunt-body warhead designs, the Army was taking a different approach in terms of both the thermal-protection technique and the test procedures used. Meteors, of which there are two major types, are the only natural objects that experience conditions comparable to those of a reentering warhead. Iron meteors bear characteristics similar to the metal heat sink designs of the Thor and early Atlas reentry vehicles. Stony meteors, as the name suggests, are similar in composition to terrestrial rocks.

The heat of entry into Earth's atmosphere affects the two types of meteors very differently. Although large iron meteorites survive reentry, they often bear deep pits and hollows, where heat penetrated the surface and burned away the iron. Stony meteorites show a different characteristic. Entry heating vaporizes their surface, and the resulting gaseous material carries away the heat while leaving the interior virtually unaffected. This allows the bulk of a large stony meteorite to survive and reach Earth's surface. Recovered stony meteorites also bear a black coating on their surface, usually without the holes seen on iron meteorites. The process of removing the surface material is called *ablation*.

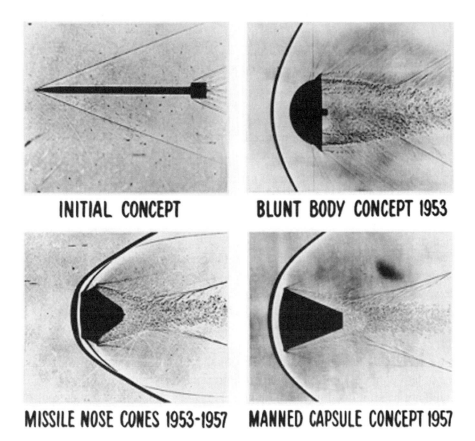

INITIAL CONCEPT　　**BLUNT BODY CONCEPT 1953**

MISSILE NOSE CONES 1953-1957　**MANNED CAPSULE CONCEPT 1957**

Fig. 2.2 Wind-tunnel tests of four reentry vehicle concepts.

The U.S. Army Ballistic Missile Agency did extensive research on ablation as part of development work on long-range rockets. Wernher von Braun and the German engineers who had built the V-2 were conscripted for service in the Army program after WWII. Like the Air Force had, Army researchers faced the problem of reentry; however, their approach focused on pragmatic engineering rather than theoretical physics. They had a great deal of experience testing high-temperature materials by placing them within a rocket's exhaust. This provided a workable simulation of reentry temperatures, with sufficient test time to produce ablation. The very limited test time of the shock tube was far too short for this purpose.

Tests began in November 1952 with materials such as ceramics being exposed to the hot exhaust of a rocket engine on a test stand at the Army's Redstone Arsenal at Huntsville, Alabama. By late 1953, the work had shifted to testing of nose-cone models. Tests of reinforced plastics followed in July 1954.

In November 1955, the test efforts took on a new urgency when the Army received government approval to begin development of the Jupiter IRBM. The missile was to have a range of 1500 n miles and a reentry speed of Mach 15. Like the Air Force Thor, this missile could strike targets in the Soviet Union from bases in Western Europe. To develop the reentry vehicle, Army engineers considered different concepts, but their primary focus remained on ablative heat shields.

A wide range of materials was tested for the Jupiter warhead, including silicones, phenolics, melamines, Teflon, epoxies, polyesters, and synthetic rubber. The materials were reinforced with soft glass, fibers of silicon dioxide and aluminum silicate, mica, quartz, asbestos, nylon, graphite, beryllium, and cotton [3, pp. 36–38, 50].

Despite the different materials being considered, the basic principal was the same: During reentry, the surface temperature on the vehicle increased, and this increase was absorbed by the ablative material. Depending on the material, the method by which the heat was eliminated differed. Some ablative materials vaporized and mixed with the boundary-layer flow. Other types of ablative material melted and became a viscous glassy liquid that was swept downstream by the viscous shear stress from the boundary layer. With reinforced plastics, the ablative material charred from the heat and formed an insulating layer [2, p. 166].

The ablative materials complicated the physics and chemistry of hypersonic reentry. Reactions occurred between the air and the gasses produced by the ablative process entering the flowfield. These reactions and their effects could not be simulated in shock tunnels. However, the gas phase reactions of the ablative products with air could be studied in shock tube experiments that used the tube as a chemical reactor rather than an aerodynamic simulation device.

The Army was ready to begin flight tests of subscale nose cones in 1957. Based on ground-test results, plastics reinforced with glass fibers were selected as the ablative materials. Unlike in the Air Force X-17 tests, the Army's simulated warheads were to make a complete reentry and be recovered from the Atlantic Ocean after splashdown. The Jupiter reentry vehicle shape was an elongated cone with a rounded tip. The subscale vehicles were one third the actual warhead's size, at 29 in. long and 20 in. in diameter. Each weighed 314 lb, of which 83 lb was the heat shield. After splashdown, a float kept the vehicle from sinking until it could be recovered [3, pp. 44–46].

The parametric relationship that controls terminal velocity during reentry of nonlifting shapes is the ballistic coefficient, W/C_DA. This parameter is the ratio of the vehicle's weight (W) divided by the product of its coefficient of drag (C_D) and a representative cross-sectional area (A), which was usually the base area. A high ballistic coefficient implies low drag for a given weight. Conversely, a low ballistic coefficient implies high drag for the same weight. The Mark 2 blunt body was a low W/C_DA shape, which underwent a rapid deceleration during reentry, followed a steep descent path, and had a subsonic terminal velocity. The Jupiter reentry vehicle was a high W/C_DA shape, with a comparatively longer range, shallower descent trajectory, slower deceleration, and a supersonic terminal velocity. However, the heat-transfer rate was higher for this shape compared to the Mark 2 blunt body shape [4].

The Army flight tests began in September 1956, with a test launch of the first Jupiter C rocket. Despite the name, it was actually a Redstone rocket with lengthened fuel tanks and two upper stages of clustered solid-fuel rockets. The launch was intended to test the modified missile, separation of the upper stages, and ignition. It did not carry a reentry vehicle, but rather a dummy stage as the payload. The missile flew successfully, reaching an altitude of 682 miles and landing 3335 miles downrange from Cape Canaveral.

The first Jupiter C nose cone launch took place in May 1957, with mixed results. The reentry vehicle came down some 480 miles short of its intended landing point, and was not recovered. Despite the loss, telemetry signals indicated the ablative heat shield had worked and the vehicle had survived reentry. The Jupiter C flight of August 8, 1957, saw the first successful recovery of an object from space. The reentry vehicle flew a distance of 1343 miles, and experienced 95 percent of the total heat load that a Jupiter IRBM warhead would undergo [3, pp. 45, 46].

The Air Force had initially focused on heat sink reentry vehicle designs for its Thor and Atlas missiles (see Fig. 2.3). The weapon was replaced by instrumentation, including a data-recovery capsule and telemetry equipment. Once the two missiles had begun their test launches, the Air Force began a series of ablative heat shield tests, with work beginning in December 1957. The booster was a "Thor Able," a Thor IRBM fitted with an Aerojet-General 1040 liquid propellant rocket modified from the Vanguard booster's second stage. As in

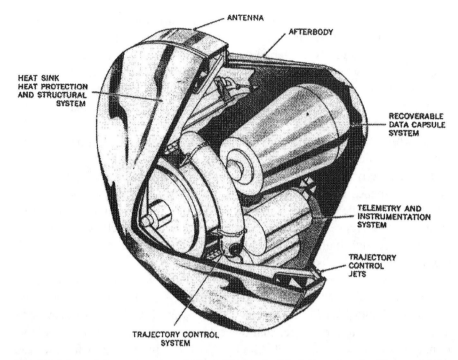

Fig. 2.3 A heat sink reentry test vehicle similar to the Mark 2 used on the Thor and Atlas D missiles.

the Army tests, the reentry vehicles had onboard instrumentation to return flight data and parachutes so they could be recovered after splashdown.

This initial effort, called Able Phase 1, involved tests of a biconic sphere shape. (This was a spherical nose on a shallow sloped cone 3 ft in diameter.) The Thor Able reentry vehicles also each carried a mouse to test the biological effects of prolonged weightlessness and the high g-forces of launch and reentry.

The first Thor Able launch was made on April 23, 1958, from Cape Canaveral (see Fig. 2.4). The launch initially appeared successful, but 146 s into the flight, the Thor's turbopump failed and the booster exploded. Both the reentry vehicle and its passenger, named "Minnie," were lost.‡

The failure delayed subsequent launches until July 9 and 23, 1958. In both cases, the launches were successful and the reentry vehicles reached altitudes of about 600 miles and traveled nearly 6000 miles downrange. The telemetry data indicated the reentry vehicles on both flights had survived the heating. The mice, "Laska" and "Wickie," survived the g-forces and weightlessness of

---

‡The Thor was deployed in England under joint British/U.S. control. The Thor later served as the first stage of the Delta satellite booster.

the flight. Unfortunately, neither reentry vehicle was recovered, making it impossible to examine either the heat shields or the mice.

These Air Force ablative tests were followed by flight tests of the Army's Jupiter IRBM with full-scale dummy warheads in late 1958 and early 1959 (see Figs. 2.5 and 2.6). In addition to undergoing reentry tests, these flights

Fig. 2.4 A Thor IRBM on the pad before a test launch. It is carrying a dummy Mark 2 heat sink warhead.

Fig. 2.5 A U.S. Army Jupiter IRBM undergoes testing.

also carried biological experiments using monkeys. The first launch was on December 13, 1958. The reentry vehicle carried a South American squirrel monkey named "Gordo." The launch was successful, and Gordo experienced 10 g's during powered flight, followed by about 9 min of weightlessness, without ill effects. The dummy warhead apparently also survived the 10,000-mph reentry. However, the recovery ship failed to pick up the recovery beacon, and no trace of the reentry vehicle was found. Like his predecessors on the Thor Able flights, Gordo was lost. Although the data collected by the Air Force and Army reentry tests indicated that high g-forces and weightlessness could be endured, recovery was proving difficult.

The second Jupiter IRBM test was launched on May 28, 1959. Aboard was a rhesus monkey named "Able" and a smaller squirrel monkey named "Baker." The Jupiter liftoff went smoothly, and, after engine shutdown, the dummy warhead separated. The two primates experienced about 9 min of weightlessness followed by very high g-loads during reentry. The monkeys endured the experience, and as the forces diminished their vital signs returned to normal. The parachute opened successfully, and the reentry vehicle splashed down in the Atlantic some 1700 miles downrange. The dummy warhead was recovered, and neither monkey exhibited any ill effects from their flight [5]. (Able died several days after the flight during a minor operation to remove an ECG electrode. This was traced to the anesthetic used and not to any injuries from the spaceflight. Baker lived another quarter century, reaching an age of 27 years.)

Although similar in performance to the Thor, the IRBM reentry vehicle bore a more advanced ablative heat shield and a low-drag cone shape. The cone shape gave the reentry vehicle a higher terminal speed and lighter weight compared to a high-drag heat sink shape. As a result, the heat sink concept had a limited service life.

By the late 1950s, as the Thor Able phase I and Jupiter reentry tests were being made, the capability to make ground tests of ablative materials had improved significantly. The advance of note was that of an *arc tunnel*, which heated the airflow using an electrical arc and provided a continuous flow inside the tunnel, unlike the brief pulse of a shock tube. An arc tunnel also did not have to reach the hypersonic speeds of a shock tube. The temperature of the flow needed to match the temperature endured by a reentering warhead. However, arc tunnels had limited pressure capability, as well as aerodynamic simulation capability, which compromised the fidelity of the ablation data under many conditions.

The first arc tunnel in the United States was operational at the NACA's Langley laboratory in May 1957, but represented only a prototype. The electrical arc heated a flow of air, nitrogen, or helium, turning it into plasma with a temperature of 5800 to 7000 K. This was a higher temperature than that reached in existing hypersonic wind tunnels.

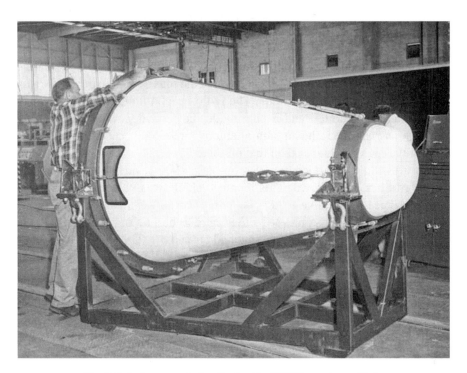

Fig. 2.6 A close-up photo of a Jupiter IRBM reentry vehicle.

Kantrowitz at Avco's laboratory recognized the arc tunnel's utility, even with its initial limitations. His first arc tunnel had a power output of only 130 kW but could produce temperatures equivalent to a reentry speed of 21,000 fps. From this data, researchers concluded that opaque-quartz ablative material could absorb 4000 BTUs per lb. Avco's next arc tunnel, at its Research and Advanced Development Division, was far more powerful, but still had a "homebuilt" quality. Its power rating was 15 MW, provided by 2000 12-V truck batteries. Before a test, these were charged in about an hour, using generators. The batteries could supply enough power to simulate reentry heating for about a minute.

The second step in the Air Force's ablative research program involved two different reentry vehicle shapes. The first of these was the RVX-1, which had a cylindrical main body, a rounded nose, and a flared base. It was 67 in. long, 28 in. in diameter, and weighed 645 lb. The RVX-1 test vehicles were built by both General Electric and Avco, and were used to test different ablative materials. The booster used for the tests was a modified Thor Able II, which featured the original design but had an Altair third stage added to increase payload and range.

The first Able phase II launch was made on January 23, 1959, but ended in failure when the Thor booster shut down early. Of the remaining five launches, another failed outright and four successfully flew the reentry profile. As with the phase I launches, recovery proved a problem. The first successful recovery was made on the third launch, on April 8, 1959, with an Avco RVX-1 coated with fused opaque quartz. The reentry vehicle reached a maximum altitude of 764 miles and traveled 4944 miles downrange before splashing down in the Atlantic. The peak speed was Mach 20, with the maximum heating occurring at Mach 16 and 60,000 ft altitude. The radio beacon failed but the vehicle's dye marker was spotted and a ship made the recovery. When the reentry vehicle was examined, the amount of ablative material burned off was found to be only 3 percent higher than that predicted.

The fifth flight of a General Electric-built RVX-1, on May 21, was also recovered after splashdown. This reentry vehicle was covered with several different types of ablative materials. Wires were also imbedded at various thicknesses within the samples. As the material burned off, the wires broke, indicating the amount of material that had burned off. The RVX-1 shape was adapted for the Mark 3 reentry vehicles, used on the Atlas D ICBMs, replacing the Mark 2 heat sink warhead. The Mark 3 reentry vehicle, like the RVX-1, had a blunt nose, a cylindrical body, and a flared tail.

The next step in the Air Force test effort was a cone-shaped low-drag reentry vehicle like that on the Jupiter IRBM. This was the RVX-2, an elongated cone 147 in. in length and 64 in. in diameter. Thermal protection was provided by phenolic nylon, and a total of 510 lb of instrumentation was carried. The RVX-2's total weight was about 2400 lb. Unlike the earlier reentry experiments, the launch vehicle was not a specialized test rocket but, rather, an Atlas ICBM. This allowed the tests to be made under reentry conditions that matched those of operational launches [3, pp. 47–49, and 6].

Three RVX-2 launches were made in the initial series. These were intended to be recovered, but only on the final launch was this successfully accomplished. The reentry vehicle flew 5047 miles downrange on July 21, 1959. This was the largest object ever recovered from space up to that time. The data from these tests led to a follow-on effort, called the RVX-2A, in 1960. These flights carried General Electric ablative materials on the first and third launches and Avco samples on the second. The first two recovery attempts failed; however, the telemetry system returned data on the reentry vehicle's performance. The third RVX-2A vehicle was successfully recovered after its flight.

The data from the flights were used by Avco to design the Mark 4 reentry vehicle carried on the Atlas E and F and the Titan I ICBMs (see Fig. 2.7). The

Fig. 2.7 A Titan I ICBM lifts off on a test flight.

Fig. 2.8 A Titan II ICBM with a cone-shaped Mark 6 reentry vehicle is pictured in its silo.

shape, a blunt cone-cylinder with a flared base, was similar to that of the Mark 3, but was larger and carried a heavier payload. General Electric used the RVX-2A as the model for the Mark 6 reentry vehicle on the Titan II. Like the test vehicle, the Mark 6 was cone-shaped but was much larger, at over 10 ft long and 7.5 ft in diameter (see Fig. 2.8) [4, pp. lxx–lxxxiii, and 7]. The Mark 6 was similar to the cone-shaped Jupiter reentry vehicle. Despite the various shapes, all these designs relied on the blunt body concept.

MANNED HYPERSONIC FLIGHT

While ground and flight tests were under way with hypersonic reentry vehicles, other researchers were looking toward manned hypersonic flight, which they envisioned as a continuation of subsonic and supersonic atmospheric flight. Since the Wright brothers' era, aircraft had been flying progressively higher, faster, and farther. Eventually, it was thought, aircraft would be capable of flying so high and fast that they could leave the atmosphere, traveling on a suborbital trajectory into the fringes of space.

Like development of ballistic missile reentry vehicles, this idea had a long history. Scientists in Nazi Germany made two test flights of V-2s fitted with swept-back wings from the rocket test site at Peenemünde. Their goal was to extend the missile's range by allowing it to make a high-speed/high-altitude glide to a target city, extending its range beyond a simple ballistic trajectory. Called the A-4b, it was the predecessor to later boost-glide concepts. The A-4b had a planned range of 465 miles, compared to 200 miles for a standard V-2 rocket.

The first A-4b flight, made on January 8, 1945, suffered a control system failure moments after launch and crashed within sight of the launch pad. The second, on January 24, was more successful. The A-4b lifted off successfully, followed a ballistic trajectory, reentered, and transitioned into a Mach 4 glide. The flight ended when high air loads caused a wing to fail. With Nazi Germany only months away from total collapse, however, no further efforts were made. Wernher von Braun and the other rocket engineers were soon forced to flee by the advancing Soviet Army and seek refuge in southern Germany. It was here that they surrendered to the U.S. Army, and were sent to White Sands, New Mexico.

Several proposals were made in the early postwar years for hypersonic winged vehicles, the first in 1949 by Hsue-shen Tsien[§] at the California Institute of Technology. His proposed design would have a maximum speed of 9140 mph, and a 3000-mile range. In the early 1950s, Bell Aircraft undertook studies of two-stage winged vehicles able to fly long distances using a boost-glide flight profile. Robert J. Wood of Bell Aircraft proposed development of a manned hypersonic aircraft at the October 4, 1951, NACA Committee on Aerodynamics meeting. Wood encountered difficulty gaining support from the other committee members. He repeated his proposal at the committee's January 1952 meeting, again without success [4, pp. i–ii, and 8].

This outcome was not surprising. Langley researcher John V. Becker called the Bell Aircraft concepts "most stimulating." He recalled, however, that there was also "such a multiplicity of enormous technical problems that these systems seemed very far in the future." As for manned spaceflight, Becker recalled, ". . . its added problems and the unanswered question of safe return to Earth was seen then as a 21st century enterprise" [4, p. 380].

Wood continued his efforts, however, and in mid-1952 the Aerodynamics Committee endorsed a proposal that the NACA "devote a modest effort" to hypersonic studies at speeds of Mach 4 to 10. Hugh L. Dryden, NACA Director of Research, realized that a modest effort would accomplish little, and reduced the recommendation to a study aimed at identifying the problems of hypersonic flight, rather than actual research on the problems [4, p. 380, and 8, p. 102].

The initial Langley response to Dryden's change in plan was to establish a three-man ad hoc study group. The members were Chairman Clinton E. Brown

---

[§]Tsien was accused of being a security risk a few years later and was deported to China. He subsequently became head of China's missile and nuclear weapons development programs. Tsien was named "Man of the Year" in 2008 by *Aviation Week and Space Technology* magazine.

of the NACA Compressibility Research Division; Charles H. Zimmerman of Stability and Control; and William J. O'Sullivan of PARD. They were asked to assess the problems, develop ideas for a research program, and define a manned research aircraft for hypersonic research.

The Brown group met for several months before issuing its report in June 1953. Report results anticipated that hypersonic flight would reveal significant differences in terms of temperatures and boundary layer conditions from those of supersonic flight. The report also questioned the usefulness of existing ground-based hypersonic tunnels, such as the 11-in. facility at Langley or the 10-by-14-in. continuous flow tunnel at Ames. These, the group believed, could not simulate the required conditions. Rather than use ground-based simulations, they recommended flight testing to produce the actual high-temperature flight conditions. This would require an expansion of PARD's rocket activities. The hypersonic testing would require test models to be flown across the Atlantic and be recovered in the Sahara [4, pp. 381–383, and 9].

The Brown group's recommendation was not the first such debate over wind tunnels vs. rocket-launched models as the proper research method for high-speed testing. Between 1947 and 1949, PARD launched 386 models. Each model was expensive, required extensive work by Langley's shop staff, had to be instrumented, and was lost at the end of the flight. The scale of the effort created friction with the wind tunnel researchers, who complained PARD's "voracious appetite" resulted in a major slowdown in the production of models for wind tunnel research. They also complained that PARD's reports discussed and analyzed the performance of different model configurations, but gave little attention to the flow processes involved. The PARD personnel countered that a single rocket flight could provide enough data to establish the flight parameters. This made the rocket research at least as productive as Langley's wind tunnel efforts.

Although the Brown group's recommendation had little direct influence, it marked the beginning of a process that led ultimately to manned hypersonic flight. In October 1953, the Air Force Science Advisory Board Aircraft Panel reported that an extremely high-performance research aircraft was needed and that its feasibility should be examined. This was an important endorsement, because the NACA lacked the money to build such an aircraft without military support. Following release of the Aircraft Panel's report, the issue of a new research aircraft program was placed on the agenda of the NACA's Interlaboratory Research Airplane Panel February 1954 meeting.

This set in motion studies at Langley, Ames, Lewis (now NASA Glenn Research Center), and the High-Speed Flight Station (HSFS, now Dryden). The major effort took place at Langley. John Becker headed the study effort, which included contributions by Maxime A. Faget (rocket propulsion), Thomas A. Toll (control), Norris F. Dow (hot structures), and James B. Whitten (test pilot). By converting speed into altitude, the aircraft could follow a parabolic trajectory and reach space. Although manned hypersonic flight was to be the aircraft's mission, the Becker group also looked at the biomedical aspects of spaceflight [3, pp. 58, 59, and 9, pp. 356, 357].

As with missile warheads, the problem was how the aircraft—and its crew—would survive the hypersonic reentry environment. When Becker and Peter F. Korycinski first considered the reentry problem during March and April of 1954, they believed reentry should be made with the aircraft's nose pointed along the flight path. However, the aircraft would still be flying at a high Mach number when it reached the denser lower atmosphere where dynamic pressure and heating on the vehicle would soon exceed its structural limits.

Becker and Korycinski used Langley's 11-in. hypersonic wind tunnel to determine that the angle of attack should be increased to between 11 and 26 deg. The higher angle was necessary to reduce the vehicle's lift-to-drag (L/D) ratio, maximum dynamic pressure, heating rates, and the amount of time the vehicle would be exposed to the heating rates. This in turn reduced the total heat load the structure would have to endure. Speed brakes could also be used to adjust the L/D to remain within heating limits. They quickly realized that this was an extension of Allen's "blunt body" principle earlier applied to warhead reentries. Becker wrote [4, pp. 386, 387]:

> As we increased angle of attack our configuration in effect became more "blunt," dissipating more of its kinetic energy through heating of the atmosphere and less in the form of frictional heating of the vehicle.... our heating analysis provided the first clear detailed insights into the reentry heating problems of winged vehicles and their probable solution by use of combined high lift and high drag.

While Becker's group at Langley defined what would become the X-15, engineers and researchers at other NACA centers offered their differing ideas about the new research aircraft. HSFS engineers proposed an enlarged, faster, and higher-flying rocket-powered vehicle similar to those of the existing X-1 series. The Ames concept focused on suborbital long-range flight, and proposed a military vehicle powered by an air-breathing engine. Lewis engineers took a contrary position. They opposed building a new manned research aircraft, arguing that hypersonic research should instead be done using rocket-boosted models. Like the Brown group, Lewis engineers proposed that the Wallops Island rocket launch facility on the Atlantic coast be expanded to support the hypersonic tests.

With the proposals of the different NACA centers in hand, the next step was to bring the Air Force and Navy into the program. This took place in a meeting on July 9, 1954, at the NACA headquarters, and resulted in an agreement that the NACA would write a document outlining details of a Mach 7 research aircraft. The final step was a meeting of the NACA

Aerodynamics Committee at the HSFS on October 5, in which a final decision on the proposal was to be reached. (NACA meetings were held at the different centers on a rotating basis.)

At the meeting, Floyd Thompson, an NACA engineer who had worked at Langley since July 1926, said hypersonic wind tunnels could not examine such facets of hypersonic flight as "the distortion of the aircraft structure by the direct or indirect effects of aerodynamic heating" and "stability and control at very high altitudes at very high speeds, and during atmospheric reentry from ballistic flight paths" [9, p. 362]. He also noted that although rocket-launched models could reach speeds of Mach 10, they could not provide the amount and quality of data that a manned research aircraft could. He therefore urged the NACA laboratories to examine the feasibility of such a vehicle.

Clarence L. "Kelly" Johnson of the Lockheed "Skunk Works"—the company's Advanced Development Projects Division—was the only person at the meeting who opposed any extension of the X-plane program. Johnson recommended an unmanned research vehicle be built to provide data on structural temperatures and the stability and control issues a hypersonic aircraft would face. If a decision was reached that the aeromedical problems of hypersonic flight were "paramount," a manned research aircraft could subsequently be built. He recommended, however, that it should be designed for use as a strategic reconnaissance aircraft. He argued that the earlier research aircraft had not provided data useful for tactical aircraft, did not have configurations like those of operational aircraft, and had taken too long to develop [9, pp. 361–363].

One of the problems facing the early research aircraft program was the rapid advance of aviation technology during the postwar period. One of the goals of the X-15 program was to build a research aircraft that performed far in advance of any tactical aircraft. At the time Kelly Johnson recommended the manned hypersonic research aircraft also be used for reconnaissance missions, he was working on the CL-282 aircraft. This had an XF-104 fuselage with long, sailplane-like wings, and was designed for very high altitude flights. President Dwight Eisenhower approved its development in December 1954, and it first flew in August 1955 as the "U-2." The A-12 Oxcart Mach 3 reconnaissance aircraft would fly in 1962.

Johnson's comments were immediately challenged. Walter Williams, head of the HSFS, said the X-1 had reached speeds of Mach 1.5 and altitudes of 55,000 ft five to seven years before tactical aircraft were able to match this performance. Gus Crowley, associate director for research at NACA Headquarters, made a wider-ranging response. He noted that the concept of the hypersonic research aircraft was "to build the simplest and soundest aircraft that could be designed on currently available knowledge and put into flight research in the shortest time possible" [9, p. 364].

On the issue of manned vs. unmanned research vehicles, Crowley observed that the X-1 and other rocket- and jet-powered research aircraft had successfully made hundreds of flights despite experiencing control problems, equipment malfunctions, and other complications. The human pilot, he argued, had saved these flights, allowing further research. Crowley said that automatic controls could not be depended upon to save the vehicle under unexpected conditions.

At the conclusion of the meeting, with Johnson casting the only "no" vote, the NACA Aerodynamics Committee endorsed a report saying that a manned hypersonic research aircraft should be built that could reach Mach 7 and altitudes of several hundred thousand feet. In June 1956, North American Aviation won the contract to build the X-15.

The three X-15 aircraft made a total of 199 flights between June 1959 and October 1968. They provided a wealth of hypersonic heating and stability data, biomedical data, experience with the effects of hypersonic flight on materials and systems, and extended manned flight to the edge of space. Thirteen of the X-15 missions exceeded an altitude of 50 statute miles, qualifying as spaceflights. Eight of the X-15 program's 12 pilots received astronaut wings for these flights. But the X-15 was only the initial step in the goal of building a manned orbital spacecraft that could land like an airplane on a runway. This proved to be a difficult goal.

DYNA-SOAR AND ASSET: WINGS INTO SPACE

The X-15 was envisioned as part of a three-round process leading to manned spaceflight. Round 1 would entail the initial series of rocket-powered research aircraft, which were to achieve speeds of up to Mach 3 and altitudes of 100,000 ft. Round 2 would be flights with the X-15, which would extend the flight envelope to hypersonic speeds and the edge of space. Round 3 would see aircraft launched by rockets into suborbital trajectories, and, finally, into orbit.

The boost-glide vehicles studied by Bell Aircraft in the early 1950s influenced rounds 2 and 3. German General Walter Dornberger developed the concept. He had been commander of the Peenemünde test site, where he oversaw development of the V-2 rocket. After the war, he came to the United States and joined Bell Aircraft. The Bell concepts called for two-stage winged vehicles, and both of the stages would be manned. The first stage would provide the initial boost, with the second stage separating and gliding to a landing on the other side of the Earth. These aircrafts' missions would be long-range bombing missions and reconnaissance.

As the X-15 program was getting under way in 1956, the Air Force began Project HYWARD (*Hy*personic *W*eapon *a*nd *R* and *D* system). This effort would define a successor to the X-15, with a maximum speed of about Mach 12. Floyd Thompson of Langley set up an interdivision study group to examine the problems the vehicle would encounter. Becker again served as chairman, and the group also included Maxime A. Faget and Peter F. Korycinski from the X-15 study group. One of the group's first decisions was to increase the speed of the proposed vehicle from the planned Mach 12 to Mach 15. This was nearly five times the maximum speed reached by a manned aircraft to that time—Mach 3.2, by the X-2 in September 1956, and from which the pilot did not survive. Becker and the other panel members saw Mach 15 as the lowest speed for which a military boost-glide mission could be defined.

When the first formal report was issued, on January 17, 1957, the maximum speed had been raised to Mach 18. Analysis indicated that at this speed the boost-glide vehicle would experience the peak heating environment. Above Mach 18, the vehicle's flight altitude would greatly increase until it was in a near vacuum, where frictional heating was minimal.

Another consequence of increasing the maximum speed was a change in the wing/fuselage design. All Bell boost-glide designs featured a mid-fuselage wing position, like that on the X-15. Heating analysis by Becker and Korycinski indicated that a delta wing on the underside of the fuselage would provide the smallest critical heating surface for a given wing loading. As a result, the heat protection system requirements were minimized. The rest of the fuselage would be in the wing's lee side and be relatively cool.

The vehicle would be kept aloft by the centrifugal force provided by its forward speed, rather than by aerodynamic lift. This meant that a low L/D ratio could be used, which greatly reduced related problems with the vehicle's configuration, weight, and heating. The study found a 40 percent reduction in wing size increased the vehicle's range from 4700 to 5600 n miles. The HYWARD's heating and aerodynamic analysis provided the basic design concepts for hypersonic lifting-body vehicles.

The Air Force formally approved development of a boost-glide research vehicle on November 25, 1957, with initial funding of $3 million. This was called the *Dyna-Soar* program (a contraction of "dynamic soaring"; see Fig. 2.9). If initial work in 1958 and 1959 demonstrated the concept to be feasible, the effort would be accelerated. By the spring of 1959, the Dyna-Soar project goals were defined as suborbital flights to a speed of 22,000 fps, with a launch by a ballistic missile. The vehicle would be manned, could maneuver during reentry, and would land like a conventional aircraft. Secondary missions called for the Dyna-Soar to fly at orbital speeds and test military subsystems. These would be undertaken only if doing so created no adverse effects on the primary mission.

In November 1959, the Dyna-Soar development plan and the contractors who would build the aircraft were selected. The development plan was a three-step effort. Step I involved drop tests of the Dyna-Soar from a B-52, followed by suborbital flights launched by modified Titan I boosters. Step II was to entail orbital flights and the initial military subsystem testing, and step III was to be development of the operational weapon system. The contract

Fig. 2.9 Astronauts with a mockup of the X-20 Dyna-Soar.

was split between Boeing Aircraft Company for the Dyna-Soar vehicle and the Martin Company for the Titan I boosters.

But the effort was brought to a halt when J.V. Charyk, Assistant Secretary of the Air Force for Research and Development, refused to release any Dyna-Soar funding. Instead, in November 1959, Charyk ordered a reevaluation, called *Project Alpha*, which examined the three-step approach, the booster, the glider size, and the flight test objectives. The Dyna-Soar emerged from Project Alpha unchanged from the original development plan. It was not until April 1960, however, that Charyk allowed contractual arrangements for the step I efforts to proceed.

Over the next two years the basic development plan underwent several changes. The Titan I booster payload was marginal for the suborbital mission, and was replaced by the Titan II. Study of potential operational weapons systems for Dyna-Soar envisioned in step III was cancelled in early 1962, but studies of possible orbital and follow-on missions did continue. "Step IIA" examined an orbital Dyna-Soar, and "step IIB" studied potential military applications in reconnaissance, satellite inspection, and logistics.

The three-step plan would take time, however. Merging steps I (the suborbital launches) and IIA (the orbital launch study), and going directly to orbital flight would save time and money. This idea had the support of both Air Force and Boeing personnel. Secretary of Defense Robert S. McNamara then weighed in on the question, asking Air Force representatives whether Dyna-Soar represented the best expenditure of the available money. Based on McNamara's comments, the Air Staff developed a policy stating that until the Air Force had demonstrated a manned orbital capability, any discussion of military applications would be premature.

As part of this, a "streamlined" program with an orbital Dyna-Soar was approved. The booster was changed to a Titan IIIC, which could put a Dyna-Soar into orbit. The booster would be launched into orbit from Florida, fire a retro rocket just shy of a full orbit, and land at Edwards Air Force Base. Two unmanned test launches would be made, followed by manned flights. The change was made official on December 11, 1961. Reflecting the experimental nature of the program, the vehicle was renamed the "X-20" the following year. (The name *Dyna-Soar* also continued to be used interchangeably, however.) [4, pp. 219, 221, 226–228, 238, 244, 245, 266]

Despite the change in plan, the future of Dyna-Soar remained in doubt. Whereas the X-15 was a partnership between the Air Force and NASA, the Dyna-Soar was an Air Force project with only limited NASA involvement. The X-15 was also solely a research aircraft. Although use of the Dyna-Soar was focused on research, it also needed to meet Air Force operational concerns and to determine what military roles in space were feasible. This meant the Air Force had little direct control over the Dyna-Soar's future. That belonged to McNamara and the small group of people from universities and think tanks, known as the "whiz kids," who had control of policy and planning in the Pentagon [10].

In February 1962, McNamara expressed interest in using NASA's two-man Gemini spacecraft to test military missions such as manned rendezvous in space. At one point, he even suggested that the Air Force take over the Gemini program from NASA. McNamara again questioned Dyna-Soar's usefulness in January 1963, and directed that a comparison be made between Dyna-Soar and Gemini to see which was more valuable in assessing military space missions. He also asked for an evaluation of the Titan IIIC compared to other launch-vehicle options.

Following a review of the Dyna-Soar project in March 1963, McNamara expanded his directions regarding the Dyna-Soar vs. Gemini study. He said that the Air Force had put too much focus on maneuvering reentry and not enough on tests for military missions such as inspection, reconnaissance, or defense of space vehicles. His comments seemed to reflect a shift away from research goals for the Dyna-Soar, and toward an operational role, even before the vehicle had flown. McNamara added that the most feasible means of doing these tests would be on a space station supported by a ferry vehicle.

This idea eventually led to cancellation of the Dyna-Soar project on December 10, 1963, and its replacement by the Manned Orbiting Laboratory (MOL). This effort called for a Gemini spacecraft attached to a laboratory module to be launched as a single unit by a Titan III. The MOL project also went through study after study, changes in mission, funding problems, and technical difficulties before it was cancelled in 1969, also without ever having been flown [4, pp. 292, 293, 296–310].

The Dyna-Soar's end did not mean a halt to research on winged reentry. The Air Force Aerothermodynamic/elastic Structural System Environmental Test (ASSET) program flew subscale gliders. Although they resembled the Dyna-Soar, there were little if any technical exchanges between the two programs. The ASSET vehicles had a 70-deg delta wing, with a tilted underside ramp for hypersonic trim and a flat underside, and a fuselage that was a sharply tapered cone/skewed cylinder. The nose cap and wing leading edges were rounded. The vehicle was built of high-temperature metals like those planned for the Dyna-Soar. The ASSET vehicle was 68.8 in. long and 32.8 in. high, with a wingspan of 54.9 in. The project was under the direction of the Air Force Flight Dynamics Laboratory. Formal approval was given on January 31, 1961, and McDonnell Aircraft was selected as the contractor that would build the vehicles.

The ASSET program's purpose was threefold: to evaluate structural designs and materials at near-orbital speeds, to provide flutter data on aircraft panels and oscillatory pressure data on a trailing edge flap, and to investigate aerodynamic and aerothermodynamic phenomena in near-vacuum conditions. The flight data would be compared to theoretical calculations in efforts to provide reliable design information for future winged spacecraft. With the capability to simulate reentry conditions on the ground still limited, the ASSET was intended to fill the gaps in knowledge and to validate or disprove theoretical assumptions and analysis.

To undertake the research activities, two different types of ASSET vehicles were built, using different flight profiles and boosters. The aerothermodynamic structure vehicles (ASVs) were to measure temperatures, heat fluxes, and pressure distribution, as well as evaluate structural concepts and materials during planned reentry speeds of 16,000 to 18,500 fps. Of the four ASV flights, one would use a Thor IRBM booster and the others would be launched by Thor-Deltas. The ASVs weighed about 1130 lb and were fitted with a flotation bag, dye marker, homing beacon, and other recovery systems.

Two aerothermoelastic vehicles (AEVs) would be launched by Thor boosters. Their roles were to test aerodynamic control-surface unsteadiness and panel flutter at velocities of 13,000 fps. At 1225 lb, the AEVs were heavier than the ASVs due to their experiment load; however, they lacked recovery systems.

The first ASSET launch of ASV-1 came on September 18, 1963 (just under three months before Dyna-Soar's cancellation). The launch of the Thor went well, as did the reentry. However, after splashdown the flotation bag failed and the only trace of the vehicle was its dye marker. Although a postrecovery examination could not be made, telemetry data showed the ASSET vehicle had successfully withstood reentry heating.

The ASV-2, launched by a Thor-Delta on March 24, 1964, was unsuccessful. After separation the Delta second stage failed to fire. The ASSET vehicle separated, but it was in a 40-deg bank angle, beyond the allowed limit, and the destruct system fired. Modifications were made to prevent a similar Delta failure in the future.

The ASV-3 launch was successfully made on July 22, 1964. The reentry went as planned and the dye marker and flotation bag were spotted by a recovery aircraft 5 min after splashdown. When the recovered ASSET vehicle was examined at the McDonnell plant, only minor water damage was found. The hypersonic reentry had caused "significant deterioration" to the columbium wing leading edges, the only damage to the vehicle. Some of the coatings on the metal panels showed signs of being affected by reentry heating, but only minor damage. The aerodynamic data showed that ground tests had accurately predicted the vehicle's flight characteristics. The only error was in calculation of the L/D, which resulted in a 5 percent error in the vehicle's range. McDonnell engineers believed the ASV-3 could have been refurbished and flown again.

The AEV-1 and -2 launches followed on October 24 and December 8, 1964. Both were successful, with the exception of the panel flutter experiment on AEV-2. Flight data indicated the techniques used to predict flutter on hypersonic vehicles were "adequate," but that thermal requirements greatly influenced stiffness. The control surface test indicated that the unsteadiness effect was so small as to be negligible.

ASV-4, the program's final flight, was made on February 13, 1965. The flight was the fastest of the program, at 19,400 fps. The data were the highest quality of the six ASSET flights. The vehicle's aerodynamic behavior was in good agreement with predictions. The only failure, unfortunately, was that of the parachute system, and the vehicle was lost.

The quality of the aerodynamic and thermodynamic data produced in the ASSET program was such that it could serve as a standard for evaluation and comparison with ground-based test facilities. It also increased confidence in theories and ground test techniques. The ASSET missions also illustrated the benefits of free-flight testing of unmanned vehicles for reentry research. The flight data showed a winged spacecraft to be a viable option [4, pp. 449–452, 460, 463–465, 471, 510, 512–523].

THE LIFTING BODIES: FLYING WITHOUT WINGS

Even as the Dyna-Soar program began, a different concept for an airplane-like spacecraft was emerging. During the early blunt-body wind tunnel tests at Ames some of the models were positioned at an angle to the airflow. The data from these tests showed the models developed a small amount of lift as a result of the misalignment. The value was small, but it implied a wingless shape could develop enough lift to be flyable. Even a hypersonic L/D as low as 1.5 would be enough to reduce the vehicle's reentry loading from 8 g's to only 1 g. The vehicle could also maneuver during reentry and land within a large area, called a *footprint*. Because the lift came from its body, rather than from wings, it was called a *lifting body*.

The task became transforming a blunt cone into a shape that could both survive reentry and be controllable at low speeds for a runway landing. Alfred Eggers's first concept, developed in early 1957, was called the *M-1*. (The "M" stood for "manned.") The cone was cut in half so it was flat on the top and round on the bottom. The flight controls were flaps around the base of the vehicle. The M-1 shape had a hypersonic L/D of 0.5. This was followed a year later by the M-2 shape, which had a hypersonic L/D of 1.4. The M-2's subsonic stability was questionable. This issue was corrected by adding two fins, a cockpit, and a boat tail to the top and bottom of the aft fuselage.

At the same time Ames researchers were working with the new concept, Langley engineers also became interested in lifting bodies. Unlike their counterparts at Ames, those at Langley were examining concepts more refined than the blunt-cone shapes. Called *horizontal landing (HL)*, these shapes went through nine configurations before the definitive version emerged. The HL-10 was the reverse of the Ames M-2 concept. The HL-10 was a delta shape, with a flat bottom and rounded top. The fins were angled to the sides, with another fin in the center [11].

The final NACA Conference on High-Speed Aerodynamics, held at Ames in March 1958, saw three different proposals presented on "manned satellites." Maxime Faget argued for a small, simple, nonlifting capsule. Thomas Wong of Ames proposed a 30-deg half-cone lifting body, which had a peak reentry load of 2 g's and could land within an area 460 miles wide and about 700 miles long. Becker made the final presentation, arguing for a small winged vehicle incorporating ideas from the HYWARD studies.

In the end, Faget's small and simple capsule became the basis for the Mercury capsule. It not only took advantage of the proven technology used in reentry vehicle designs, but also was the lightest-weight design. Both the

Redstone and Atlas boosters had limited payload weights. Wong challenged Faget's capsule, but he did not press the lifting-body concept strongly. The Ames engineers were reluctant to be involved in a hardware development program, or in launching either manned or unmanned spacecraft. The Flight Research Center would bring the lifting-body concept to reality, in its own, particular manner [9, pp. 377–389, and 12].

By 1962, there were a number of different lifting-body concepts on various drawing boards, but no official support for any of them. NASA was going to the Moon, while the Air Force was struggling to keep the Dyna-Soar alive. Additionally, an Air Force scientific panel had concluded that lifting bodies would have controllability problems. This created a dilemma. To prove the concept, a lifting body would have to be flown. But until the concept was proven, no lifting body would fly.

R. Dale Reed, an engineer at the NASA Flight Research Center (FRC), became interested in lifting bodies in 1962. He knew the idea was viewed with skepticism, so he decided to start with flight demonstrations on a very small scale. The first tests used paper lifting bodies, flown down the hallways of the FRC main building. Reed recalled that other engineers rolled their eyes at these flights. He followed this effort with another involving a balsa wood half-cone shape similar to that of the M-2. Reed launched the model by hand into tall grass, and later from the roof of the main building. These flights proved the shape was stable. Reed next flew the model like a kite on the end of a long string. Following this, the model was towed aloft behind a gas-powered model airplane. The model was released by a timer, and made a steep descent to the ground. To show the stability of the lifting-body model, Reed's wife Donna filmed it with a movie camera.

Reed also began to assemble a small team to help him plan and "sell" a manned lifting body project to FRC management. The engineer was Dick Eldredge, who had built three gliders himself, and the research pilot was Milton O. Thompson. Additionally, Eggers, now Ames deputy center director, encouraged Reed's efforts and offered support. Reed and Eldredge originally planned to test three lifting-body designs. The plywood and fiberglass fuselage shells were to be attached to a common internal structural frame with the landing gear, control system, ejector seat, and rollover structure.

The designs featured the M2-F1 shape, the M1-L half-cone, and a Langley lenticular design. The M2-F1 was a 13-deg half-cone that used body flaps to improve stability. The M2-L was a 40-deg half-cone that had an inflatable boat tail to improve its landing L/D performance. The lenticular vehicle (which Reed's wife called the "powder puff") would extend wings after reentry for control. Thompson suggested they build the M2-F1, because this shape seemed the most flight worthy; he saw the M1-L and lenticular shapes as being too radical [11, p. 59, and 12, pp. 11–15]. One possible flaw in the lenticular lifting body concept was at high angles of attack, the airflow would separate from the blunt leading edge, causing a loss of lift. The extendable wings also would have been heavy and a complicated engineering problem.

Paul Bikle, the FRC director, gave approval for additional work. As with the model lifting body tests, this was a step-by-step process. First was a six-month feasibility study, which was completed in the fall of 1962, followed by Bikle's approval to build the M2-F1. The money came from the director's discretionary fund. Center personnel working part-time would do much of the work in-house at the FRC. The exception was the fuselage shell, because FRC personnel lacked woodworking experience. An El Mirage, California, sailplane manufacturer was issued a contract for this work. Officially, the M2-F1 was only a full-scale wind-tunnel model for testing at Ames. If it just happened to be airworthy, that was out of management's control.

The M2-F1's initial flights were done towed behind a souped-up 1963 Pontiac Catalina convertible fitted with the largest available engine and racing modifications. The car would tow the M2-F1 on a 1000-ft tow-rope across the Rogers dry lakebed at about 100 kt. The initial car tow tests, on March 1, 1963, indicated control problems. The lifting body bounced back and forth on its main wheels. The M2-F1 was shipped to Ames for extensive wind-tunnel tests to analyze the instability. After these were completed on March 15, and the control system modified, it was cleared for more car tow tests.

Car tow flights resumed on April 5, 1963. After several tries, Thompson found he could remain airborne for the full duration of the run. In all, 11 tow flights were made that day. This confirmed that the vehicle was controllable, and cleared the way for envelope expansion through the spring and into the summer that included the first free flights. Thompson climbed to about 200 ft, released the towline, and descended to a landing. These brief flights provided a limited amount of M2-F1 handling data. This amount was sufficient, however, to clear the M2-F1 to be towed aloft behind a C-47 transport. The lifting body would then be released from the towline, and glide to a landing on the lakebed.

The first M2-F1 free flight was made on August 16, 1963. The two aircraft climbed to about 5000 ft above the lakebed. Thompson released the towline and descended to the lakebed for a landing. The descent path required only minimal maneuvering in case the M2-F1 had control difficulties. The descent took under 2 min, at a 4000-ft/min descent rate. Thompson made several test maneuvers to assess the M2-F1's handling. The flight was successful, and cleared the way for further research flights and pilot checkouts.

Over the next 3 yr, between 71 and 90 air tow flights were made, along with some 400 car tows. The M2-F1 showed that a lifting body could fly, but it was far from an operational spacecraft. Before the M2-F1 was retired, the next step in the lifting-body saga was under way [11, Chap. 5].

In January 1964, the FRC issued a request to aerospace contractors for bids on the design and construction of two heavyweight lifting bodies. The first would be the M2-F2, a refined version of the original M-2 design. The "elephant ears" control surfaces were eliminated, with roll and pitch control provided by flaps on the aft fuselage. The other heavyweight lifting body would be the Langley-designed HL-10. Whereas the M2-F2 had a half-cone shape, the HL-10 had a much more refined design. It had a delta-shaped fuselage and was flat on the bottom with a rounded top. Two outboard fins and a tall center fin provided stability.

Both of the lifting bodies were powered by an XLR11 rocket engine like that used on the X-1 and D-558-II. Rather than being towed aloft, they would be dropped from a B-52 mothership. They were designed to fly at speeds under Mach 2 and at altitudes of less than 100,000 ft. By limiting the vehicles' performance, costs were kept down. This approach eliminated the need for high-temperature materials, and allowed use of existing components. The focus of the research was on the vehicles' transonic and supersonic characteristics. Hypersonic testing would require a further step.

When research flights began, the M2-F2 and HL-10 both exhibited major control problems. During the first M2-F2 glide flight, on July 12, 1966, the vehicle rolled 90 deg from side to side. Thompson was able to get the vehicle back under control, and made a safe landing. Control problems reoccurred on several later M2-F2 flights. On the HL-10's first glide flight, on December 22, 1966, NASA pilot Bruce Peterson found it to be almost uncontrollable. Peterson had to fly the HL-10 faster than planned in order to keep it under control. His touchdown speed was 322 mph, but it landed safely. Not surprisingly, the aircraft was grounded for an analysis of the control problem.

On May 10, 1967, Peterson was making the 16th M2-F2 glide flight when the vehicle began to roll from side to side at a rate of 200 deg/s. He was able to regain control, but was distracted by a helicopter in his landing path. Peterson flared and dropped the landing gear, but he was too low; the M2-F2 touched down on the lakebed before the landing gear had fully extended. The vehicle bounced into the air, rolled, and tumbled six times, ripping off one fin, the canopy, and the landing gear, before coming to rest on its back and remaining fin. Peterson was pulled from the wreckage with a fractured skull, severe facial injuries, a broken wrist, and damage to his right eye. He underwent reconstructive surgery, but lost his eye due to an infection [12, pp. 106–108].

The lifting body program was now in disarray. The M2-F2 had suffered major damage in the crash, and the HL-10 was deemed unsafe to fly. This was a major disappointment, because it was expected that the HL-10 would have the best handling characteristics of the various lifting bodies.

Both Northrop and FRC engineers wanted to rebuild the vehicle, but NASA Headquarters was very reluctant. The FRC management sent the M2-F2 fuselage to Northrop, where it was placed in the assembly jigs. The reason given was so the damaged vehicle could be "inspected." In reality, NASA and Northrop engineers were slowly rebuilding the lifting body. NASA Headquarters did not formally announce the M2-F2 would be rebuilt until January 28, 1969. By that time, significant "inspection" work had been done.

Meanwhile, to understand the HL-10's control problem, an extensive series of wind-tunnel tests was conducted. The problem was finally traced to flow separation on the upper aft part of the fuselage when at high angles of attack. The solution was a small change in the camber of the outboard fins' leading edges. This was accomplished by adding a fiberglass "glove."

The HL-10 flew again on March 15, 1968, and exhibited none of the problems seen in the initial flight. Eight more glide flights followed before the XLR11 rocket engine was added. The first successful powered flight was made on November 13, 1968. The following spring, on May 9, 1969, the HL-10 made the first supersonic flight of a manned lifting body. The speed and altitude build-up continued for nearly a year. On February 18, 1970, Major Pete Hoag reached a maximum speed of Mach 1.86. NASA research pilot Bill Dana reached an altitude of 90,303 ft in the HL-10 nine days later. These were the highest and fastest flights made by a manned lifting body. These high-speed/high-altitude flights simulated the final part of a reentry from Earth orbit [12, pp. 114–117]. There was also another element in the lifting body effort: the SV-5J jet-powered lifting body trainers. The Aerospace Research Pilot School commandant, Colonel Charles Yeager, had been impressed with his flight in the M2-F1, and asked for a lifting body trainer. Although two fuselage shells were built by Martin, the aircraft had a large "dead-man zone," where if the engine failed immediately after takeoff there was no way for the pilot to recover.

This period also saw the debut of a third lifting-body design. This was developed by Martin for the Air Force and resembled a potato with fins. The Air Force had several areas of interest in regard to lifting bodies. They could be used to return film from reconnaissance satellites to a precise location for recovery. A lifting body could also serve as a resupply vehicle for a manned military space station, which might follow the MOL. The Air Force was a partner with NASA in the testing of the M2-F2 and the HL-10. The Martin lifting body served both the Air Force/NASA partnership and specific Air Force needs.

The first element of a wide-ranging effort to test the new lifting-body shape was the X-23A PRIME (Precise Recovery Including Maneuvering Entry) vehicle. Also called the SV-5D, this was a subscale vehicle to be boosted to near-orbital speeds by an Atlas rocket to test its reentry and hypersonic characteristics—as the earlier ASSET flights had been designed to do—as well as its ability to maneuver to a specific landing site. The second program element involved a rocket-powered manned lifting body that became the X-24A. This vehicle would cover the subsonic, transonic, and low supersonic flight envelope.

The first X-23A launch was made on December 21, 1966, from Vandenberg Air Force Base. The flight plan called for a simulated reentry from low orbit with no cross-range maneuvering. The vehicle's launch, reentry, and initial parachute deployment went as planned. However, the main parachute did not deploy properly, and the X-23A was destroyed when it hit the Pacific Ocean.

The second launch followed on March 5, 1967, and successfully demonstrated a 499-n-mile cross-range maneuver. This meant that the X-23's control system would maneuver the vehicle using its aerodynamic control surfaces from a ballistic reentry toward a specific point over the Pacific, which would facilitate a reconnaissance satellite returning film to Earth even though the satellite did not pass directly over the recovery zone. The reentry and parachute deployment were successful, but the planned midair recovery by a JC-130B aircraft could not be attempted. The X-23A splashed down, but its flotation equipment tore free and the vehicle sank.

This string of recovery failures ended with the third and final flight. Launched on April 19, 1967, the flight simulated a low-orbit reentry with a maximum cross-range maneuver. During the reentry, the vehicle generated 710 miles of cross-range. A recovery aircraft caught the parachute at 12,000 ft and reeled the X-23A aboard. This brought the X-23A program to a close. Attention then shifted to manned X-24A flight tests [4, pp. 691–703].

Not until two years later did the X-24A fly its first missions. Like the other manned lifting bodies, the X-24A was dropped from the B-52 mothership. Its first glide flight was made on April 17, 1969, during which the control system indicated a need for modifications. A total of nine glide flights were made. The XLR11 engine was then installed for the powered flights. The first of 19 powered X-24A flights was flown on March 19, 1970. The vehicle's first supersonic flight was made on October 14, 1970—the 23rd anniversary of the first supersonic flight by Captain Charles E. Yeager in the X-1. Several of the X-24A flights suffered XLR11 engine problems, however, including its last flight, on June 4, 1971.

The last of the initial generation of lifting bodies was also the first to fly. Now fully "inspected," repaired, and modified, the M2-F3 made its first glide flight on June 2, 1970, with Bill Dana as pilot. The most visible change was the addition of a small center fin. The effect was readily apparent to the pilots. The unmodified M2-F2 required constant pilot inputs. After the fin was added, handling was greatly improved. Jerauld Gentry, who had flown the M2-F1,

M2-F2, HL-10, and X-24A, made a glide flight on February 9, 1971, to assess the modifications. He had been highly critical of the M2-F2's original handling, but after his flight in the M2-F3, Gentry said the vehicle flew as well as the HL-10, considered the best-handling of the lifting bodies. The first powered M2-F3 flight was made on November 25, 1970, also by Dana. The vehicle was retired in December 1972, after a total of 27 flights (plus 16 as the M2-F2) [8, pp. 154–158].

SUMMARY

When long-range ballistic missiles were first considered in the mid-1940s, reentry was judged the most difficult technical roadblock to their development. Solving the problem required not only the conceptual breakthrough of a blunt-body reentry vehicle, but also the heat-sink and ablative heat shields, and ground-test techniques such as the shock tube, shock tunnels, arc tunnel, and related instrumentation. The task also had room for quick and dirty approaches such as using a rocket exhaust to test ablative heat shield materials.

A similar pattern can be seen in the reentry tests. The initial tests with the X-17 looked at the basic physics of reentry using different shapes, rather than direct simulations of operational reentry profiles. Theoretical calculations, shock tube and other ground-test data, and rocket tests were combined to develop the operational heat-sink reentry vehicles used on the Thor and early Atlas missiles.

With ablative reentry vehicles, a similar step-by-step approach was also used. When the U.S. Army was developing the Jupiter missile, they first used subscale models launched by modified Redstone rockets, followed by full-scale dummy warheads launched by Jupiter missiles. The Air Force used a similar approach with the testing of the RVX-1, RVX-2 and RVX-2A shapes. The data from the test flights were then used in the design of the Mark 4 and Mark 6 reentry vehicles.

Different approaches were necessary for manned hypersonic flight. The X-15 constituted a major leap over previous research aircraft, yet the research was conducted in a systematic step-by-step manner. The vehicle represented a three-fold jump in existing aircraft speed and altitude performance. At the same time, however, the X-15 also represented a continuation of the rocket-powered X-planes. Like the earlier X-1 series, the D-558-II, and the X-2, it was launched from a mothership, used a rocket engine, and glided down to a lakebed landing.

The boost-glide vehicles and the X-20 Dyna-Soar represented departures from past practice. These entailed ground launches, rather than air launches, and the increases in technical demands and vehicle performance of these vehicles were far greater than those of the X-15. After the Dyna-Soar was approved, its development was made more difficult by a review of the design and the change from a suborbital profile to a single-orbit mission. More serious were the changes in the program's goals and McNamara's management theories.

The manned lifting-body program showed the advantages of step-by-step development, a concept very few believed in at the time. Although a winged spacecraft seemed technologically feasible, an orbital vehicle looking like a bathtub or a finned potato was another matter. To overcome doubts, more than wind-tunnel testing was required. Flight tests were needed, and they started with a paper airplane. Each subsequent step—from a wooden model tossed into the grass and towed aloft behind a model airplane, to a feasibility study, to the car and airplane tow tests of the M2-F1, to the heavyweight lifting bodies, and, finally, the X-23A suborbital launches—was necessary to show that the concept was valid.

In the process, a body of knowledge was assembled from the widely disparate elements. Taken together, the theoretical studies, ground tests, low-speed atmospheric flights, and suborbital reentry tests cleared up many of the unknowns of hypersonic flight and atmospheric reentry. The flight behavior of a hypersonic vehicle design could now be accurately predicted using theoretical calculations and ground testing. Stable hypersonic, supersonic, transonic, and low-speed stability were demonstrated in actual flight. A vehicle could also make cross-range maneuvers during reentry and reach a specific recovery area, allowing a glide to landing on a runway. The potential was that a winged or lifting body could provide access to space on a lower-cost basis than could expendable rockets and capsules.

The knowledge base developed in the research effort was applied to the design of the space shuttle. Although the orbiter was a winged-body vehicle rather than a lifting body, the experience gained during the previous decades made its design and construction possible. Unlike the Bell boost glide vehicles and some of the early round 3 concepts, the shuttle was not a full two-stage vehicle (i.e., an orbital airplane carried aloft by a larger reusable winged booster or an expendable rocket). An expendable external tank and two solid-rocket boosters mated to the orbiter served as the first stage [4, pp. 695, 696]. Because the space shuttle main engines continued to burn, using the liquid oxygen and liquid hydrogen from the external tank, after the solid rocket boosters burned out and were jettisoned, this configuration was known as a "stage-and-a-half."

The space shuttle's arrival, although a landmark event, still fell short of realizing the ultimate step—a fully reusable single-stage-to-orbit (SSTO) vehicle. This as-yet-unrealized vehicle would see rockets replaced with air-breathing engines. The result would be the ultimate expression of aircraft development: a plane that could roll down a runway, take off, and fly into orbit.

REFERENCES

[1] Bonestell, C., and Ley, W., *The Conquest of Space*, Viking Press, New York, 1958.

[2] Allen, J. E., *Aerodynamics: The Science of Air in Motion*, 2nd ed., Granada, London, 1982, pp. 165–166.

[3] Heppenheimer, T. A., *Facing the Heat Barrier: A History of Hypersonics*, NASA SP-2007-4332, Washington, DC, 2007, pp. 29–30.

[4] Hallion, R. P. (ed.), *The Hypersonic Revolution: Case Studies in the History of Hypersonic Technology*, Vol. I, Air Force History and Museums Program, Washington, DC, 1998, pp. lxvii–lxxi.

[5] Burgess, C., and Dobbs, C., *Animals in Space: From Research Rockets to the Space Shuttle*, Praxis, Chichester, UK, 2007, Chap. 5.

[6] Encyclopedia Astronautica, Thor Able, http://www.astronautix.com/lvs/thorable.htm.

[7] Stumpf, D. K., *Titan II: A History of a Cold War Missile Program*, University of Arkansas Press, Fayetteville, 2000, pp. 58–61.

[8] Hallion, R. P., and Gorn, M. H., *On the Frontier: Experimental Flight at NASA Dryden*, Smithsonian Books, Washington, DC, 2002, pp. 101–102.

[9] Hansen, J. R., *Engineer in Charge: A History of the Langley Aeronautical Laboratory, 1917–1958*, NASA SP-4305, Washington, DC, 1987, pp. 269, 270, 353.

[10] Byrne, J. A., *The Whiz Kids: Ten Founding Fathers of American Business—And the Legacy They Left Us*, Current Doubleday, New York, 1993, Chap. 26.

[11] Thompson, M. O., and Peebles, D. *Flying Without Wings: NASA Lifting Bodies and the Birth of the Space Shuttle*, Smithsonian Institution Press, Washington, DC, 1999, pp. 60–62.

[12] Reed, D. R., and Lister, D., *Wingless Flight: The Lifting Body Story*, NASA SP-4220, Washington, DC, 1997, pp. 8–10.

Chapter 3

HYPERSONIC AIR-BREATHING PROPULSION

> The simplest, however, of all conceivable flying machines would be a cylinder
> blowing out gas in the rear and driving itself along on the principle of the rocket. . . .
>
> *Scientific American, September 8, 1860*

Rocket engines had two basic limitations when it came to hypersonic flight. First, their burn time was short—measured in minutes. Second, and more important, rockets require both the fuel and the oxidizer to be carried internally. Although this makes them independent of the atmosphere, it greatly increases vehicle weight, thus rendering them inefficient.

Specific impulse is the measurement of the efficiency of a reaction motor such as a jet, ramjet, or rocket. It is defined as the thrust produced by the engine per unit weight of propellant burned per unit time. Specific impulse is thrust in pounds divided by propellant flow in pounds-mass/seconds. The units are pounds-second/pounds-mass, which is abbreviated as "seconds" as the unit of measure [1].*

Conventional jet engines have a specific impulse of 5980 seconds. In contrast, the space shuttle main engine, which burns hydrogen and oxygen, has a specific impulse at sea level of 363 seconds. The high specific impulse of a jet aircraft is due to its use of oxygen from the atmosphere to burn the kerosene, rather than carrying the supply internally. Additionally, the thrust of a jet engine per unit flow rate of fuel is some 20 times higher than that of a rocket engine.

On the negative side, a jet engine's chamber pressure is typically 20 atm. A rocket, in contrast, may have a chamber pressure of 200 atm. Another shortcoming of a jet engine is its internal heat limits. The airflow velocity into a jet engine must be below the speed of sound. Aircraft-inlet shock waves compress the air and slow it to subsonic velocities before it reaches the turbine blades. During this compression, the kinetic energy from the velocity of the air's mass is converted into pressure. This is referred to as *pressure recovery*. In the process, the air also becomes hotter. The engine will be damaged if its compressor inlet temperature becomes too great. After the air is further compressed by the

---

*Additional information from Charles Rogers.

turbine blades, fuel is mixed into the compressed air and ignited. The increased heat from the combustion raises the temperature of the engine structure. This establishes another limit, the exhaust gas temperature. Again, too high a temperature will result in engine damage. Typically, the maximum value is around 1700 K.

The situation with a rocket engine is different. Propellant flowing through small tubes inside the combustion chamber and nozzle structure keeps the rocket's temperature below the melting point of the vehicle's structural materials. Many liquid-fuel rockets use propellants that are super-cooled gases stored on board in a liquid state; circulating them around a rocket engine's nozzle cools the nozzle even as combustion occurs. Even though the rocket exhaust temperature is typically about 4000 K, the rocket engine's structure remains below temperatures that would weaken its structure. The high temperatures and pressures of a rocket produce very high exhaust velocities for a given exhaust flow rate. A jet engine's exhaust velocity, due to its lower temperature and pressure limitations, is greatly inferior to that of a rocket.

A final shortcoming of a jet engine is that as the aircraft increases altitude its thrust drops due to lower ambient air pressure. The air density finally becomes too low for the combustion of the fuel to continue, and the jet engine flames out. In contrast, a rocket engine's specific impulse increases as it climbs. The highest specific impulse occurs in a vacuum. The original main engine of the space shuttle had a specific impulse at sea level of only 363 seconds, but reached 453 seconds in the vacuum of space, where, the rocket's exhaust expanded unimpeded. Within the atmosphere, energy is lost due to the surrounding atmosphere slowing the velocity of the rocket's exhaust gases. A later version of the space shuttle's main engine used an improved Pratt & Whitney turbopump and a larger throat area. This engine had a sea-level specific impulse of 366 seconds and a vacuum specific impulse of 452.5 seconds. The sea-level specific impulse was increased from the original version due to the increased throat area, which decreased the engine's expansion ratio [2, 3].

THE DEVELOPMENT OF JET PROPULSION

Although rockets were invented in 10th-century China, air-breathing "jet" propulsion had its origins in the early 20th century. The first air-breathing jet engine was conceived by Frenchman René Lorin in his work on the theoretical calculations for a ramjet, made before World War I. Albert Fono, a German researcher, patented several ramjet designs in 1928 for propelling supersonic aircraft. His designs were the first to show all the basic details of the modern ramjet engine. These included the inlet and spike, combustion chamber, fuel injectors, flameholder, and nozzle. René Leduc of France undertook extensive research on ramjet designs in the 1930s. This included refining the inlet for better performance and conducting ground tests, which replicated the airflow speeds and temperatures of actual flight [4].

The simplicity of a ramjet was part of its appeal. Indeed, it is often described as the simplest of engines, because there are no moving parts. The ramjet lacks the spinning compressor blades and other mechanical components of a turbojet engine. Without these components, the ramjet can withstand greater temperatures than a turbojet; however, this lack of a turbine also imposed a major operational constraint on a ramjet. A jet engine's rotating fan blades pull in air to begin the combustion process. A ramjet, however, cannot develop thrust when standing still. It must be moving through the air before combustion can begin.

Some ramjets are designed with an inlet that has a cone-shaped centerbody. When supersonic airflow strikes the centerbody, oblique (i.e., sloping) and normal (i.e., 90-deg) shock waves are formed. This slows and compresses the airflow as it passes through the subsonic diffuser and into the combustion chamber. Injectors spray fuel into the combustion chamber, and the fuel mixes with the subsonic airflow and is ignited, generating hot exhaust gases. To maintain stable combustion, flameholders are used. These sometimes take the form of several rings of different diameters, ensuring a flame front that covers as much of the combustion chamber diameter as possible. The combustion chamber's diameter becomes smaller and then expands to form the nozzle. This speeds the velocity of exhaust gases, and generates thrust [5].[†]

The basic design of the ramjet was understood by the end of the 1930s. Yet it was a propulsion system far ahead of its time. This was an era when all-metal propeller aircraft represented the cutting edge of aviation technology. Ramjets were most efficient at supersonic speeds, which were far beyond the performance of existing airplanes. Many engineers and scientists also thought jet propulsion was not feasible.

THE RAMJET IN AMERICA

Despite these doubts, initial ramjet studies and tests were done in the United States beginning in the 1940s, which provided basic data on their operation. Stewart Way of Westinghouse was the first U.S. researcher to note the potential of the ramjet. He published an analysis of the performance of ramjet-powered subsonic aircraft in July 1941. As part of this work, Way used

---

[†]Although the word *jet* is now associated only with turbojet engines, the term was originally applied to both air-breathing and rocket engines that used hot exhaust gases to produce thrust. Turbojets, ducted fans, ramjets, pulsejets, and rockets were all once called "jets." The Jet Propulsion Laboratory was originally established at the California Institute of Technology for research on liquid fuel rockets, not air-breathing engines. However, because rockets were considered too "Buck Rogers," the (slightly) more respectable word *jet* was used.

a ramjet that employed electrically heated air, rather than burning fuel, to generate positive thrust.

John V. Becker and D. D. Baals of the NACA's Langley laboratory analyzed ramjet efficiency and tested a ramjet in a wind tunnel. Their September 1942 report focused on subsonic aircraft. NACA research on turbojet engines also had application to ramjets. This included work on flame stabilization in high velocity airflows of up to 120 fps using spray and vaporization chambers [6].

By this time, the person most associated with early ramjet technology in the United States was Roy Marquardt, Director of Aeronautical Research at the University of Southern California. In 1944, the Navy Bureau of Aeronautics issued a contract for development of a 20-in.-diameter subsonic ramjet engine. Marquardt won the contract, but because USC lacked manufacturing and test facilities, he established the Marquardt Aircraft Company as a subcontractor to the university. Marquardt and nine friends obtained $1000 as start-up money and opened the company on November 3, 1944.

The firm had minimal production capability, tooling, and workspace, and was competing with other industries in wartime for skilled craftsmen and technicians. Test facilities were also a problem, because it was necessary to produce airflow that would simulate both the high speeds and the high temperatures of flight. To test the 20-in. ramjet, Marquardt went to the Kaiser Steel mill in Fontana, California. The plant had a 13,000-hp blast furnace compressor, which could supply the high-speed air for the tests. A "tee" connector was added to the plant's main air duct, which allowed the airflow to pass through the ramjet engine during test runs.

Marquardt and his team were able to deliver the 20-in. ramjet to the Navy in 1945. The Army Air Forces (AAF) was also interested, and ordered several 20-in.- and 30-in.-diameter subsonic ramjets. The AAF was the first to fly the 20-in. ramjets. In early 1946, two of the ramjets were fitted to the wingtips of a P-51D Mustang. The P-51D took off under prop power and climbed to altitude. The ramjets were then ignited, creating a flame longer than the Mustang's fuselage. The aircraft's speed increased by 40 mph.

Soon after, the Navy began a more ambitious ramjet effort. Initial tests in 1946 involved a 20-in. ramjet attached to an F7F Tigercat, a twin-propeller carrier fighter. The ramjet weighed just over 100 lb but produced as much thrust as a turbojet engine weighing 10 times more. This was followed in 1947 by additional Navy ramjet flight tests using a Bell P-83 (an early jet prototype) and an F-82 Twin Mustang.

The Air Force continued its early ramjet efforts, and in 1948 a pair of 30-in. engines was mounted on the wing tips of an F-80 jet fighter. The aircraft took off and accelerated to 400 mph under the thrust of its jet engine. The pilot ignited the ramjets and idled the jet engine. The F-80 was now propelled solely by the 4000 lb of thrust produced by the ramjets, allowing it to accelerate under their power. The same year, the Air Force contracted for a 48-in. ramjet. The engine was carried under a B-26 on 17 test flights.

NACA researchers at Langley were also undertaking subscale ramjet free-flight tests. In 1947, a NACA research ramjet was ground-launched by a small rocket booster. After separating from the rocket, the ramjet ignited and accelerated to Mach 1.45. This test was followed in 1949 and 1950 by tests of the more advanced F23 ramjet, which burned ethylene. As in the 1947 test, the F23 was launched by a solid rocket. The ramjet reached a burnout speed of Mach 3.12 at an altitude of 67,200 ft on June 6, 1950. It coasted upward to a peak altitude of 159,000 ft before descending into the Atlantic Ocean.

Later ramjet launches were made from F-82 Twin Mustang and F2H-2 Banshee fighters. One type of ramjet tested in this manner featured a Lewis Laboratory design with a 9.75-in. diameter and a length of 118.72 in. The ramjet was fueled by pentaborane. Rather than a large rocket booster attached to the vehicle's tail, the Lewis ramjet used a small rocket fitted into its nozzle. Once the rocket completed its burn, the ramjet fired and ejected the spent rocket casing. The ramjet was launched on February 23, 1956, from an F2H-2 flying at 42,000 ft. The engine reached a speed of Mach 3.02.

The highest speed reached by a NACA-sponsored ramjet was a ground-launched F29 engine, fired from Wallop Island in 1958. This burned a slurry mixture of 50 percent JP-4 jet fuel and 50 percent magnesium–magnesium oxide–aluminum blend. The maximum speed reached by this ramjet was Mach 3.84—significantly faster than the Mach 3.2 reached in September 1956 by Capt. Mel Apt in the Bell X-2 rocket plane.

The Air Force recognized the potential for ramjet applications in both air defense and strike missions, and requested Marquardt develop a 20-in. supersonic ramjet capable of reaching Mach 2.5. Although the basic design of the new engines (inlet with cone, subsonic airflow, combustor, flameholder, and nozzle) was similar to that of earlier ramjets, their improved performance meant greater demands on the components.

One necessary alteration would involve the test vehicle carrying the ramjet. Although earlier ramjets could be attached to conventional propeller or jet aircraft, the new ramjet would fly for sustained periods at Mach numbers well above those of any existing manned aircraft. A specially built unmanned test vehicle, the Lockheed X-7, had to be built to carry the new high-speed ramjets [7].‡

The X-7 had a cylindrical fuselage with a sharply pointed nose (see Fig. 3.1). The thin wings were straight, with a low aspect ratio similar to those of the

---

‡Given the NACA ramjet research and the preference for a jet engine over rockets as a propulsion system, the failure to develop a ramjet-powered research aircraft might be surprising. At the same time as the X-1 was being developed, Langley did studies of a ramjet research aircraft and recommended one be built. This proposal did not gain favor with the Air Force or Navy, and no action was taken. The X-7 was the only significant ramjet research vehicle to be built in the 1950s.

Fig. 3.1 An X-7 being prepared for launch from a B-29 mothership.

X-1 series. The horizontal stabilizers were positioned low on the vertical fin. The vehicle was made of steel, for protection against aerodynamic heating. The X-7 flew a preprogrammed mission profile. The ramjet was mounted beneath the fuselage, rather than internally. This allowed engines of different sizes to be tested without modifying the X-7. The vehicle was designed to test ramjet engines at speeds near Mach 3; a modified version reached test speeds of Mach 4.

Although the vehicle was conceptually simple, it had an unusual design feature. Test data were transmitted to the ground during flight, but engineers also wanted to recover the vehicle to examine the ramjet after the flight, so the X-7 was built with a parachute that deployed from the aft fuselage. The vehicle descended until the long nose spike hit the ground and stuck; when recovery crews arrived, the X-7 would be vertical, its nose buried in the ground and the heat-scorched fuselage and wings wrapped in the ribbon parachute. But after refurbishment, an X-7 could be flown again.

Like the manned rocket planes, the X-7 was air-launched. The drop plane was a modified B-29 flown from Holloman Air Force Base in New Mexico. A large rocket booster attached to the X-7's tail ignited after the drop, and boosted the vehicle to separation speed. The length of the booster, however, which was nearly as long as the X-7 itself, as well as the size of its large fins, prevented the assembly from being carried under the B-29's fuselage. Instead, it was mounted under the bomber's left wing, between the number 1 and 2 piston engines.

Although there were problems with the X-7's rocket boosters during initial flights, it soon proved successful. Two flights were made, one with a 20-in. ramjet built by the Marquardt Company and the other with a 20-in. ramjet built by Wright Aeronautical. Tests of the Marquardt 28-in. supersonic ramjets were then begun with the X-7. These were to provide data for production of the 28-in.-diameter RJ43-MA-3 ramjet for the Bomarc A surface-to-air missile, an unmanned winged missile with a top speed of Mach 2.8 at 70,000 ft. The Bomarc was launched vertically from the ground by a rocket in its fuselage.

Once at the necessary speed, a pair of ramjets mounted under the fuselage was ignited, powering the missile toward the target aircraft.

The first flight of the 28-in. test ramjet on the X-7 was made December 17, 1952. This resulted in a burn time of 10 s, followed by a blowout that caused the combustion to stop, causing minor damage to the engine. The next flight, on February 2, 1953, also ended prematurely. The April 8, 1953, mission had a burn time of 20 s and reached a top speed of Mach 2.6—faster than any rocket plane had flown until that time. However, the engine's fuel system failed, causing the thrust to decay, cutting short the flight. Another X-7 flight on September 12 also ended prematurely. The flight on December 8, 1953, failed due to problems with the ramjet's graphite exit nozzle, leading to its replacement with a metal nozzle [7, p. 3, and 8].

The difficulties reflected the change from basic propulsion research to development of a ramjet that could be used operationally. The research ramjets either operated at a specific point design of speed and altitude or had very narrow operating envelopes. Flight outside these speeds was referred to as off design performance. An operational system would have to operate in a much wider range of Mach numbers, altitudes, and maneuvers. As a result, there would be considerable variations in the ramjet's aerodynamic and stagnation temperatures, the internal pressures, structural loads, and other flight conditions. These would affect the engine's operation, rendering it inefficient or creating malfunctions such as blowouts. The challenge for the engine's designer was to develop systems that functioned properly despite these fluctuating conditions [7, pp. 6–8].

The original X-7 was designed for speeds of Mach 3. In 1954, after the initial tests of the 28-in. Marquardt ramjet, development of a modified X-7 design began. This iteration had a slightly longer fuselage and shorter-span wings. The most important change was elimination of the large booster mounted on the tail, which was replaced by two smaller boosters positioned under the wings. The recovery parachute and nose spike were unaltered. The new X-7 could be carried under the fuselage of a B-50 bomber, rather than under a wing as with the B-29. The X-7 was placed in a pit, the B-50 was rolled over it, and the X-7 was raised into position and attached to the mothership.

The new X-7 configuration was used to set several ramjet speed records. The flight of August 29, 1957, powered by a 28-in. Marquardt ramjet engine, was one such flight. Following launch from the B-50 at 33,700 ft, the two solid rocket boosters ignited and accelerated the vehicle to a separation speed of Mach 2.25. The ramjet ignited and the X-7 climbed to 54,500 ft, leveled off, entered a shallow descent, and began the speed run. The ramjet burned for 91 s, shutting down when fuel was exhausted. The peak speed, reached at engine shutdown, was Mach 3.95. The X-7 recovery system worked properly. Postflight examination indicated the skin temperatures were above 600°F, high enough to cause internal damage to the ramjet.

The fastest X-7 flight was made on April 27, 1958. The X-7 carried a 36-in.-diameter Marquardt ramjet engine. The B-50's drop altitude was lower, at 28,500 ft, and the two booster rockets' burnout speed was Mach 1.99. The ramjet ignited and accelerated the X-7 until it reached a top speed of Mach 4.31, or 2881 mph. Just three seconds later the thrust dropped suddenly, and the X-7 broke up in flight.

Work was also under way on a ramjet-powered cruise missile called the Navaho. This was a reflection of the missile development plan formulated by Theodore von Kármán. It was to be the final step between pilotless aircraft and long-range ballistic missiles. The Navaho had delta wings, two delta canards at the nose, and two vertical fins. The vehicle was launched by a liquid-fueled booster. Once the Navaho separated, its two XRJ47-W-5 ramjets ignited. They were some 40 in. in diameter and built by Wright Aeronautical, Marquardt's chief competition in the ramjet field.

The contract was awarded in October 1951, with the first engine mockup delivered in March 1952. The engine design began flight testing in April 1952 using a 20-in. subscale ramjet on X-7s. A total of seven X-7 flights were made with the engine. Three were failures, three were partially successful, and one was a completely successful flight. The test results were indicative of problems to come.

The first Navaho launch was made on November 6, 1956, from Cape Canaveral. It lifted off the pad, but a guidance system failure caused the vehicle to go out of control after only 26 s. Technical problems delayed the second Navaho launch until March 22, 1957. The booster was unable to reach ramjet-ignition speed, and the vehicle crashed into the Atlantic.

By this time several Thor and Jupiter IRBM launches had been made. Although all the launches were failures, they represented the potential for much more capable weapons. This potential was underscored on June 11, 1957, when the first Atlas A launch was made. It also failed shortly after launch, but Navaho's fate was sealed. A high-speed cruise missile was now an outmoded concept. On July 13, 1957, the Air Force cancelled the Navaho program [9].

SUPERSONIC COMBUSTION RAMJETS

Although the X-7 Mach 4.31 flight was the fastest made by a ramjet-powered vehicle, it also illustrated the performance limitations of this type of engine. Within seconds of reaching the peak speed, the ramjet suffered a failure and the X-7 broke up. A speed of Mach 6 or 7 was considered the maximum attainable by a ramjet. This was due to a basic design component—like turbojet engines, ramjets required that incoming air be slowed to subsonic speed.

The problem with a conventional ramjet operating at hypersonic speeds was that slowing the air to subsonic speeds resulted not only in significant drag (potentially greater than the thrust produced), but also in raising the air temperature within the ramjet. The temperature became so high that when the hydrogen fuel was injected into the burner, the atoms dissociated. This process actually *reduced* the temperature of the air/fuel mixture. Further cooling occurred as heat was absorbed into the engine structure. To prevent the engine structure from overheating, a lean fuel mixture would have to be used, reducing thrust. Additionally, because the atoms in the air/fuel mixture were dissociating, combustion reactions would not be completed. The combustion took place in an expanding flow within the nozzle. As the volume expanded, exhaust gas temperature dropped—a reversal of the compression raising the air's temperature. As a result, a point was reached at which the burner no longer performed any useful function, because no heat was added to the exhaust gases [10, 11].

These shortcomings could potentially be eliminated, however, if the combustion occurred in a supersonic airflow. The pressure and temperature within the burner would be great enough to allow ignition and burning of the fuel, although the temperatures would not be so high that fuel-molecule dissociation occurred. The supersonic airflow and reduced internal temperature also would ease the heat transfer to the engine's structure. Such an engine's potential was awe-inspiring. In theory, it could power an airplane into Earth orbit [10, pp. 32, 33].

The quest to build such an engine did not start with goals so lofty, however. Rather, it sprung from efforts to research the possibility of burning fuel in the supersonic airflow on a wing to generate thrust or added lift. Irving Pinskel and John Serafini at the NACA Lewis Flight Propulsion Laboratory made the first theoretical calculations in 1950 using airflow over simplified two-dimensional shapes, such as a wing or axisymmetric body. In 1952, the researchers analyzed the effects of heat being added to the supersonic airflow under a wing. The results suggested that burning the fuel externally would add more lift or thrust than a similar amount of fuel being burned in a jet engine afterburner [11, p. 100].

Their calculations did not answer the fundamental question of whether supersonic combustion was even possible; it was considered impractical by many due to the difficulty of stabilizing the flame front. Edward A. Fletcher, Robert G. Dorsch, and Melvin Gerstein, also at Lewis, decided to test the possibility in 1953. They selected aluminum borohydride as the fuel for the wind-tunnel tests, because it had been shown to easily ignite in air, even in extreme conditions. The goal of their tests was to determine the feasibility of supersonic combustion and to measure the resulting disruption to the airflow within the tunnel.

The transient runs were made at speeds of Mach 1.5, 2, 3, and 4, and were often explosive in nature. A few of the runs showed subsonic flow due to shocks generated by the combustion. In contrast, all the steady-state runs, made at speeds of Mach 1.5, 2, and 3, were successful, with the aluminum borohydride burning smoothly for as long as several seconds. This was accomplished without a flameholder. There was no evidence that the combustion caused shock waves to form, and in all of the steady-state runs the airflow remained supersonic throughout the burn [12].

The next major step came when NACA researchers Richard J. Weber and John B. MacKay produced a fundamental design analysis of what they called a "supersonic combustion ramjet." This was later abbreviated as "scramjet." (Weber and MacKay used the acronym *SCRJ* for supersonic combustion ramjet. Scramjet came later.) The Weber and MacKay study, NACA TN 4386, published in 1958, concluded that at speeds of about Mach 5 to 7, the ramjet and scramjet were about equal in performance. At speeds above Mach 7, their calculations indicated that the scramjet's performance would be superior.

Weber and MacKay's study was a landmark in the history of scramjet development because it defined problems designers would struggle with for the rest of the 20th century. These included combustion temperatures, combustor gas dynamics effects, wall cooling, frictional losses, and nozzle performance. Performance losses due to molecular dissociation and the failure of the atoms to recombine were considered. Design issues examined included inlet configurations and their efficiencies, the need for a divergent combustor shape, and the wall pressure forces on such a design. Despite their efforts, Weber and MacKay still faced a basic unknown. They noted: "The present study starts with the basic assumption that stable supersonic combustion in an engine is possible" [13, 14].

What had been a minor effort by a few individuals was now attracting the attention of major institutions. In 1957, William Avery and Gordon Dugger at the Applied Physics Laboratory of the Johns Hopkins University began a study of scramjets. They first focused on external burning. Their initial engine bore the shape of a broad inverted triangle, with the shallow forward ramp compressing airflow. Fuel was injected at the peak. To ease problems of ignition and combustion, a very short cowling was placed at the apex. Once the fuel was ignited, hot exhaust gas flowed over the raised rear slope and expanded to produce both thrust and lift. The tests were run in a Mach 5 wind tunnel.

To run the tests, a young researcher named Frederick Billig was hired. His first task was to prove the basic assumption that supersonic combustion was really possible. The initial fuel Billig tried in 1958 was hydrogen, with an electrical spark as the ignition source. This was unsuccessful, and he switched to triethyl aluminum, which spontaneously ignited on contact with air. A successful test was made on March 5, 1959, when the triethyl aluminum ignited with a large white flame. Subsequent runs showed that much of the combustion was occurring within the supersonic airflow.

The following year, Billig published a study comparing kerosene-fueled ramjet and scramjet performance. The results largely confirmed the results of the Weber and MacKay study, which had assumed the use of hydrogen. Billig found that a ramjet was limited to a Mach-6-to-8 speed range, and that a scramjet would be superior above this range.

Another milestone in the early history of scramjet development occurred in September 1958, the same time the Weber and MacKay study was published. This was the First International Congress in the Aeronautical Sciences, held in Madrid, Spain. At this meeting the major figure in the first generation of scramjet researchers made a brief appearance. Antonio Ferri was on the faculty of the Brooklyn Polytechnic Institute. During the meeting, he reported on work being done at the school, and confirmed that steady combustion without strong shocks had been achieved in a Mach 3 airflow.

The scientist who would become synonymous with early scramjet research had an unusual background. He held PhDs in electrical engineering (1934) and aeronautical engineering (1936) from the University of Rome. He subsequently directed work on the supersonic wind tunnel at Guidonia, Italy. With the Italian surrender to the Allies in 1943, the Office of Strategic Services, the wartime U.S. intelligence agency, arranged for him to leave Italy and work at the NACA. Ferri became director of the Gas Dynamics Branch at Langley in 1949. But he wanted an academic life, so he joined the Brooklyn Polytechnic Institute in 1951. He worked as a consultant while there, and, in 1956, established the General Applied Science Laboratory (GASL), which specialized in scramjet development [14, 15].

During the late 1950s and early 1960s, Ferri developed an elegant but technically demanding scramjet concept. It combined high performance with a small size, in order to reduce the propellant and tank weight, while using few or no moving parts. To cope with a wide range of Mach numbers, the inlet would normally require a variable geometry system to keep the shock properly positioned. Ferri proposed a concept called *thermal compression* be used, rather than a complex mechanical inlet system. Thermal compression was accomplished by adjusting the fuel flow into the burners. The burning fuel would create an oblique shock wave within the inlet, which would slow and compress the airflow, replicating the effect of a movable inlet.

This, he argued, prevented the formation of internal shocks and the resulting pressure loss. Because the combustion was controlled by the mixing of the fuel, the Mach number entering the burner could vary according to the Mach number of the vehicle itself. This ability to compensate for off-design performance eliminated the need for the area of the inlet and nozzle to vary. Such a fixed-geometry design was much simpler, lighter, and more reliable. This also allowed the combustor to be shorter than that of a conventional ramjet engine, but with about the same cross-section [10, pp. 32, 33].

Ferri noted several potential trouble areas had to be solved before such an engine design could be built. The first was whether a fixed-geometry inlet/ nozzle combined with the thermal compression would actually work, due to difficulties in adjusting the fuel injection process rapidly enough to meet rapidly changing flight conditions. The technology to build digital electronic fuel controls did not yet exist. Any system would be large, complex, and have to operate under extreme conditions. Another issue was the feasibility of the scramjet's proposed structural design. Perhaps the two biggest questions were whether the fuel and air would mix and ignite within the short time period, despite the high-velocity airflow, and whether shock formation occurred during the combustion process [10, p. 33, and 14, p. 1139].

Despite the numerous unknowns, scramjet studies, both theoretical and those entailing ground testing of subscale engine designs, were undertaken in the late 1950s and early 1960s. The next step was scramjet flight tests. It should be remembered that this was a technologically optimistic period. Between 1945 and 1956, aircraft speeds had gone from subsonic to Mach 3, and the X-15 program soon expanded this to Mach 6. The Dyna-Soar aircraft was to continue this speed advance right up to high suborbital velocities. Operational Mach 2 military aircraft were in service, and development of jet airliners had cut travel times in half. Planned supersonic transports would reduce this even further.

Development of a scramjet-powered vehicle would be the ultimate expression of the era, even if just how to accomplish this remained an open question. The X-7 and the Navaho represented the two different approaches used in flight testing of ramjets. The X-7 was for small-scale flight testing of the basic technology. With the Navaho, only minimal subscale flight testing was done before the all-up testing of the vehicle and full-scale ramjet took place. During later Navajo Fly Five and Project RISE flights, made after the program had been formally cancelled, several Navahos were successfully launched and their ramjets ignited. A Fly Five launch, made on January 10, 1958, was the longest Navaho flight, lasting 42 min and 28 s; it flew at a sustained speed of Mach 2.8 and travelled 1075 miles. However, when the Navaho's guidance system made a course change, the airflow into one of the ramjets was disrupted, leading to a blowout. As the vehicle slowed, the other ramjet also lost power, and the Navaho crashed into the Atlantic. Several other launches met the same fate before the project ended [9, pp. 73–76].[§]

The first U.S. attempt to build an air-breathing hypersonic vehicle used the "all-up" approach. Called the Aerospaceplane, this was a scramjet-powered single-stage-to-orbit aircraft. The vehicle was to be fueled with liquid hydrogen, which, due to its low density, meant the vehicle would be very large in order to carry the amount of fuel needed to reach orbit. Despite the technical

---

[§]Blowouts were a concern during the development of the Lockheed D-21 reconnaissance drone in the 1960s. This vehicle used the Bomarc ramjet as its engine. The engineers discovered that the ramjet would blow out during the test flights, but the engine structure was hot enough to relight once the airflow was reestablished. Blowouts were never a problem.

demands, aircraft contractors began studies of Aerospaceplane concepts between 1957 and 1959, using both Air Force and in-house funding. Republic Aviation, General Dynamics, Lockheed, Douglas, Boeing, Goodyear, and North American Aviation eventually became involved with the Aerospaceplane. Republic teamed with Ferri, of GASL, who proposed a vehicle using scramjet engines from Mach 8 on up to orbit.

General Dynamics, Douglas, and North American each received $500,000 study contracts from the Air Force in June 1963 to prepare detailed development plans. Martin built a full-scale wing-fuselage structure akin to that necessary for an operational Aerospaceplane. Research was also done on various air-breathing engine concepts, structures, and materials.

The Aerospaceplane offered the potential of airline-like access to space. With this, however, came a daunting series of technical challenges. The scramjet was still little more than a concept that had undergone only a limited amount of ground testing using subscale models. There were no flight data at all. The project's expense dwarfed that of the X-15 and other X-planes. Given the earlier reluctance of both President Eisenhower and his scientific advisers to fund the Dyna-Soar, it is not surprising the Aerospaceplane lacked support.

As early as December 1960, the Air Force's Scientific Advisory Board (SAB) warned that the Air Force was neglecting the more conventional problems the vehicle faced. The incoming Kennedy administration held the same negative attitude toward the Aerospaceplane that it did toward the Dyna-Soar. By 1962, the technical problems had reached the point that the single-stage-to-orbit concept was abandoned, replaced with a more feasible two-stage concept. Even this less ambitious goal was not sufficient to satisfy the SAB's doubts, and in October 1963, an SAB report enumerated the program's shortcomings. These included failure to establish a requirement for a fully reusable launch vehicle, the fact that the necessary hardware development eclipsed what available technology could support, and the program's high cost. The report continued: "While these factors dominate the picture, the Air Force must focus on advancing the important technical fields involved and prepare themselves for the time when the projected total payloads into orbit per year will increase to the point where such recoverable launching systems are competitive" [5, p. 951].

A few months later, the Dyna-Soar was canceled and funding for the Aerospaceplane was cut. Very soon, all work on the project ground to a halt. Studies continued on a wide range of reusable vehicle configurations, but the focus in U.S. scramjet research shifted to small-scale efforts [5, pp. 948–952].

THE HYPERSONIC RESEARCH ENGINE

NASA's initial scramjet research project was with the Hypersonic Research Engine (HRE). This was to be a small liquid hydrogen–fueled scramjet attached to the lower fin of the X-15A-2. Among the reasons a flight-test program was undertaken was because hypersonic conditions could not be simulated in ground-test facilities. To attain speeds of Mach 7 to 8, combustion-heated air would be needed to simulate the temperatures the engine would experience. As a result of the combustion heating, the air was contaminated with water vapor and, depending on the fuel used to power the tunnel, other products of combustion.

A second issue driving flight testing was more basic: some Langley researchers believed that a $30 million scramjet ground-based research effort had little to no chance of being funded, whereas a scramjet flight-test program would be. Indeed, one NASA researcher argued engine research and development "is better done on the ground," saying that he thought "there is no genuine need" for flight tests [5, pp. 757, 759, 760, 763].

Four contractors submitted HRE scramjet designs—Garrett, General Electric, Marquardt, and Pratt & Whitney. The first three progressed to the next phase of the bidding process; the Pratt & Whitney proposal was rejected by the NASA selection board. Garrett's concept soon emerged as the clear favorite. Its axisymmetric design had a barrel-shaped outer shell and a center body with a sharp inlet spike using variable geometry and a rounded aft body. A step on the center body served as the flameholder.

The Garrett engine was seen by the selection board as having the smallest and simplest as well as the overall best structural design. They thought it would be the easiest to cool and the cheapest to develop. It was also seen as having the best research potential, because its shallow annular combustor would produce a quasi-two-dimensional flow. This, board members felt, would be easier to analyze, and the data would be applicable to a wide variety of scramjet designs with similar flow, including designs featuring combustors with rectangular cross-sections. The selection board was also impressed by the amount of work performed by Anthony duPont, the engine's designer, and Garrett in developing the technical details of their proposal.

In contrast, the General Electric scramjet was considered too large (at 129 in. long, versus the Garrett design's 86-in. length), too heavy, and too hard to cool. Additionally, its combustor was thought to be too deep to enable production of quasi-two-dimensional flow. Finally, GE's cost estimates and development schedule were twice those of government estimates. Marquardt produced the only nonaxisymmetric design proposal. Designed by Ferri, its scramjet had a rectangular front view and incorporated thermal compression to avoid use of a variable geometry inlet. The Marquardt engine was rejected by the selection board because it had complex three-dimensional flows, an inferior engineering design, and a lack of data that would support the feasibility of Ferri's concepts [5, pp. 773–776].

Garrett's design was also favored due to the overall purpose envisioned for the HRE. Although often pictured as a prototype scramjet, the reality was much different. Earl Andrews, who worked on the HRE, later recalled:

> The axisymmetric [design] was chosen because it was thought that it would be less risky because it used more of the state-of-the-art at the time. The engine was not to be a prototype concept but for bringing an inlet, combustor, and nozzle together for research study of scramjet operation. We were always having to remind everyone that this was a research engine and not meant to be a prototype engine.[¶]

The HRE program ran into problems nearly as soon as it began. Costs rose, projected engine performance dropped, and the development phase dragged on. The combustor design proved unworkable, and the axisymmetric design's problems included thick wall boundary layers that led to internal flow separation, focused shock waves, and complex interactions. Although the selection board had been impressed by the effort Garrett made in developing its proposal, the company had no experience in scramjet design, which created many of the later problems.

With the actual HRE not ready, X-15A-2 flight tests began, using a dummy engine built from 3/8-in. plate steel. It had the same shape and size as an actual scramjet, but lacked internal components or flow path. Two different inlet spikes were built, one with a 20-deg cone and another with a 40-deg cone. The X-15A-2's ventral fin was also modified to accommodate the dummy ramjet. The lower fin was cut back, making it shorter, and was fitted with a vertical leading edge, which replaced the original sloped edge. The new fin also had impact rakes on the leading edge for pressure measurements. The aircraft was covered with an ablative heat shield to protect it from Mach 7 heating. The heat shield could be refurbished for a subsequent flight. The aircraft also carried two large drop tanks for additional fuel (see Fig. 3.2). These separated at a speed of about Mach 2.

To measure its effects on the aircraft's stability and control, Maj. William Knight made two speed build-up flights in 1967 in the X-15A-2 fitted with the boilerplate dummy ramjet. The first flight took place on May 8, without the two drop tanks or ablative heat shield. Maximum speed attained during the flight was Mach 4.75. The X-15A-2 was subsequently covered with the ablative material and the tanks were fitted. A speed of Mach 4.94 was reached during the second dummy scramjet flight, made on August 21. A postlanding inspection showed most of the ablative coating had survived the flight with only minor damage. The exception was the lower fin. It was heavily charred along

---

[¶]Earl Andrews email to Curtis Peebles, "FW: Airframe Integration Questions," July 9, 2008, 11:28 A.M. Dryden History Office files.

Fig. 3.2 The X-15A-2 being prepared for a test flight.

its entire length; very little char remained on the leading edge, and the lower portion was completely eroded. Similar damage, although not as extensive, had occurred due to sand being kicked up from the lakebed on an earlier landing.

On October 3, 1967, Knight reached a top speed of Mach 6.7 with the dummy scramjet. The multiple strong shock waves generated from the engine created localized heating, which burned through the aircraft's ablative coating, melted the skin of the lower fin, and damaged the hydraulic systems, preventing the flaps from being extended (see Figs. 3.3–3.5). The damage

Fig. 3.3 The X-15A-2 following its October 3, 1967, flight, during which it reached Mach 6.7.

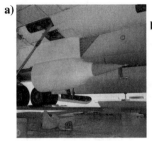 

Fig 3.4 a) The dummy HRE and the modified lower ventral fin prior to the flight, b) shock-on-shock heating during the Mach 6.7 flight damaged the lower fin, and the dummy HRE fell off. It was recovered from the desert and examined.

also caused the dummy scramjet to fall off the fin. Although the X-15A-2 landed successfully, it never flew again [16, 17].

The mishap led to ill feelings among the HRE researchers at Langley. They had been studying the problem of shock heating. The actual pylon planned for the HRE/X-15 flight tests would have had shielding and possibly active cooling. Langley personnel considered the X-15A-2 pylon to have been poorly engineered and thrown together for the stability and control testing. Another objection was that the Langley staff that had been assigned to the NASA Flight Research Center (now Dryden) had not been included in the planning of the dummy HRE testing, nor did they contribute to the analysis and resolution. The experience was viewed as a poor example of teamwork between two NASA centers on a joint project.**

When the X-15 program ended the following year, the HRE program was reoriented for ground testing. The first part of the reorientation involved the Structural Assembly Model (SAM). This was a full-scale, flight-weight model of the HRE built of Hasteloy-X, which was cooled using hydrogen. The model was tested in the Langley 8-ft High Temperature Tunnel during 1971 and 1972. The SAM tests provided thermal fatigue data, as well as thermal and mechanical stress data on the engine structure. These showed changes would be needed to increase engine life through lower coolant flow rates. Specific areas needing attention included the need to reduce the range of temperature changes the vehicle would encounter as well as a need for higher design-surface temperatures and a modified cooling jacket design.

The second element of the ground tests involved the Aerothermodynamic Integration Model (AIM), a boilerplate model built of Nickel-200 alloy, which was water-cooled and burned hydrogen. The tests were made in the NASA Lewis Plumbrook Hypersonic Test Facility. The goals of the AIM tests included determining the effects on the engine's internal thrust at speeds between Mach 5 and 7 due to fuel/air ratio, angle of attack, simulated altitudes, and varying inlet-contraction ratios. The determination of the inlet-combustor

---

**Jerry Reedy history interview, April 4, 2001, NASA Dryden History Office, and Earl Andrews email to Curtis Peebles, "RE: Airframe Integration questions," July 9, 2008, 6:42 P.M.

Fig. 3.5 Excessive heating during the Mach 6.7 flight also burned through the lower ventral fin, damaging the X-15A-2's internal structure.

interaction limits, inlet and combustor performance, and the combustion mixing length were also among the test goals. Still another was showing that the HRE engine could shift efficiently between supersonic and subsonic combustion. This was referred to as dual-mode operation, and was an important capability for an operational scramjet engine. The AIM tests began in October 1973 and continued through April 1974. With their completion, the HRE project was ended [5, pp. 773–778, and 18].[††]

Simultaneously with the HRE program, the Air Force initiated several ground-based scramjet research programs. These included work with an 18-in.-diameter variable geometry scramjet built by United Aircraft Research

[††]Chuck McClinton email to Curtis Peebles, "RE: Airframe Integration questions," July 9, 2008 9:21 P.M., and Chuck McClinton e-mail to Curtis Peebles, "Re: Hyper-X Question 2," July 2, 2008, 8:37 A.M.

Laboratory and tested at Mach 5; a 9-in. General Electric Mach 7 scramjet that featured a component-integrated model design; a GASL low-speed, fixed-geometry testbed scramjet incorporating Ferri's ideas; a Marquardt 18-in.-diameter flight-weight scramjet; and another Marquardt "dual mode" design. These proposed models, like the HRE, were to be attached to the vehicle's airframe.

Beyond these ground tests was an Air Force program to fly a scramjet-powered vehicle. The Scramjet Incremental Flight Test Program called for use of ground-launched rocket-boosted vehicles. The planned booster was a Castor solid-fuel rocket. The test vehicle had a cylindrical center section, a pointed nose, and four fins at the tail. Four Marquardt/GASL modular scramjets were mounted just behind the nose section. Inside the test vehicle were a liquid hydrogen tank and a smaller gaseous-hydrogen tank to fuel the scramjets. To protect the vehicle from aerodynamic heating during hypersonic flight, the nose cone and edges of the fins were covered with thermal protective materials. The Air Force Aero Propulsion Laboratory managed the effort.

The scramjet modules were designed by Ferri and encompassed his thermal-compression and fuel-mixing concepts. The box-shaped modules with their "sugar-scoop" inlets represented a very different engine concept than did the HRE's axisymmetric design. (The Aero Propulsion Laboratory engine had been the basis for Marquardt's HRE proposal.) Ground tests of the engine at speeds of Mach 5 began in April 1965. The data indicated the airflow remained supersonic during the combustion, and without harmful combustor-inlet interactions. Despite being only a preliminary design, the GASL scramjet engine reached 80 percent of design thrust level during ground tests.

The flight program envisioned four launches for the GASL flight program. The first was to be a test check of the vehicle/booster combination. This was to be followed by three "hot" scramjet launches to test their operation, the first actual flight tests ever made of a scramjet engine. The engines would ignite at a speed of Mach 5.4 at an altitude of 58,740 ft. The scramjets would have to operate at a speed of about 6000 fps, and were to have a burn time of over 5 s. The program goal was to demonstrate positive thrust from the scramjets. The planned velocity increase was 600 fps. The data from the flights were to be used in the design of subsequent scramjets.

Although the GASL ground-test data proved the scramjet design valid, further development in the project proved difficult. The modules required five different structural designs to address projected heating problems. Aerothermodynamic issues with the modules remained throughout the program. The test vehicle also underwent extensive wind-tunnel testing at the Arnold Engineering Development Center and at Lockheed's Rye Canyon Research Laboratory. The Rye Canyon tests covered a speed range between Mach 0.4 and Mach 4.4. These tests showed that the test vehicle's drag was a problem, and major efforts at reducing this were made between April 1965 and July 1966. The test vehicle's weight was another problem facing engineers. The original weight target was 400 lb, which soon grew to 600 lb. The dummy test vehicle's final weight was 680 lb.

The dummy test vehicle was flown January 11, 1967. Despite all the wind-tunnel efforts, it showed higher-than-expected drag. This reduced the test vehicle's separation speed from Mach 5.4 to 5.18, and the separation altitude was reduced by almost 1000 ft. Peak altitude was only 177,712 ft, compared to the estimated 228,500 ft. This was traced to unexpectedly high "spillage" drag from the dummy scramjet's inlets.

Beyond the technical issues with the Scramjet Incremental Flight Test Program, there were also fiscal problems. The Vietnam War put pressure on military research and development funding. Program costs had ballooned from the original estimate of $3.5 million to $12 million. Traditionally, when military funding became tight, the first area to be cut was research and development. Beyond developmental difficulties with the scramjet modules, the research goals had been modified extensively to enable collection of more elaborate data. With no near-term scramjet application at hand, the program became expendable. In August 1967, it was cancelled, and Air Force attention shifted to scramjets using storable hydrocarbon fuels.

Because the Scramjet Incremental Flight Test Program ended prematurely, its contributions to scramjet development can never be known. On one level, the program could have represented the best chance for the United States to gain actual data from a scramjet in flight. However, Langley researchers working on the HRE in early 1967 believed the Scramjet Incremental Flight Test Program was far more risky as well as much less cost effective than that of the HRE [5, pp. VI-xvi–VI-xxi, 780, 781; and 19].

Looking at the larger picture, there was a more significant issue. Scramjets represented an immature technology on several levels. An understanding of the physics of supersonic combustion was lacking. Computing capabilities in the 1960s were limited, meaning that theoretical models were essentially a mixture of limited solid data and high hopes. Wind-tunnel test limitations restricted the data obtainable to a narrow range of Mach numbers. Most of the scramjet designs tested by the Air Force during this period entailed performance as theoretically calculated, but the tunnel results could not establish the real-world advantages and flaws of scramjet designs.

When wind tunnels proved unable to accurately simulate supersonic flight in the 1940s, the alternative was to build research aircraft. With scramjets, this was not possible, due to the uncertainties already cited. The HRE and Scramjet Incremental Flight Test Program both encountered major design and operational issues. As problems multiplied costs rose, schedules stretched, and performance declined. Eventually, the programs were cancelled when the results were deemed no longer worth the price.

There were also management issues regarding plans, scale, and research goals. The Aerospaceplane was a totally impractical concept—it required a huge airframe based on questionable structural designs, had as its goal the demanding single-stage-to-orbit profile, and was to be powered by scramjet concepts that existed exclusively as ideas and subscale test models. The Scramjet Incremental Flight Test Program was also hampered by a lack of data on scramjet operation and the inability of wind tunnels to simulate flight conditions. Additionally, expansion of research goals increased costs at a time when research funding was tight and research activities were falling out of favor.

SECOND-GENERATION SCRAMJET RESEARCH

With the end of the HRE and Scramjet Incremental Flight Test programs, the high hopes and great dreams seen in the field during the early 1960s faded. The Air Force's follow-on scramjet effort was soon abandoned. A Marquardt scramjet was wind-tunnel tested in 1970 and 1971 at speeds between Mach 3 and 6, but Air Force leadership decided scramjets were outside of mission requirements at a time of diminishing resources and higher-priority demands. The program was cancelled in 1972, ending Air Force scramjet activities for over a decade [4, p. 12-8].

Efforts to restart NASA's scramjet flight-test effort also ended in failure. Even so, before the X-15 program ended, some at the Flight Research Center were urging development of a follow-on vehicle. By the early 1970s, several studies of hypersonic air-breathing vehicles were under way at Langley and FRC. At Langley, two concepts were being considered. The more advanced vehicle was called the Hypersonic Research Facilities study (HYFAC) and was to be a Mach 12 vehicle, doubling the performance of the X-15. Less ambitious was the High Speed Research Aircraft (HSRA), which was to have a top speed of Mach 8.

Paralleling the NASA efforts were those with Air Force hypersonic concepts. Among the latter, the first was for a Mach-3-to-5 vehicle. The more advanced was the Incremental Growth Vehicle, which was to fly at Mach 4.5, then Mach 6, and finally to Mach 9. In the climate of limited funding, the logical approach was for a joint NASA/Air Force development program leading to hypersonic air-breathing vehicles.

Such a program was established in July 1974, when the last lifting body, the X-24B, was undergoing flight tests (see Fig. 3.6). This aircraft was actually the X-24A with a new fuselage grafted over the original "finned potato" shape. The new shape was intended to represent capabilities for a high-speed cruise vehicle. Thus it was an ideal starting point for the new hypersonic vehicle. Two "X-24C" concepts emerged, one with a pair of cheek inlets for air-breathing engines, and another powered by an XLR-99 rocket engine.

In December 1975, NASA and the Air Force set up an X-24C Joint Steering Committee. By July 1976, a completely new concept emerged establishing the National Hypersonic Flight Research Facility (NHFRF, but pronounced "nerf"). The vehicle to be built in the NHFRF program differed significantly from the original X-24C concept. It was to undertake not only high-speed research and testing of air-breathing engines, but also material, systems, and weapon-separation testing. It was to have a 40-second cruise capability and inter-changeable modular fuselage sections. The NHFRF enjoyed strong support at Dryden, while at Langley, both the propulsion and aerodynamics researchers saw it as the culmination of over three decades of work. The NHFRF also offered an opportunity to maintain the hypersonic expertise that had been built during the X-15 program, and which would otherwise soon be lost. Finally, the program could revitalize classic aeronautics research at NASA.

The NHFRF's wide range of research goals, however, carried a significant price tag. The cost estimate for construction of two vehicles and a 10-year/200-flight program was $200 million, a major expense at a time of shrinking NASA budgets and the demands of building the space shuttle. In the wider society, the technological optimism of the late 1950s and early 1960s was long gone. This era saw reduced federal spending for, and hostility toward, science and technology as well as the end of the Vietnam War, a decline in the national space-related euphoria that had characterized the Apollo era, unwillingness to fund the final three Apollo missions, high inflation, a gas

Fig. 3.6 Maj. Mike Love and the X-24B on Rogers Dry Lake.

crisis, and sundry other factors that collectively had the effect of dooming prospects for such expensive programs.

NASA and the Air Force continued discussions about the NHFRF into 1977, but program costs kept rising. Although the Dryden and Langley centers backed the effort, NASA Headquarters did not, and canceled the NHFRF in September 1977. James J. Kramer, the agency's acting associate administrator for aeronautics and space technology, gave the reasons as "the combination of a tight budget and the inability to identify a pressing near-term need for the flight facility had led to a decision by NASA not to proceed to a flight test vehicle at this time" [20].

The failure of the NHFRF showed the inability of the aeronautical research community to adapt to the new political and social environment of the late 1970s. The original X-24C concept would have been expensive and difficult. It was, however, a logical extension of the previous work with lifting bodies and the X-15, and offered an opportunity to continue hypersonic research at reasonable cost. Instead, the joint NASA/Air Force committee opted for a completely new, unproven, and very large vehicle. Some of the models being considered would barely fit under the B-52 mothership. The NHFRF would be all things to all researchers. Its goals, such as testing high-speed reconnaissance equipment and weapons separation, lacked any near- or long-term operational requirements. The project had a very large price tag, at a time of economic difficulties and diminished expectations.

This was a problem not limited to the aeronautics community. Planetary probes became fewer in number beginning in the 1970s. As the number of missions declined, scientists attempted to pack more experiments onto the few launches being made. This increased payload, requiring larger boosters. In turn, the missions' costs increased, making it even more difficult to gain approval. Mars exploration effectively ended between the late 1970s and the early 1990s because of this spiral. A way out of this dilemma was not yet apparent.

Declining interest in scramjets is reflected in the fate of the Marquardt Company. During the early 1950s, it was the third-largest employer in the west San Fernando Valley. The company continued to prosper into the mid-1960s. In August 1967, following cancellation of the Scramjet Incremental Flight Test Program, Marquardt dropped two-thirds of its employees and closed its Ordnance Aerophysical Laboratory in Texas. The company continued research work on ramjets and scramjets from the late 1960s into the 1980s with the ASALM missile propulsion effort [7, pp. 6, 14–16, and 11, p. 202].

The major figures that dominated the first generation of scramjet research passed from the scene. Antonio Ferri, who had been involved with early scramjet development, died in 1975. Roy Marquardt, who had created the ramjet industry in the 1950s, died in 1982 [15, pp. 57–59].

In the broader view, scramjet technology was still not yet a viable propulsion system. The difficulty of the task, the limitations of existing ground-test facilities, and the failure to successfully undertake a flight-test program left too many unknowns. After the end of the NHFRF, research activities were limited to wind-tunnel studies for a decade.

REFERENCES

[1] Adams, F. D., *National Aeronautics and Space Administration Aeronautical Dictionary*, NASA, Washington, DC, 1959, p. 157.

[2] Fry, M. R., "Liquid Propellant Rockets," *Space Education*, September 1982, pp. 155, 156.

[3] Marbarger, J. P., *Space Medicine: The Human Factor in Flights Beyond the Earth*, University of Illinois Press, Urbana, 1951, p. 15.

[4] Waltrup, P. J., "Hypersonic Airbreathing Propulsion: Evolution and Opportunities," *Conference Proceedings on the Aerodynamics of Hypersonic Lifting Vehicles*, AGARD, April 1987, pp. 12-4, 12-5.

[5] Hallion, R. P. (ed.), *The Hypersonic Revolution: Case Studies in the History of Hypersonic Technology*, Vol. II, Air Force History and Museums Program, Washington, DC, 1998, pp. VI-iii, VI-v-vii.

[6] Avery, W. H., "Twenty-Five Years of Ramjet Development," *Jet Propulsion*, November 1955, pp. 605–608.

[7] Steelman, R. C., and Allen, R. C., "History of Ramjet Propulsion at the Marquardt Company—1944 to 1970," AIAA 2004-3538, pp. 1–3, 5.

[8] Miller, J. *The X-Planes X-1 to X-45*, Midland, Hinckley, England, 2001, pp. 113–117.

[9] Gibson, J. N., *The Navaho Missile Project*, Shiffer, Atglen, PA, 1996, pp. 37, 38, 63–77.

[10] Ferri, A., "Supersonic Combustion Progress," *Astronautics & Aeronautics*, August 1964, pp. 32, 33.

[11] Heppenheimer, T. A., *Facing the Heat Barrier: A History of Hypersonics*, NASA SP-2007-4332, Washington, DC, pp. 98, 99.

[12] Fletcher, E. A., Dorsch, R. G., and Gerstein, M. "Combustion of Aluminum Borohydride in a Supersonic Wind Tunnel," NACA Research Memorandum RM E55D07a, June 20, 1955.

[13] Weber, R. J., and MacKay, J. S., "An Analysis of Ramjet Engines Using Supersonic Combustion," NACA Technical Note TN 4386, September 1958.

[14] Curran, E. T., "Scramjet Engines: The First Forty Years," *Journal of Propulsion and Power*, November-December 2001, p. 1138.

[15] Busemann, A., "Antonio Ferri, 1912–1975," Memorial Tributes: National Academy of Engineering, Volume I, 1979, pp. 57–59.

[16] Jenkins, D. R., and Landis, T. R., *Hypersonic: The Story of the North American X-15*, Specialty Press, North Branch, MN, 2003, pp. 143–145.

[17] Andrews, E. H., and Mackley, E. A., "Review of NASA's Hypersonic Research Engine Project," AIAA 93-2323, pp. 1, 2.

[18] Kumar, A., Drummond, J. P., McClinton, C. R., and Hunt, J. L., "Research in Hypersonic Airbreathing Propulsion at the NASA Langley Research Center," ISABE-2001 Invited Lecture 4, pp. 943–949.

[19] Peschke, W. O. T., "Approach to In Situ Analysis of Scramjet Combustor Behavior," *Journal of Propulsion and Power*, September-October 1995, pp. 943–949.

[20] Hallion, R. P., and Gorn, M. H., *On the Frontier: Experimental Flight at NASA Dryden*, Smithsonian Books, Washington, DC, 2003, pp. 169, 170.

Chapter 4

DEVELOPMENT OF AIRFRAME-INTEGRATED SCRAMJETS

When we walk to the edge of all the light we have and take the step into the darkness
of the unknown, we must believe that one of two things will happen. There will be
something solid for us to stand on or we will be taught to fly.

Patrick Overton, educator, poet, playwright, and speaker

Despite the disappointments of the 1960s and 1970s, these years saw a conceptual breakthrough in scramjet design. Ironically, the breakthrough came as an outgrowth of the problems with the HRE. The pod-type axisymmetric Garrett HRE design was selected to ensure success in meeting program goals, specifically, understanding the engine's performance. The analytical tools available in the mid-1960s were very limited, so the simplest configuration was required.*

Although this type of engine was good for use in research-based assessment of scramjets, it was known at the start of the HRE program that a podded scramjet engine, particularly one with a forward-facing throat like that used in the Garrett HRE, had severe performance shortcomings. Chuck McClinton, a Langley researcher who was involved with scramjet development during this period, later noted the HRE "would not provide sufficient thrust to accelerate anything, including itself."†

A second shortcoming of the pod-type scramjet came to light as a result of shock heating that occurred on the final X-15A-2 flight. Although data in wind-tunnel testing could be acquired with the HRE, putting pod-type scramjets on a real hypersonic vehicle required careful integration work to prevent such damage. This was not an issue with conventional jet engines, which were attached to an airliner's wings or fuselage by a pylon, or with ramjets because the lower speeds involved did not result in the shock heating found with the HRE on the X-15.

The most significant problem with pod-type engines was their operating environment. The hypersonic speeds of a scramjet-powered vehicle, even at very high altitudes, dictate that it must operate in a high-dynamic pressure

---

*Chuck McClinton e-mail to Curtis Peebles, "Re: Hyper-X Question 2," July 2, 2008, 8:37 A.M.
†Chuck McClinton e-mail to Curtis Peebles, "Re: Hyper-X Question," July 9, 2008.

flight regime. In simple terms, the faster the engine went, the more air had to be captured by the inlet to generate the necessary thrust. Further complicating the issue was that the faster the vehicle flew and the higher its altitude, the thinner the air it traveled through.[‡]

The effect of these factors is apparent in the specific impulse of a hydrogen-fueled scramjet at speeds between Mach 6 and Mach 25. At the lower end of the speed range the specific impulse value is about 3000 s. At Mach 8 a steady decline begins so that by Mach 20, the specific impulse is between 400 and 1000 s. The lower value is less than the specific impulse of a hydrogen-fueled rocket [1].

As a result, in order to capture the necessary airflow at higher speeds the scramjet's inlet had to constitute a larger percentage of the total vehicle frontal area. On a conventional subsonic aircraft, the inlet makes up only a small percentage of its frontal area. For an SR-71 flying at Mach 3, the ratio is significantly higher. For a hypothetical scramjet-powered vehicle flying at Mach 15 or greater, the inlet would have to comprise nearly the entire frontal area to provide sufficient air to the engine.

For these reasons, the HRE, with its nose spike and narrow annular inlet, never had a chance of being adopted as a viable scramjet design. Pod-type engines were "out," as McClinton later remarked. "This was well known, except in some of the hypersonic fringes, as long as I was in the [scramjet] business"[§]

THE AIRFRAME-INTEGRATED SCRAMJET

Although the airframe-integrated scramjet concept originated in 1968, the idea of integrating an aircraft's propulsion system with its airframe to increase performance was not new. In July 1957, Roger W. Luidens, a researcher at the Lewis Flight Propulsion Laboratory (now the NASA Glenn Research Center), published a report, "Arrangements of Jet Engines and Airframe for Increased Range." The report detailed how changing the location of a ramjet-powered Mach 5 aircraft's wings, fuselage, inlets, and engines would improve its range.

Among the possibilities Luidens offered was locating the inlet in order to increase pressure recovery. If the Mach number of the airflow ahead of an inlet is reduced, an increase in total inlet pressure recovery would occur. He suggested two things: putting the inlet under a wing or putting it under

---

[‡]Chuck McClinton e-mail to Curtis Peebles, "Re: Airframe Integration question," July 9, 2008, 9:21 P.M.; Earl Andrews e-mail to Curtis Peebles, "Re: Airframe Integration question," July 9, 2008, 6:41 P.M.; and Lawrence D. Huebner e-mail to Curtis Peebles, "Re: Airframe Integration question," July 9, 2008, 3:23 P.M.

[§]Chuck McClinton e-mail to Curtis Peebles, "Re: Airframe Integration question," July 9, 2008, 9:21 P.M.

the nose of the fuselage, both of which offered other advantages. Luidens wrote [2, p. 5]:

> One of these is that the direction of the inlet airflow is independent of airplane angle of attack. A change in airplane angle of attack results only in a change in local Mach number ahead of the inlet. The inlet can usually more readily adapt itself to a Mach number change than to a change in flow direction. Also, by locating the inlet in a region of favorable compression, the maximum Mach number for which the inlet must be designed may be reduced. This is important to the design of the inlet for engine-inlet matching.

Improvement in range, Luidens found, could also be achieved by canting the exhaust jet downward at a small angle to the flight direction. Doing so results in an increased increment of lift, but practically no loss of thrust. The increment of lift created by the canting of the exhaust no longer had to be carried by the wings, and the drag created by the lift was slightly decreased. The engines could be throttled back slightly, reducing the amount of fuel burned and increasing the range. By optimizing the configuration, a theoretical range improvement of 25.6 percent at Mach 4 could be achieved [2, 3].

The idea of making the inlet part of the airframe appeared in early scramjet designs. Earl Andrews, who had worked on scramjets at Langley in the early days, recalled seeing in the 1958–60 time period cutaway diagrams of one of the Aerospaceplane concepts with scramjets that used the vehicle's underside for inlet compression and nozzle expansion. During the HRE program, he recalled attending a meeting with other Langley researchers at which details on the four HRE concepts were presented. This included the Marquardt engine with its "alligator jaw" concept with side-spillage open areas. When the Langley researchers returned from the conference, John Henry and John Weidner looked at applying several of the HRE design concepts. Among those they considered were the Marquardt alligator jaw and the HRE axisymmetrical design combined with a sidewall compression inlet, which used swept leading edges and downward spillage to allow the engine to start.

Weidner and Carl Trexler designed and tested inlet models in Langley's Mach 4 Blowdown Tunnel (see Fig. 4.1), while Andrews laid out a streamline traced inlet (by hand) and wrote up the results in a memo. Later, he and Clay Rogers suggested a study be made of a streamline traced inlet with a throat plug to vary the contraction. John Henry rejected the idea because he thought valid analysis could be done only at design conditions and wasn't doable due to "the fact that the branch [the Langley division responsible for carrying out the work] had too much on its plate already."[1]

---

[1]Earl Andrews e-mail to Curtis Peebles, "Re: Airframe Integration question," July 9, 2008, 6:41 P.M.

Fig. 4.1 A model of a large high-speed air-breathing vehicle being prepared for testing in a Langley wind tunnel in 1959. The flat underside of the forward fuselage gave a hint of things to come.

Andrews was assigned to analyze a Ferri scramjet concept at GASL. He determined that the forebody underside started with a 10-deg cone a short distance from the nose, with the compression delayed as the airflow moved around the cone from the bottom centerline on either side. This caused the cross-section to become elliptical, with its long axis in the horizontal plane, followed by the forebody turning inward after a short swept throat. The cowl had swept leading and trailing edges. Andrews had a model built and tested in the 20-in. tunnel at Mach 6 to survey the flowfield. When Andrews was put on loan to the HRE project office, his Langley colleague Marvin Torrence picked up the Ferri inlet work. An inlet model was made and Torrence wrote a report on the test results.**

By the late 1960s and early 1970s, the airframe-integrated scramjet was well established as the preferred design. Langley's John V. Becker and Frank S. Kirkham published a paper in 1971 on hypersonic transports. After examining (and rejecting) boost glide and vertically launched shuttles for

---

**Earl Andrews e-mail to Curtis Peebles, July 9, 2008, 11:28 A.M. The *Aviation Week and Space Technology* magazine of June 22, 1959, showed a boost glide vehicle wind tunnel model with a sloped underside and an underslung rectangular engine being prepared for a test at the Lewis Research Center.

long-range hypersonic transports, due to concerns about noise, g-forces, and high operating costs, they focused on hypersonic air-breathing cruise vehicles. These, in their view, would be much more economical to operate and had the advantage of low sonic boom overpressure, allowing high-Mach overland flights. The illustration in their paper was the iconic delta wing vehicle with the modular scramjet engine under the fuselage [4].

Turning the concept of an airframe-integrated vehicle into reality involved a set of interconnected tasks. Integrating a scramjet aerodynamically and geometrically into the airframe provided both maximum inlet capture area and maximum nozzle-expansion area, while maintaining minimum cowl drag. The factors that directly affect inlet and engine performance are the mass flow ingested by the scramjet, static pressure, local Mach number, and flow angularity. These factors define the state of the flow, and if they are uniform, the remaining flow variables are also uniform. If the precompression flow at the inlet face were uniform, the need for a complex inlet design that would provide efficient operations in variable flow conditions would be eliminated. Should the flow mass be reduced, there would be a corresponding reduction of the thrust developed by the scramjet. Maximum performance would require a complex fuel schedule.

The underside of the aircraft's forward fuselage would serve as the inlet ramp. Oblique shock waves formed by the ramp began the compression process before the air reached the inlet. This reduced both the compression the inlet would have to produce and the physical size of the scramjet engine modules. One estimate indicated that an airframe-integrated scramjet would reduce the engine's size by a factor of three at Mach 10, as compared with conventional designs. The cowl drag is also reduced in this process.

To maximize performance, the engines would be mounted on the vehicle's underside and positioned far enough back so the bow shock from the vehicle's nose almost hit the cowl lip, for maximum mass capture.[††] This encompassed an area several times wider than it was high and allowed several small rectangular scramjet modules to be positioned side by side across the vehicle's underside. An individual engine module was relatively small, so it could be tested in existing ground facilities. A single, large scramjet would have to be ground tested as a subscale model, which could introduce errors into the data.

Although this approach would offer a major advantage in reducing the scramjet modules' size and weight, a new problem appeared. The "capture area" of the shock wave would diminish as the vehicle's Mach number increased. This would put great demands on the inlet, because engine performance was dependent on capturing the maximum amount of flow between the fuselage and bow shock across the width of the fuselage.

[††]If the bow shock was ingested by the engine, an "overspeed" condition would occur. If the bow shock hit the cowl, it would cause high heating.

The scramjet would also have to be integrated with the aft fuselage. With the HRE and the Scramjet Incremental Flight Test programs, hot exhaust gases exited the scramjet engine without interacting with the vehicle itself. In an airframe-integrated scramjet, the curved aft fuselage would serve as half of a bell nozzle. This would greatly increase the engine's effective exhaust velocities and reduce its length and cooling requirements while creating only a small increase in the engine's external drag. The ambient air pressures would shape the exhaust plume, acting as the missing half of the nozzle and maintaining the optimum shape for maximum efficiency. (A form of an aerospike nozzle.)

The scramjet nozzle design would be determined by the test vehicle's thrust and stability requirements. These involved such factors as the location of the scramjet modules, orientation of the thrust vector, and trim penalties due to misalignment of the thrust and the vehicle's center of gravity. Depending on conditions, the vehicle might fly nose high or low. To correct this, the control surfaces would have to be deflected, adding drag. There was also the potential for strong interactions between the nozzle exhaust and the nonuniform flowfields around the fuselage afterbody and external cowl.

Although the airframe-integrated scramjet offered an elegant solution for building a workable engine, the advantages the concept offered would be achieved only if the integration of the airframe and scramjet was begun early in the design process. This task was made difficult by the interactive constraints the design entailed—the engine modules' size and number must meet the mission requirements, and the flowfield must meet performance requirements. Directly affecting these would be a good forebody design; aerodynamic, structural, and internal volume requirements; and other constraints [5].

SCRAMJET RESEARCH AT NASA LANGLEY

The winding down of HRE research efforts and the emergence of the airframe-integrated scramjet overlapped. The year 1968 saw the end of the X-15 program, and the shift to ground testing of the HRE. That same year, the individual design elements of the various Ferri/Marquardt/GASL engine concepts merged to become the airframe-integrated scramjet. Aspects of scramjets studied at Langley during the 1960s and early 1970s included fuel injection and mixing, component testing, direct connect combustor, injector combustor tests, engine tests, fuel ignition flameholding, and mixing recipes [3].

As the SAM and AIM ground tests of the HRE were being planned, analytical studies of airframe-integrated scramjet concepts were being undertaken at Langley by John Henry and Griffin Y. Anderson. The focus of their study was a basic question: If the underside of the vehicle is used effectively to perform inlet and exhaust nozzle functions, can the relatively small fixed-geometry engine module provide the desired levels of thrust and low cooling characteristics?

Their results were presented at the First International Symposium on Air Breathing Engines, held in Marseille, France, in June 1972. In designing the airframe-integrated engine concept, they used baseline performance values, such as the engine's specific impulse, from previous studies. As a result, its projected performance was about midway between the goal and minimum values for the HRE scramjet. Their engine concept also reflected three design requirements: fixed inlet geometry, low cooling, and the shape of the space between the vehicle bow shock and vehicle undersurface available for the scramjet modules.

Fixed geometry required that the sidewall leading edges be swept to allow airflow to spill downward upstream from the cowl leading edge during engine startup. The initial Langley engine design used a leading-edge sweep angle of about 48 deg for the sidewall compression surfaces. The sidewalls had a wedge angle of 6 deg in the flow direction to compress incoming air. The inlet had a starting speed of Mach 3.

One of the study's goals was to measure performance losses caused by such features. Using a fixed geometry posed difficulties in establishing a supersonic flow at the low end of the engine's Mach number range. At high speeds, the inlet had to maintain combustor airspeed velocities at levels that minimized momentum losses and generated high thrust levels.

The study had mixed conclusions about fixed vs. variable geometry inlets for an airframe-integrated scramjet. The study showed variable geometry would increase thrust by only 16 percent. This limited improvement would come at the cost of increased complexity as well as seal and joint problems. The scramjet's internal pressures would also be increased by an amount that could exceed the heat-sink limitations of hydrogen fuel. The added weight would tend to negate the improved performance.

Use of a fixed-geometry inlet had design consequences for other parts of the scramjet as well. The internal compression was created by the inlet sidewalls and the leading edges of the three fuel injector struts. These were placed horizontally in the narrowest part of the scramjet, just ahead of the combustor. Hydrogen injected through the struts mixed with the air, ignited, and expanded out the nozzle.

Constant flow properties tended to be parallel to the swept leading edges. As a result, the sweep angle of the three fuel injector struts had to be the same as that of the leading edges. The design of the fuel injector struts was a critical issue. Tests of a similar inlet design indicated the struts' leading edges would have attached shock waves at just under Mach 4. Additionally, the strength of these shock waves would be low, and no boundary-layer separation was expected to occur. However, with the sidewall/strut design, it was impossible to prevent the sidewall shocks from merging with the strut shocks at some flight speeds. This was dealt with by changing the struts' wall slope, which would either cancel a shock or weaken its strength.

Engine cooling requirements were the next issue. The cooling system for the scramjet would be similar to that found in rocket motors. In what was called a "regenerative" system, liquid hydrogen would be circulated through cooling jackets or tubes, which formed the engine's internal walls. The fuel would absorb the heat, be injected into the combustor airstream, and ignited. The engine would remain at a constant temperature because the same amount of heat absorbed by the cold fuel would be added by the fuel's combustion. However, the heat exchanger technology available in the late 1960s and early 1970s could not meet the engine's cooling requirements.

The solution to the cooling-system problem was to reduce the amount of heat the engine produced. One means was using three injector struts. Because the mixing and combustor length were proportional to the gap between the struts, the total area subject to high heating rates was reduced. The supersonic combustion also reduced both the pressure and the heat transfer in the combustor. As with other elements of the airframe-integrated scramjet, various constraints were imposed on the combustor design.

For an engine design in which combustion occurs close to the fuel injectors, a 20 percent reduction in the cooling requirement results in only a 3 percent drop in specific impulse. In an engine with the same geometry, but where the combustion occurred over a longer distance, there was a further reduction in the cooling requirement, but also a greater loss of impulse.

The combustor concept involved an array of uniformly spaced multiple fuel injectors. The mixing and merging of fuel patterns for the different injectors became important in engine operations. Calculations indicated that the merging of the different mixing patterns increased the time needed to complete air/fuel mixing.

The combustor was not the only engine/airframe element contributing to the cooling requirements. The nozzle also put a significant heat load on the test vehicle. As with the combustor, shortening the nozzle's length was the primary means of reducing its cooling requirements. One of the study's goals was to find ways to accomplish this. The initial design in the study called for a relatively long nozzle. Henry and Anderson noted that an optimization of different factors, such as its thrust performance over the vehicle's speed range, cooling requirements, and aerodynamic factors such as lift and trim drag, would result in a reduction in size over the initial configuration.

Their comparison of the baseline and predicted kinetic energy efficiencies showed a difference in specific impulse of only 20 seconds. The loss of thrust was caused by the growth of the boundary layer and the assumption that the nozzle entrance would experience frozen flow. The limitations of the analysis and of existing technology also complicated estimates of nozzle performance. More analyses were needed to determine the effects of the propulsion system on the vehicle's aerodynamics. Additionally, the engine cooling system had yet to be developed, and new insulation techniques or coatings could result in major reductions in cooling requirements [6].

By the time the HRE office closed down in 1974, research at Langley had shifted to the airframe-integrated scramjet concept. These tests used subscale scramjet models, which employed a design similar to that of the Henry and Anderson study. These included a fixed geometry inlet with a 48-deg sweep surface, mixing controlled combustion to regulate the burning process as a function of the flight Mach number, and a cutback cowl design for starting. Compression of the airflow in the vertical direction was achieved by the forebody, while horizontal compression was produced by the inlet sidewalls. The compression process was completed by the three 48-deg swept struts, on which the fuel injectors were mounted. Earlier inlet component testing had confirmed that this configuration exhibited performance rivaling that of variable geometry inlets, but without the weight and design complications.

The strut fuel injector arrays were arranged both perpendicular and parallel to the airflow. This allowed multi-plane sprays, which shortened the combustor's mixing length. At Mach 5 or below, most of the fuel spray would be parallel to the airflow. This would slow the mixing/burning process, preventing "thermal choking" of the engine. At higher speeds (i.e., Mach 7) fuel was injected at a right angle to the airflow for rapid mixing and burning. The injectors were located behind steps in the struts, which acted as flameholders to allow sustained combustion in the supersonic airflow. The steps, like rocks in a stream, created an eddy allowing stable combustion.

Because the engine model tests were conducted as research into the airframe-integrated scramjet concept rather than as demonstrations of a completed engine, the models were built with only the initial portion of the nozzle. As a result, the nozzle exit area was about equal to the inlet area. Independent research would be required to investigate the unknowns of an integrated half-nozzle.

The supersonic nozzle flow was complicated by the nonuniform entrance flow, complex chemistry of the combustion process, internal shock and viscous effects, interactions at the nozzle exits between the adjoining engine modules, and interactions between the exhaust and the air spilled by the inlet. These unknowns were analyzed, including with simulations of the nozzle flow using gasses other than air.

The ground testing of the new concept was made using a step-by-step approach. The initial tests were made at Mach 7, where fuel mixing and burning represented the unknowns, and at Mach 4, where unstart problems were expected. Two separate engines were tested using research engines rather than flight-worthy designs. They lacked the cooling system of an operational engine. More important, they featured a modular design, which allowed the inlet, combustor, nozzle, and fuel injection struts to be replaced without affecting the other components. This allowed major changes to be made without building a new engine each time.

Both the Mach 4 and Mach 7 engine designs were 8.13 in. high and 6.50 in. wide. The Mach 4 engine model was 56.88 in. in length; the Mach 7 design was longer, at 60.44 in.‡‡ Because the models would experience different heating conditions, each was built of different materials. The Mach 4 engine model was constructed from Nickel-200 and was a heat-sink design without an active cooling system. The Mach 7 engine was built of copper. Though also a heat-sink design, the engine had water-cooling on high-heat areas such as the inlet leading edges. The Mach 4 engine was instrumented to measure data on pressures, temperatures, and thrust. Instrumentation on the Mach 7 engine model also recorded these measurements, along with data on the engine's heating rates.

Another difference between the two models was in the Mach 4 engine's inlet design, which had a "drooped" interior cowling surface. This feature reduced the amount of local flow compression in the cowl region. The Mach 4 model also had an updated combustor design and a different internal geometry. This resulted in combustor expansion beginning earlier with less expansion occurring on the sidewalls and the top surface.

The two engine models were tested at different facilities that could meet test requirements posed by airframe-integrated scramjets. The Mach 4 engine tests were conducted in GASL's Westbury, New York, facility. The facility simulated the Mach 4 enthalpy§§ level of 890 K by burning hydrogen and oxygen to heat the air. This simulated the compression heating of the air as it was slowed by the ramp and inlet. Oxygen was added to replace the oxygen burned. As a result, the oxygen and nitrogen airflow was contaminated by water vapor. This difference in composition from that of atmospheric air (referred to as "real air") had to be allowed for in data analysis. Altitudes of from 71,410 ft to 86,300 ft and a test time of 13 s could be attained in the GASL tunnel.

The Mach 7 engine tests were done using the NASA Langley Arc-Heated Scramjet Test Facility (AHSTF). This was the first of several new or upgraded hypersonic test facilities to enter service during the 1970s and 1980s. The arc-heated tunnel could produce the Mach 7 enthalpy level of 2220 K. Unlike the GASL tunnel, which used combustion heating, the Langley tunnel used a 10-MW DC electrical arc to provide proper test conditions. Beyond higher

---

‡‡The report on these tests was published at a time when the United States was attempting to switch to the metric system, and the report used centimeters and meters as the units of measure. The two engines were both 20.32 cm high and 16.32 cm wide; the Mach 4 engine was 142.2 cm long and the Mach 7 engine was 151.1 cm long. The Mach 4 engine was tested at simulated altitudes of 21,640 m and 26,150 m, and the Mach 7 engine was tested at 35,350 m.

§§*Enthalpy*, or total heat, is the amount of energy in a system capable of doing mechanical work. Enthalpy (H) is defined as the sum of the internal energy (E) + pressure (p) × volume (v). The equation is $H = E + p \times v$.

speeds, the arc-heated tunnel could also provide a higher simulated altitude—116,660 ft—and a longer test time of 20 s.

Mach 4 and Mach 7 speeds were selected because the engine would be tested in radically different operating conditions, including how the hydrogen fuel was injected and how airflow was captured. Other reasons for testing at those speeds involved predictions for ignition and inlet-combustor interactions. All these issues needed verification. A major research objective was achieving the highest-possible combustor efficiency by testing different fuel injector designs and modifying the engines' internal geometry.

The Mach 7 engine tests were the first to begin, with 90 tests conducted between April 1977 and February 1979. These were divided into several different research areas. The initial tests were to verify that inlet starting and performance matched the results of earlier inlet component tests. These showed excellent agreement. The inlet pressure measurements also agreed with theoretical calculations, with the exception of those near the cowl, which differed due to a shock wave being generated by downflow from the swept inlet shocks.

The primary focus of the tests was on achieving ignition and sustained combustion, which proved difficult. The problems appeared immediately once Mach 7 tests began, and were traced to several factors. The low combustor-entrance pressure level occurred due to the tests being conducted at higher simulated altitudes than those for which the engine had been designed. Other factors affecting ignition and sustained combustion included low fuel and engine wall temperatures.

Several test approaches were taken in attempts to resolve the difficulties. Some were not intended for use in an actual flight-worthy engine, but rather as quick and easy methods to ignite and burn the fuel so other problem areas could be examined. One of these involved modifying the center strut to increase static pressure and temperature in the area just downstream of the fuel injectors. Also part of this redesign was injecting a small amount of air or nitrogen from tubes in the sidewalls roughly a third of the combustor length downstream of the combustor entrance. The result simulated the effects on the airflow of a geometry change in the combustor.

Zirconium oxide rods were also fitted to the modified center strut. These extended about one third of the distance to the side struts, and acted as igniters/flameholders. A combustor geometry modification was also used to sustain the reaction. Sidewall plates were added inside the combustor to decrease the flow expansion immediately downstream of the struts. This maintained higher pressure, which aided the reaction.

As a result of ignition problems, a research effort was begun to find operational ignition methods. Additionally, plans were made for Mach 7 tests at dynamic pressures up to 2.5 times higher than those of initial tests. These conditions would be similar to the design pressures of the engine and, as a result, the ignition and combustion problems were expected to be significantly reduced.

This was not the only problem area to appear during the Mach 7 tests. Inlet-combustor interactions occurred that varied in severity and were a function of combustor geometry, the amount of fuel burned, and the amount of the surface boundary layer ingested by the test engine. The results ranged from increased inlet pressure and flow spillage to inlet unstarts. A partial solution for the boundary-layer ingestion was to lower the engine's position in the wind-tunnel flow. This did not remedy the situation completely, however, and several other approaches were evaluated to correct this issue. These included fuel injector modifications and various splits between the parallel and perpendicular fuel injectors. To define the causes of the inlet-combustor interaction, tests were done in a conventional wind tunnel. To simulate fuel combustion, air was injected into the engine. The subsequent report noted that the ignition/reaction and inlet-combustor problems could not be dealt with in isolation. Changes needed to correct the ignition/reaction problems at low pressures could make the inlet more sensitive to a pressure rise in the combustor.

Overall performance results of the engine at Mach 7 were mixed. During the tests, the engine was able to produce just enough thrust to overcome its own drag (something the HRE was incapable of doing). The highest engine performance occurred with either the strut modifications and nitrogen injection or the zirconium-oxide rods and the sidewall modifications. Both of these produced higher compression. This, in turn, caused faster ignition and higher combustion efficiency.

Despite the engine being able to produce more thrust than drag, its performance was still well below the theoretical estimates for complete combustion. This difference was thought to be due to a combination of factors: low combustion efficiency, inlet-combustor interactions, and ignition delay. Theoretical analysis supported the hypothesis that the poor performance was due to ignition delays.

While the Mach 7 tests explored the engine's hypersonic behavior, the GASL Mach 4 wind-tunnel studies looked at the low end of the engine's performance spectrum. During 1978, more than 70 tests runs were made. Because the Mach number at the inlet throat was 3.4, researchers expected the combustor pressure rise to cause inlet-combustor interactions. As with the Mach 7 tests, lowering the model engine's position in the wind tunnel lessened the effects of these interactions.

Another problem was the appearance of detached shock waves from the inlet struts caused by the combination of the 48-deg inlet sweep and the inlet itself. Efforts to correct the interaction in the strut area involved moving the fuel injection away from the top, cowl, and sidewalls and adding steps in the walls ahead of the normal fuel injection station. The changes produced small improvements, but did not solve the interaction problem.

Igniting the fuel and maintaining stable combustion also proved difficult during the Mach 4 tests. Because of the low enthalpy level, the static temperature of incoming air was only 550 K at the engine section where the fuel was injected—not high enough to cause autoignition within a reasonable distance. Researchers realized that an ignition system was necessary to correct the problem. For the purposes of the wind-tunnel tests, they decided to briefly increase the airflow's total temperature to 1390 K to ignite the hydrogen, and then reduce the temperature to steady-state run conditions.

The procedure used to accomplish this was to add extra hydrogen to the normal air, oxygen, and hydrogen mixture. Once engine ignition had occurred, the extra hydrogen flow was cut off, and the lower airflow temperature and correct oxygen levels were established. Further temperature reductions were made by increasing the heater airflow above the initial conditions.

This high initial temperature, followed by a significant drop, produced transients in the airstream total pressure and airflow rates. This was not the only combustion issue affecting the Mach 4 engine. The data indicated only a third of the fuel injected parallel to the airflow from the side strut was actually burned entirely. Further analysis supported this conclusion, but the reasons were not clear. Overall, researchers concluded that the engine had performed well in some tests, the remaining problem areas were defined, and, with additional research, the desired performance at Mach 4 would be achieved.

As with the Mach 7 tests at Langley, most of the problems were interactive, meaning that their solutions were interdependent. Several research activities were begun involving the igniter/ignition and inlet/combustor interactions, and scramjet configuration-related supersonic combustor issues. Once these problems were solved, researchers expected a fixed-geometry airframe-integrated scramjet would be efficient at speeds above Mach 4.

The Mach 4 engine design was modified by moving the fuel injectors to the walls near the downstream end of the combustor. This allowed the engine to operate in a "mixed combustion mode," as both a scramjet and a conventional (i.e., subsonic combustion) ramjet. Part of the fuel would be burned in a supersonic airflow in the forward part of the combustor, and the remainder, farther downstream, was burned subsonically. The researchers believed this would produce higher performance while the engine operated at speeds below hypersonic [7].

NEW TESTS, NEW ENGINE DESIGNS

At this point, the Mach 4 and Mach 7 scramjet designs had undergone two years of ground testing. The Mach 4 engine had been tested at GASL, and the Mach 7 tests were done at Langley's arc-heated scramjet test facility. Now the two scramjets were transferred. The Mach 7 engine was sent to GASL, because its Mach 7 test facility had the advantage of higher pressure capability

(80 atm, versus 30 atm at Langley). A range of facility pressure capabilities was seen as necessary for finding a pressure limit for combustor operations.

The Mach 4 scramjet was also transferred, from GASL to a new Langley test facility called the Langley Hypersonic Propulsion Test Cell No. 1. At GASL, only one or two Mach 4 tests could be made per day. The Hypersonic Propulsion Test Cell could make as many as 10 runs per day (see Fig. 4.2).

One area of GASL's Mach 7 follow-on research was fuel-ignition problems. This effort used the three-strut engine design with the fuel injector arrays both perpendicular and parallel to the airflow. Calculations indicated the combustion problems would vanish if the combustor-entrance pressure was raised by a factor of two and an ignition source was added. The initial GASL Mach 7 tests, which involved the higher pressures the test facility could produce but with the original engine geometry and autoignition of the hydrogen used in the Langley tests, did not show dramatic improvements over the original low-pressure tests. Because the problem seemed related to ignition, researchers considered the use of chemical additives as an ignition source to be a viable alternative.

Silane, or, more properly, silicon tetrahydride (SiH_4), was identified as a potential igniter fuel. Silane is a colorless gas available from commercial gas suppliers. Silicon is a Group 4 element located directly below carbon on the periodic table, so the silane molecule is analogous to methane. As a result,

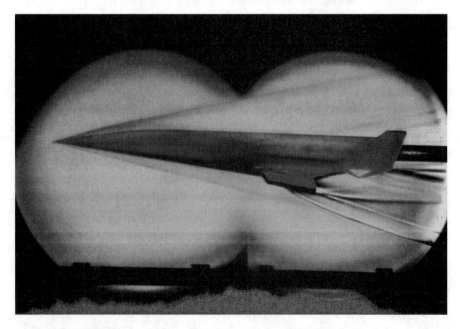

Fig. 4.2   An airframe-integrated scramjet undergoing a high-Mach wind-tunnel test. The shock waves are clearly visible.

silane is compatible with hydrogen fuel systems. Silane had been considered as a potential rocket fuel but had never been used. Its principal industrial use was in the production of semiconductor devices for the computer industry.

Silane did pose several problems, however. It was pyrophoric, meaning the gas ignited on contact with oxygen, giving off a bright bluish-white flame. It had to be handled with care, because any leaks would ignite a fire. A purge of the piping with an inert gas was required both before and after test runs. However, similar care had to be taken while handling hydrogen gas. And unlike silane, hydrogen burns with an invisible flame, posing a significant risk to personnel [8]. (A Lockheed engineer who worked on the Saturn V's S-II second stage recalled the dangers of working around liquid hydrogen. While on a catwalk, he noticed heat haze in front of him. A line had a pinhole leak and was spraying liquid hydrogen across the catwalk. The hydrogen ignited on contact with air and was burning with an invisible flame. To detect such leaks, the engineers held a broom in front of them. A hydrogen leak would be indicated by the broom bursting into flame.)

When a sufficient concentration of silane and hydrogen comes in contact with oxygen, autoignition occurs, producing sufficient heat to ignite the hydrogen gas (about 800 K). The reaction goes to completion, producing SiO_2 and H_2O. (The exhaust products are water and a white powder with a particle size of 1 μ.)

The initial tests used a 20/80 ratio silane/hydrogen mixture from the perpendicular and parallel injectors for 1 s, followed by hydrogen only once a flame front was established. These proved disappointing, because once the silane was shut off, combustion ceased. Curing the problem required a modification to the injector, which was based on an HRE design. A second set of perpendicular fuel injectors was added downstream of the existing set. This procedure was called *staged injection*. The two sets of fuel injectors created overlapping "separation zones," which served as a flameholder.

After a transition stage lasting about 2 s, the silane/hydrogen mixture was replaced by hydrogen. Unlike in the earlier attempt, both stable combustion and flameholding continued after the switchover. This also proved the effectiveness of silane as an igniter.

As the Mach 7 tests got under way, research activities began with the Mach 4 engine at the Langley Hypersonic Propulsion Test Cell No. 1. The facility had been newly modified for engine research, and the Mach 4 scramjet work was the first project undertaken following its checkout. Eight to ten engine tests, lasting about 20 s each, could be made over a single day. Initial tests used the original strut configuration rather than the modified staged design of the Mach 7 engine. As with the Mach 7 tests, a silane/hydrogen mixture was injected to ignite the air/fuel mixture, and was switched to hydrogen only when combustion began. Despite the lack of staged injection, the silane established the flameholding pattern.

The next step was a systematic study to determine the critical parameters for effective flameholding using staged fuel injection, and the most effective use of silane. The decision to use a 20/80 mix of silane and hydrogen in most of the Mach 7 and 4 tests had been arbitrary. No information was available on the optimum and minimum ratios. Although the 20/80 mix gave good results, the cost was about 25 times that of using hydrogen alone. Subsequent tests tentatively showed that a 5/95 mixture of silane and hydrogen gave nearly the same results. A 1980 report noted: "These results are very positive from a cost standpoint," but cautioned, "they are preliminary at this point ... having covered only a narrow range of test conditions" [8, pp. 642–646].

The scale of the three-strut engine testing was an impressive 268 runs. At Langley, 90 runs were done in the AHSTF, and the Combustion-Heated Scramjet Test Facility (CHSTF) was used for 178 test runs of the engine.

These tunnels were part of wide-ranging improvements to Langley's ground test facilities made following the end of the HRE project. The first of the new engine test facilities to come on line was the AHSTF, which opened in 1976. Test gas was produced by heating air with an electrical arc and then adding unheated air to produce the required stagnation pressure and enthalpy conditions. The AHSTF speed-range capability was Mach 4 to 8, maximum stagnation pressure was 40 atm, and it could sustain a relatively long test period. At Mach 4, a run could last up to 60 s; at Mach 8 the test duration was 30 s.

The next facility to become available, in 1978, was the CHSTF. As the name suggests, air was heated in this facility by combustion of hydrogen. Air was added to the test gas to bring the oxygen content to 21 percent. The test gas could replicate stagnation enthalpy flight speeds between Mach 3.5 and Mach 6, with a maximum stagnation pressure of 34 atm. Test time in the CHSTF was 25 s.

The third of the engine test facilities was the 8-Foot High Temperature Tunnel (8-Ft. HTT), originally designed in the late 1950s and put into service in the mid-1960s. Its original research activities involved aerothermal loads and aerothermostructures testing. After some two decades of activity, the 8-Ft. HTT was upgraded in 1993 for propulsion testing. As in the CHSTF, test gas in the facility was heated using combustion, but with methane rather than hydrogen being burned. Oxygen was added to bring the test gas composition to the proper value. Maximum stagnation pressure in the 8-Ft. HTT was about 136 atm at Mach 7, and the run time was about 30 s at the test point.

Although these three facilities were used for testing of all scramjet engine components, beginning in the late 1960s the Direct-Connect Supersonic Combustion Test Facility (DCSCTF) was used for tests of scramjet fuel injectors, combustor configurations, inlet isolators, and nozzle expansions. The DCSCTF was not an independent facility, but rather was a parallel test cell attached to the CHSTF. The test gas was heated by burning hydrogen, followed by the addition of oxygen to bring the composition up to 21 percent.

The DCSCTF capabilities were more limited than those of the engine test facilities. Stagnation conditions were only about 40 atm and 2100 K. These were equivalent to flight speeds of about Mach 7.5.

All of these test facilities were located at the Langley Research Center and operated in the high-supersonic to low-hypersonic speed range, from Mach 4 up to Mach 7 or 7.5. Scramjet testing at speeds of Mach 12 to 17 required a much more technologically advanced facility, one capable of producing true flight stagnation enthalpy. This was accomplished by using a shock wave to heat and accelerate test gas to the conditions of the flow as it entered the combustor at the planned flight Mach number.

This capability was met with the NASA HYPULSE (*hypersonicpulse*) shock-expansion tube, which was located at the GASL facility. HYPULSE, originally built at Langley as a tool for investigating atmospheric entry at super-orbital velocities, had been closed down until 1988, when it was moved to GASL and reactivated. The HYPULSE could simulate speeds as high as Mach 15 for a few milliseconds at a time, at dynamic pressures as high as 1000 psf. The data collected included wall pressure and heat flux, flow visualization, and fuel plume imaging. Some of the results achieved with the HYPULSE included developing a procedure for analyzing scramjet data from a pulse facility taken in hypersonic flow conditions, data to verify scramjet design methodology, imaging of the fuel plume and quantifying air/fuel mixing, and measuring water vapor to determine combustion performance [3, pp. 4, 5, 10; and 9].

Despite the availability of improved ground-based research facilities, however, the three-strut scramjet was still far from being realized as an operational engine. At Mach 4 it had suffered from combustor-inlet interactions, which were resolved by adding injectors and placing them farther back in the combustor. At Mach 7, performance was below the level predicted until silane was used to aid ignition. Mach 7 tests also saw flameholding problems that proved more intractable because they were caused by the small size of the struts and fuel injectors.

To correct the problems caused by the struts' scale, researchers considered different remedies. One approach was use of a single, larger strut, but shock wave problems were inherent in this configuration. The solution was the "reverse sweep inlet." In this design, the leading edge of the strut is swept in the opposite direction from the sidewalls. The oblique shock from the strut intersects the sidewall shock along a line, and the combined shocks strike the wall at a point, preventing boundary-layer separation.

A new version of the three-strut scramjet was designed, called the *strutless parametric engine*. This engine design was longer and narrower than the three-strut engine. The sweep was retained at the inlet but eliminated for the rest of the engine. The inlet was V-shaped, but much narrower than that of the three-strut engine. Beyond this was a short, straight section called the *isolator.* This section would prevent unstart by creating oblique shock waves in the inlet flow that would prevent the high pressure in the combustor from propagating upstream.

Following this was the combustor, which had the perpendicular and parallel fuel injectors. The combustor faired out, merging with the nozzle, and used a rectangular fixed geometry with wedge sidewalls for compression of the airflow, as on the three-strut engine, but, as the name makes clear, the struts were eliminated from the design.

The engine was designed in separate sections to allow easy changes of the inlet sweep without having to pull the whole engine from the wind tunnel. The initial goal was to eliminate the combustor-inlet interactions that showed up in Mach 4 tests. In the mid-1980s, the parametric engine underwent 238 tests in the CHSTF and another 212 in the AHSTF. The 450 total runs were a significantly greater number than had been achieved with the three-strut engine.

The third in the series of Langley scramjet engine designs, the *step-strut engine*, was a refinement of the parametric engine. It had a single long strut down the center of the engine. The leading edge was a series of unswept wedges that staggered the shock waves striking the unswept sidewalls. A series of fuel injectors lined the walls, in a configuration similar to that of the parametric engine. This inlet-strut design had several advantages. The step configuration behaved as if it were swept for engine starting. Once the engine was running, however, the downflow due to sweep was minimized. The step-strut engine was tested in the CHSTF 245 times. Tests at Mach 4 showed that the design prevented combustor-inlet interactions [3, pp. 4, 15].

The inlet designs tested all had similar mass capture, total pressure recovery, and other features. This represented an important breakthrough in scramjet development, because the sweep angle, and the number and design of the struts, are determined by the system or combustor requirements, not by the inlet itself.

Beyond the engine tests, there were other advancements in scramjet designs. Research was conducted at Langley on mixing-controlled combustion, which Ferri had proposed in the early 1960s. Most of the efforts involved injecting hydrogen into air, to measure the mixing rate of fuel from the perpendicular and parallel injectors in the supersonic flow. The ability to model the mixing process in the turbulent flow was the first step in describing the supersonic combustion process. Cold-flow mixing studies were used to produce correlations for perpendicular injection mixing, including identifying the effects on mixing efficiency caused by the strut and injector spacing, as well as the diameter of the injectors. These results were important because the flow field within the combustor of a scramjet engine is very complex, particularly with regard to supersonic fuel-air mixing and flame propagation along the combustor at various Mach numbers. At lower Mach numbers, rapid burning caused "thermal choking" of the engine. To avoid this, the fuel was injected from the struts along the airflow. This slowed the fuel/air mixing and burning, and spread the combustion over a greater length and area. At high Mach numbers, the fuel was injected at right angles to the supersonic airflow so it would mix and burn quickly.

From this research, a "combustor design recipe" was developed at Langley that defined the length needed for complete combustion, the injector spacing, and the injector angle. For injection into a supersonic airflow, a single injector fueled an area about twice as wide as it was high. As a result, with injectors on both sides of the interior, the spacing between injectors had to be about equal to the height of the gap between the struts. The length required for complete mixing increased as the fuel equivalence ratio approached 1. In contrast, both rich and lean mixtures had shorter mixing lengths, because either the fuel or the oxidizer was excessive.

With additional calculations, the design recipe also allowed good predictions of the engine's thrust and cooling requirements. These were tested to estimate the performance of the small test scramjets. Efforts were also made to develop two- and three-dimensional Navier-Stokes codes for internal engine performance calculations. A three-dimensional Navier-Stokes code was also developed to analyze several inlet flow fields.

This computational fluid dynamics (CFD) program was used for the original Langley three-strut engine, the reverse sweep inlet, and two-strut models. An algebraic two-layer eddy viscosity model was also developed for turbulent-flow calculations. When compared to the data from the two-strut inlet, the turbulent flow program calculations showed a very good match for the pressure distribution for two points on the inlet sidewall, as well as for the inlet's mass capture and pressure recovery data.

The scale of the scramjet research undertaken in the mid-1980s was impressive. Various models had undergone more than a thousand test runs, each lasting 10 to 20 seconds. This added up to nearly 5 hrs of total scramjet operating time. Given the average speed, this was equivalent to a nonstop flight around the world.

Perhaps the tests' most important achievement was summed up in a brief statement in a 1985 technical paper by Langley researchers G. Burton Northam and G. Y. Anderson: "At both Mach 4 and Mach 7 flight conditions, there is ample thrust for acceleration and cruise. These tests demonstrated the integration at the component technology to construct an engine design that delivers useful thrust levels over the Mach 4 to 7 flight speed range" [10, 11].

In the same paper, there was another measure of how much had been learned, expressed in a way both subtle and telling. On the next-to-last page of their report, "Supersonic Combustion Ramjet Research at Langley," was Figure 32, "Summary of Scramjet Test Results." The vertical axis showed thrust, and the horizontal axis showed fuel flow. The amount of thrust at Mach 4 and 7 was plotted. Also marked was the vehicle's coefficient of drag

(C_D), which indicated that the scramjet produced more thrust than the vehicle's drag. The figure had no measurements for the amount of thrust or fuel flow. A practical scramjet engine had been developed, capable of propelling a vehicle at low hypersonic speeds. How it was capable of doing this was classified [10, p. 24].

NATIONAL AERO-SPACE PLANE

These accomplishments came at a time of technological optimism. As in the late 1950s and early 1960s, the potential of a scramjet-powered, single-stage-to-orbit vehicle beckoned. Following the end of the X-24C research vehicle, interest in hypersonic vehicles remained within the Air Force. This eventually led to studies of Transatmospheric Vehicle (TAV) concepts, which in turn led to a Defense Advanced Research Projects Agency (DARPA) 1984 study called Copper Canyon. This showed that the idea of a hypersonic vehicle had promise, and generated support for a development program. This became the National Aero-Space Plane (NASP), also referred to as the X-30.

A scramjet-powered single-stage-to-orbit offered the possibility of routine and low-cost access to space. The proposed Strategic Defense Initiative (SDI) missile defense program would require the orbiting of numerous payloads. Additionally, the loss of Challenger showed the limitations of the shuttle program. Unfortunately, building the NASP would be difficult and, after a decade of effort, ultimately unsuccessful.

The original NASP scramjet concept was by HRE designer Anthony duPont, who headed a small consulting firm. Unlike standard scramjet designs, duPont's did not use a booster rocket to accelerate the engine to ignition speed, but a "combined-cycle" design (see Fig. 4.3). This allowed the propulsion

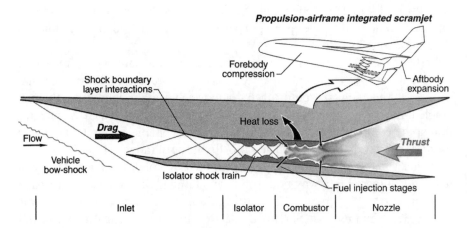

Fig. 4.3   A simplified drawing of an airframe-integrated scramjet's internal flow path. The flow path concept represented the culmination of nearly a half-century of research.

system to operate from a standing start up to orbital speed. (As previously noted, conventional ramjets and scramjets alike had to be accelerated to ignition speed by an auxiliary propulsion system before producing thrust.)

The NASP's engine used a series of injector nozzles in a secondary inlet ahead of the combustor. The injectors' exhaust generated airflow through the inlet to produce thrust, even while standing still. The result was a unified propulsion system. The injectors would be used for takeoff thrust. The engine would operate as a ramjet until reaching hypersonic speeds, and then switch to a scramjet mode. Finally, as the vehicle neared orbital speed, and the scramjet's specific impulse dropped as the atmosphere became too thin, the engine would operate as a rocket, propelling the vehicle into orbit.

The duPont NASP engine concept also used a system called the Liquid Air Cycle Engine (LACE). Air was collected in flight and cooled by liquid hydrogen until the oxygen was liquefied (LOX); the LOX was then used in the injectors as the oxidizer for the hydrogen propellant. This eliminated the need for a separate supply of liquid oxygen to be carried on the vehicle. (The LOX was required for propelling the vehicle into orbit, with the engine operating as a rocket.) The LACE concept had originated in the 1950s, and its use had been proposed in some of the Aerospaceplane concepts in the 1950s and early 1960s.

duPont made optimistic estimates of the engine's inlet, combustor, and nozzle performance; the vehicle's air drag; the ratio between fuel and structural weight; and the amount of effort developing the actual vehicle would require. His optimism was noted, but reviewers believed that, though not all the goals were probably achievable, the effort could still be worthwhile. The technical challenges of the vehicle and engine were seen in a similar light. Several potential solutions were believed to be available for each of the unknowns [11, pp. 215–217].

DuPont's engine, called the *government baseline* engine because it was a design standard defined for contractors by the government, was first tested at GASL. The small version, which used his injector design but not the LACE system, produced static thrust. The Government Baseline engine also underwent extensive wind-tunnel tests in the Langley CHSTF. Starting in 1987, a total of 114 runs were made using different engine configurations at Mach 4 flight conditions (see Figs. 4.4, 4.5, and 4.6).

The government baseline engine used in these tests had variable geometry, and operated as a mechanically throated ramjet. Unlike in many other test engines, the 20/80 percent silane/hydrogen mixture was not used for ignition; spark plugs were used instead. The objective of the government baseline engine tests was to identify major problems with the dual mode ramjet/scramjet that might render them unusable in the NASP program.

The duPont engine was only the first of several X-30 scramjet designs tested in Langley's CHSTF and AHSTF tunnels between 1987 and 1990.

Fig. 4.4   An X-30 NASP subscale engine module awaits a test run in a NASA Lewis Research Center wind tunnel.

Others were developed at Rocketdyne, Pratt & Whitney, and the Johns Hopkins University Applied Physics Laboratory. Rocketdyne produced four different engine configurations. The Rocketdyne A engine underwent 69 tests in the AHSTF, followed by another 55 AHSTF runs of the Rocketdyne A1. Of the

Fig. 4.5   The upper part of the exhaust plume is shaped by the nozzle, the lower plume by ambient pressure and airflow.

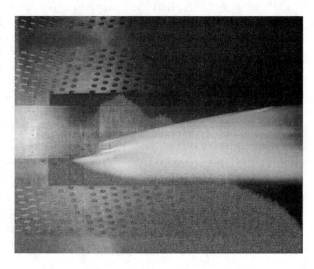

Fig. 4.6   The NASP scramjet engine is prepared for a wind-tunnel test.

four, the Rocketdyne A2 engine underwent the most extensive test series, with 321 total wind-tunnel runs (177 tests in the CHSTF and another 144 in the AHSTF). The Rocketdyne A2+ engine underwent another 72 tests in the AHSTF, bringing the total for the A series to 517 tests.

Although Rocketdyne engines underwent the largest number of tests overall, it was Johns Hopkins's Generic High Speed Engine (GHSE) Model B-1 that was given the most individual tests, with 359 in the CHSTF. The GHSE B-1 was a triangular engine with an opposed dual-ramp inlet, an isolator, and a combustor with an initial constant area section and a divergent section. The GHSE B-1 engine was tested at simulated flight speeds of Mach 4.3 and 5. The Pratt & Whitney C engine came in a distant third with 233 tests (31 in the CHSTF and another 202 in the AHSTF). All these engines were designed with flow paths optimized for transatmospheric flight and mid-range flight speeds. The tests were made to determine whether the engines could achieve high performance and had good operability characteristics.

A second series of scramjet tests was conducted at Langley starting in 1990 and continuing through 1994. These activities were the outgrowth of a NASP management decision that teamed the engine contractors. The different scramjets developed by each were used to produce a series of revised designs. The first of the team engines was the SX-20 (Subscale eXperimental-engine 20) design, which was modified into different configurations during the test series in the AHSTF. The 160 wind-tunnel runs of the SX-20 were made at simulated flight conditions of Mach 7 to 8.

The SX-20 data, along with data from other engine and component testing, were used to develop the final NASP engine design. Called the E22A, this was the scramjet that would be built to power the X-30 research vehicle. The final design called for a total length of 60 ft, which included the inlet ramp, cowled inlet, isolator, combustor, and nozzle. To test the E22A design, a 12.5

percent scale version of the engine's flow path was built, called the SXPE (Subscale eXperimental Parametric Engine). A total of 142 SXPE tests were made in the AHSTF, at flight conditions of Mach 5 to 8.

The final NASP engine design tested at Langley was the Concept Demonstration Engine (CDE; see Fig. 4.7). The engine had a flow path designed to operate at Mach numbers as high as 25, which is orbital velocity. The CDE was a 30 percent scale version of the middle scramjet module of the X-30 flow path. With a heat sink copper structure, the CDE gave an indication of the scale of the NASP research. Many of the scramjets tested in wind tunnels were small. At 16 ft long, the CDE was the largest integrated scramjet ever to

Fig. 4.7   NASP CDE being prepared for wind-tunnel testing.

be tested. Unlike the other NASP engines, the CDE was tested in the Langley 8-Ft. HTT. The 24 CDE tests were made at simulated speeds of Mach 7; the dynamic pressures, however, were only 60 percent of those of actual flight conditions.

This reduced amount was a problem with all 166 SXPE and CDE tests. Although the test facilities could simulate flight stagnation enthalpy, the dynamic pressures that could be simulated were too low. Additionally, both facilities had problems with contamination of the test gas. The 8-Ft. HTT test gas had water vapor and carbon dioxide contamination, and the AHSTF test gas contained small amounts of nitrous oxide.

The SXPE and CDE tests addressed the issues of ground simulation accuracy, parameters, and contaminants. The tests allowed a direct comparison of nearly identical supersonic combustor flow paths, at two different geometric scales and in two different facilities. As a result, the SXPE and CDE tests underscored the need to understand the effects of geometric scale, dynamic pressure, test gas composition, and viscous effects in the design and testing of scramjet engines [3, pp. 3–7, 15, and 11, p. 238].

The Langley efforts were not limited to engine tests, but also included testing of scramjet combustor components. These tests were conducted in the DCSCTF and entailed use of a plasma torch at simulated Mach 4 flight conditions. The torch was fueled with an argon/hydrogen mix that was turned into plasma by a 1-kW current. The plasma served as an ignition source within a hydrogen fuel flowfield. The plasma-torch ignition data were compared with data derived using a silane/hydrogen mix. This revealed that the silane mixture was unable to ignite the hydrogen at stagnation temperatures below 990 K. In contrast, the plasma torch could successfully ignite the hydrogen gas at stagnation temperatures as low as 290 K.

Other activities included tests of swept-ramp injectors at simulated Mach 5 to 7 speeds. These were made to improve the efficiency of air/fuel mixing, with the goal of reducing the scramjet combustor length. These tests used both swept and unswept wall-mounted ramps. The data were used to determine combustion efficiency and mixing performances, which were compared to theoretical models. Among the results: the swept injectors' performance was not affected by the stagnation temperature. In contrast, the performance of unswept injectors dropped as stagnation temperatures increased.

A related injector research activity involved testing expansion-compression ramps. The results of these tests were mixed. Tests of the compression ramp injectors showed the hydrogen fuel would auto-ignite without an external source. The reverse was true for the expansion ramp injectors, which required a 20/80 silane/hydrogen mix before ignition occurred. The expansion ramp also showed a lack of base flameholding ability. Expansion ramp injectors exhibited better combustion efficiency—the fuel mixed with the test gas and burned vigorously in a short distance. The compression ramp injectors did not perform as well. The fuel immediately ignited after injection, causing lower combustion efficiency over the same distance.

Although much of the research effort was focused on the inlet/engine combination, the integrated nozzle was also a subject of testing. One set of tests examined the effects of nozzle geometry on exhaust-flow behavior. Another tested injecting a thin film of hydrogen gas in order to reduce the nozzle wall heat flux and skin friction. The exhaust-flow tests used a scramjet with four swept ramp injectors, a combustor, and an expansion ramp nozzle. These were assembled in different configurations. The test gas velocity in the combustor was Mach 2.7, to simulate Mach-6-to-7 flight enthalpies. Three 7-in.-long nozzle throat sections were built, each with a different curvature.

Calculations had predicted the exhaust flow would become laminar as it expanded through the nozzle; however, the flow remained turbulent throughout the nozzle expansion. For the hydrogen gas film tests, a 4-in.-long film injector was installed in the scramjet. When the tests were completed, heating rates were reduced by up to 70 percent for all three nozzle designs, due to the film injection.

Although these efforts were focused on scramjet operations in the low-hypersonic speed range, work was also done at near-orbital velocities. Shock tubes had long been used to simulate such high speeds, but these test facilities had a major limitation. Although low hypersonic speeds could be achieved in wind tunnels for relatively long periods, the HYPULSE shock-expansion tube located at the GASL facility, but used in the NASA tests, operated for only milliseconds. A snapshot image of the internal processes resulted from each run.

The HYPULSE tests were of scramjet fuel injector performance at Mach 15 and dynamic pressures of 1000 psf. Data collected included wall pressure and heat flux measurements. A significant effort was made in flow visualization, which used an early attempt at laser imaging. A 50-μs-duration laser pulse was fired, causing the hydroxyl radicals to fluoresce, which were used to create an image of an 8-in. "slug" of the high-speed flow. Other imaging was done of the fuel plume's density and spread in the supersonic flow. This was accomplished by measuring the scattering of the laser light by the hydrogen gas, which was recorded by an imaging camera.

Initially, the hydrogen gas was "seeded" with silane to produce the 1-μ silicon dioxide particles. Later, uniform diameter particles were used, which were fed into the fuel stream. Measurements were also made of water vapor to measure combustor performance. Beyond producing the test data, the HYPULSE tests also served as the basis for a procedure developed to analyze shock tube data, and provided data to verify a scramjet design methodology [3, pp. 7–11].

But despite the success of the wind tunnel tests, the X-30 faced insurmountable technical problems. These would doom the effort, and threatened to bring an end to hypersonic research for years to come.

THE FALL OF THE NASP

One example of the NASP's technical problems involved a steady increase in the proposed vehicle's weight as the project progressed. In 1984, with initial studies under way, duPont had said the NASP would be a 52,650-lb vehicle. It would carry a 2500-lb payload into polar orbit, and be built in 5 years at a cost of $5 billion. His initial design was greatly simplified, however, lacking not only specifications for landing gear, orbital maneuvering engines, and retrorockets, but also any reserve for weight increases.

The concept was revised to a more practical design, with a gross weight of 80,000 lb. The scramjet engine had an average specific impulse of 1400 s, compared to 452.5 s vacuum specific impulse for the later version of the space shuttle's main engine. The vehicle itself had a cylindrical pointed fuselage, small delta wings, and a small vertical fin. This configuration became the government baseline for the NASP vehicle, and was sent to potential contractors for additional study.

Rockwell International, McDonnell Douglas, General Dynamics, Boeing, and Lockheed submitted conceptual designs, using their own estimates of engine thrust and vehicle drag. The contractors' design weights were far greater than those of duPont's initial or revised estimates. Rockwell International's design was the lightest, at a gross weight of 175,000 lb. This was over twice that of the government baseline X-30 design. The other weight estimates went far higher: McDonnell Douglas, 245,000 lb; General Dynamics, 280,000 lb; Boeing, 340,000 lb; and Lockheed, 375,000 lb. In October 1987 Boeing and Lockheed, submitters of the two heaviest designs, were dropped from the NASP competition.

As comparison, a Boeing 747 weighed some 800,000 lb fully loaded. However, the percentage of payload for the 747 was far higher; with the NASP, around 75 percent of the total vehicle weight was fuel. Even with the higher-density "slush" hydrogen used as fuel, the result was a huge airframe.

Two years later, the NASP had again grown in size, to a gross weight of 400,000 lb. Far from bearing the sleek shape of the government baseline vehicle, the new X-30 concept was a high mass capture inlet that had low drag. The vehicle had a "platypus" nose shape, a flat upper surface with two fins, flat sides, two small delta wings, and a sloping lower fuselage for the inlet ramp and integrated nozzle (see Figs. 4.8 and 4.9). Three scramjet modules were located on its underside. A 60,000-lb-thrust rocket engine was also part of the design.

The increase in the NASP's estimated weight was due to the twin banes of aircraft design—reduced engine performance and increased drag. The average specific impulse in du Pont's optimistic original estimates was 1400 s. But by 1990, the expected average specific impulse was only 1070 s, a decrease of nearly 25 percent. The estimated drag had gone up as well. To reach orbit, the NASP would have to reach a velocity of 25,000 fps, plus the

Fig. 4.8   The final NASP configuration.

velocity needed to overcome atmospheric drag during the ascent. For a conventional rocket, this was an additional 5000 fps of velocity. This brought the
total velocity requirement to 30,000 fps.

The government baseline X-30 vehicle required a total velocity of 47,000
fps. This was due to its higher drag compared to a rocket. By 1990, refined
drag estimates meant the estimated required velocity was 52,000 fps, an 11
percent increase. The combination of less efficient engines and the higher total
velocity needed to reach orbit translated into more fuel that had to be carried.
This meant a larger vehicle structure to contain the necessary volume.

Fig. 4.9   Artist's conception of the final NASP configuration in flight.

With the increase in the NASP's size, its capabilities were sharply reduced. By 1990, a polar orbit was no longer possible. Launch would have to be made due east from the Kennedy Space Center in Florida, to take advantage of Earth's rotation. Even with this assist, it was estimated that another 3000 fps of velocity would be needed to overcome drag and actually reach orbit. To do this, still another design improvement was required, one with a minimum vehicle weight of 450,000 lb. Development costs were now estimated at $15 billion.

Even if the money was forthcoming, major technological unknowns remained. Specific milestones, including vehicle design, propulsion, structures, and materials, had been established that were to be met before a commitment was made to development of a full-scale vehicle. By 1992, of 38 milestones defined, only 17 had been met. Of particular importance were the unknowns related to materials; there were 19 goals in this area, and only 6 had been met.

As support for the project evaporated, the NASP budget was cut. The bulk of NASP funding came from the Air Force, with NASA providing a lesser amount. From a peak of $320 million in fiscal year 1989, NASP spending dropped to around $250 million during fiscal years 1990 and 1991. With fiscal year 1992, and failure to meet the requirements needed for full-scale development to begin, this dropped to $205 million. In the next budget, fiscal year 1993, the amount dropped to $140 million. At that funding level, development of the NASP was out of the question [11, pp. 214–223].

The "all-up" approach to the NASP, rather than an incremental series of research efforts and test vehicles, was due to the project personnel's belief that the data from a small scramjet could not be scaled up for use in a full-size engine. As a result, only a prototype vehicle with a speed of over Mach 20 was seen as being capable of providing the "flying test bed" needed to test the ground data's validity.

This belief was widely held within the hypersonic research community, due to the limitations of ground-based wind tunnels and test facilities. Speeds of Mach 10 and above could only be simulated for a few milliseconds at a time. No existing facilities could come close to simulating Mach 25 conditions. Additionally, the scramjet, inlet, exhaust, and airframe all had to be integrated properly in order to function. Without suitable ground facilities, the only way to test the integrated system was in flight.

An incremental approach, such as one using a research aircraft for data at Mach 10 to 15, was rejected for the same reason. Researchers believed the true unknowns were at speeds well above this range. As a result, the argument was made that it would be cheaper and faster to build the full-scale X-30 than it would be to develop the system one element at a time. With this approach now eliminated, new options were sought [12].

THE SEARCH FOR ALTERNATIVES

Vincent L. Rausch was among those who attempted to shift the NASP effort toward an incremental approach. From 1986 to 1988, Rausch served in the NASP Joint Program Office at Wright-Patterson Air Force Base. His positions included acting program manager, Director of Program Operations, and Director of Systems. He also served as the first director of the NASP Inter-Agency Office at the Pentagon. His responsibilities included oversight of the Department of Defense portion of the NASP budget, congressional liaison, and day-to-day coordination with NASA. Rausch joined the NASA Headquarters Office of Aeronautics in 1991, following his retirement from the Air Force, in the position of Assistant Director for Aeronautics (High-Performance Aircraft). In this capacity, he coordinated NASA's programs with the Department of Defense. He was appointed director for the NASP program in 1992, and served as the agency focal point for NASA involvement in the program.¶¶

Rausch developed an incremental development plan for the NASP. The initial version of his plan envisioned use of wind-tunnel testing to speeds of Mach 8, CFD calculations for speeds between Mach 8 and Mach 15, and finally, rocket-boosted models for tests above Mach 15. He had located rocket

---

¶¶Vincent L. Rausch biography, ca. 2004, supplied by author.

boosters for use in the high-speed flight tests. This approach addressed the objections of NASA scientists who believed that a Mach 15 vehicle would supply most of the data needed to understand the technical requirements of a NASP-type vehicle. Finally, the incremental approach would involve hardware being flown, allowing the project to exhibit visible progress [12, pp. 189, 190].

In November 1992, the NASP management announced a finalized development plan, which took incremental steps to producing an operational vehicle. These began with rocket-launched subscale tests, a preliminary hypersonic research aircraft followed by an improved version of the research aircraft, then a prototype single-stage-to-orbit vehicle, and finally the operational vehicle. The program would begin in fiscal year 1995, and be completed in fiscal year 2012.

The first step in the revised plan was called *HYFLITE* (HYpersonic Flight Test Experiment). This incremental approach represented a break with the NASP program's past. It was intended to be a fast-paced effort overseen by a small management team. Initially, it met this requirement. The HYFLITE project office had the manufacturing flow plan for the test vehicle completed before December 1992. After years of problems, a positive mood emerged gradually among team members.

The HYFLITE I test vehicle would be wedge-shaped and simulate the forebody of the NASP; it would be used to determine boundary-layer transition. The HYFLITE II vehicle would also be wedge-shaped, but also include an inlet and engine flow path. It would demonstrate inlet operability, airflow characteristics, and scramjet performance. This last component meant the scramjet would actually produce thrust. The test vehicles would be ground launched, using a Minuteman II ICBM as a booster.

Although HYFLITE I and II were intended as low-cost alternatives to the NASP, they still involved a complex set of technical challenges. The Minuteman II, like all long-range missiles, flew a high-arching trajectory several hundred miles into space before descending toward the target. For the HYFLITE, the booster would have to lift off vertically and turn to fly horizontally at a relatively low altitude, as compared with its normal ballistic trajectory. During the launch, the test vehicle would be covered with a shroud to protect it from atmospheric heating. Before the test began, this shroud would have to separate, posing control problems and causing a brief delay in data acquisition.

The original cost estimate for the HYFLITE was $135 million. This soon increased to $200 million, and then $300 million. When a detailed estimate of program costs was made, the total for the two HYFLITE launches became $579.8 million. The issue soon worsened. HYFLITE I and II were to be tests of the scramjet components; they would not test the NASP vehicle aerodynamics. As a result, a third mission was proposed, one more akin to that of a traditional X-plane.

The HYFLITE III was to demonstrate a "free flyer." The vehicle would be launched by a Minuteman missile and was to provide stability and control data at high speed, as well as demonstrate the integrated scramjet's ability to produce more thrust than the vehicle's drag. These goals came with a very high price—$1.5 billion for the single flight. Not surprisingly, although HYFLITE I and II received tentative approval, HYFLITE III remained "pending."

NASP personnel estimated that the HYFLITE launches would meet 50 percent of the program's research goals for scramjet performance in aerothermal loads testing, just under 50 percent for boundary-layer and CFD validation for local flow, 40 percent for airframe/engine integration, 25 percent for scramjet operability and performance, and 20 percent for engine mode integration.

The HYFLITE was not the only alternative considered by NASP management. A more advanced proposal was that of the X-30X, a full-scale vehicle that would fly in the Mach-12-to-15 speed range. It would not attempt to reach orbit, nor have a low-speed propulsion system like the original X-30. As a result, it would have to either be air-launched or use a ground-launched booster or turbojet engines. Costs for the X-30X were estimated to be $3 to $5 billion [12, pp. 294, 295, 298, 299, 307, 308].

The NASP management was not the only party searching for alternatives. Several concepts were studied at the NASA Dryden Flight Research Facility. Starting around 1990, pilot Milton O. Thompson, who had flown the X-15 and the M2-F1 and M2-F2 lifting bodies, and Dr. Kenneth W. Iliff, an engineer on these projects, began looking at possible manned research aircraft that could test air-breathing propulsion systems. Called the HALO (Hypersonic Air-Launch Option), a B-52-launched vehicle was initially considered, but would have been limited to a top speed of Mach 4.5. They felt this too slow to be of value.

Their attention shifted to a larger vehicle launched from the back of the 747 Shuttle Carrier Aircraft. Two NASP-like research vehicle designs were proposed. The first was one attached to an expendable rocket. The second was a large vehicle that used onboard rockets to accelerate to the test speed. Mach 10 to 14 was the favored speed range. This was faster than the X-15 had traveled, but would avoid the severe heating and material problems of speeds approaching orbital velocities.

By late 1991, Thompson and Iliff began studying an SR-71 as the HALO launch aircraft. Although the B-52 and 747 would have launched the HALO at subsonic/low altitude conditions, the SR-71 would air-launch the vehicle at Mach 3 and 70,000 ft. The second scenario would also result in design of a lighter research vehicle. A Mach 12–14 vehicle launched from the 747 would weigh 80,000 to 100,000 lb. By contrast, an SR-71-launched HALO with the same performance would be 60 ft long, weigh about 17,000 to 18,000 lb, and have a wingspan of 20 ft.

This option was not without technical risk. The D-21 unmanned reconnaissance drone had used a similar launch profile. Like HALO, it was designed to be launched from the back of an M-21 aircraft at Mach 3. (An M-21 was a two-seat aircraft similar to the SR-71.) During an early launch attempt, the D-21 collided with the M-21, causing the death of one of the two crewmen.

Also under consideration was the launching of test vehicles by Pegasus rockets to speeds as high as Mach 15, a concept Iliff and another Dryden engineer, Henry Arnaiz, worked on. These flights would focus on gathering hypersonic aerodynamic and aerothermodynamic data, as well as boundary-layer and shock-impingement measurements. The Pegasus could also carry a larger test vehicle bearing deployable scramjet experiments. Pegasus satellite launches had already been conducted using Dryden's B-52, so center personnel were familiar with the booster. Between the HALO and the Pegasus test vehicles, most of the speed range of potential hypersonic vehicles could be covered with such a plan.

Iliff briefed various hypersonic experts within NASA, and later recalled being baffled by the different receptions with which the Dryden proposals were met. On one hand, NASP advocates said research with the Pegasus-launched vehicles would not add any information of significance, because its parameters were fairly well understood. In contrast, wind-tunnel researchers said very little was known in these areas. They needed high-quality aerodynamic and aerothermodynamic data. To accomplish this, the Pegasus flights would have to be made at exact values of angle of attack, angle of side slip, Mach number, and altitude. Researchers in this camp would only support the Pegasus tests if these conditions were met. Otherwise, they said, the results would not be useful to them.

The HALO proposal had been briefed to management at Ames, of which Dryden was an adjunct facility at the time, and also to Langley, the NASP program, and the Department of Defense. On September 3, 1992, NASA administrator Daniel S. Goldin visited Dryden and was briefed on activities there, which included the HALO and Pegasus proposals. Goldin bombarded Iliff with questions on single-stage launch vehicles, orbital debris, and rockets vs. air-breathing engines as the briefing progressed. Finally, Iliff was able to complete the hypersonic briefing [13].

Goldin apparently liked what he heard. In comments to a reporter with the *Antelope Valley Press*, Goldin said he favored an alternative to the NASP, and considered the HALO concept as a possibility. He later said the comment was opinion and not policy, and when pressed, Goldin said he did not want to abandon the NASP. Despite this, NASP personnel believed, "The vultures are circling" [12, pp. 284–286].

The end of the NASP came when Congress included language in the Air Force budget directing that the NASP "be phased out in an orderly fashion in Fiscal Year 1994," because the United States could not afford an X-plane program. Congress also directed the Air Force to continue basic hypersonic research, but provided only $60 million to support the effort. The HYFLITE was too expensive to be supported at this level of funding.

NASP management eventually decided to abandon the HYFLITE boundary-layer testing, and focus instead on one basic issue: would a scramjet actually work in flight? The new program, called HySTP, for HYpersonic Scramjet Technology Project, would launch a 25 percent scale scramjet engine to a speed of Mach 12 to 15 using a Minuteman or Peacekeeper ICBM as the booster.

In retrospect, the HySTP goals were very limited. The test would not directly lead to an X-30 engine, and might not even be valid for a scaled-up engine. Indeed, the goal simply was for the scramjet to produce "useful" thrust. This was defined as thrust from combustion beyond that produced from simply expelling hydrogen out the nozzle. Use of an ICBM as the HySTP booster meant that, as with the HYFLITE, there were potential technical problems with flying a depressed trajectory.

The Air Force approved the HySTP, budgeting $450 million over 5 years. The initial fiscal year 1995 funding was $45 million from the Air Force and another $22.5 million provided by NASA. The Air Force issued a contract for the project with existing NASP contractors, effectively transitioning the NASP to the HySTP. Finally, a scramjet flight test seemed to be on a solid footing. As before, however, events intervened.

U.S. military forces had been deployed in the African country of Somalia. To pay for the operations, the Department of Defense transferred money from other programs. NASP/HySTP lost a total of $40 million annually from the fiscal year 1996–2000 budgets. Because this level of cuts would end the program, other money was transferred from another source to sustain it. This funding, however, was also eventually committed to the Somalia operations. Secretary of the Air Force Sheila E. Widnall searched in vain for other funding with which to continue the HySTP, and the contract was terminated at a cost of $10 million. The NASP program was closed out in June 1995 when the last of the contractors ended their activities.

THE REASONS WHY

The NASP's problems were complex, but its basic failing was that, as with the Aerospaceplane program of the 1960s, technology that would meet the single-stage-to-orbit goal did not exist. As a result, the NASP program never came close to its goal of building a single-stage-to-orbit vehicle. Indeed, the project was unable to produce even a credible vehicle design. [One type of research tool that hadn't been available to hypersonics researchers prior to the NASP program was the Computational Fluid Dynamics (CFD) computer program (see Fig. 4.10). Such programs could calculate airflow around a vehicle.]

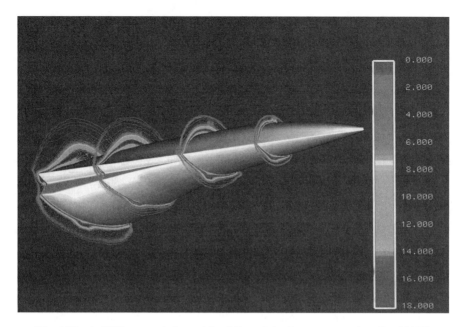

Fig. 4.10   A CFD-generated graphic of the original government baseline NASP.

The original NASP design was that of the duPont vehicle. This was a lightweight aircraft (~50,000 lb) that could be ready in a relatively short time frame (5 years) and at relatively low cost ($5 billion). This concept "sold" the idea of a single-stage-to-orbit vehicle to necessary supporters and set the NASP effort in motion. Yet it was soon apparent the design was not remotely realistic.

But because the NASP had been sold as easy, low-cost, and a single-stage-to-orbit vehicle, based on the original (and soon abandoned) duPont design, it became impossible to successfully promote the incremental approach until it was too late. Costs for the HYFLITE and HALO programs (the latter estimated at $2 billion) were too high to gain political support for the amount of data they would provide. The NASP program was tied to a goal researchers lacked the data to achieve, was not structured so as to collect the data needed, and lacked direction on how to proceed. Indeed, despite repeated reviews that showed the duPont vehicle was completely impractical, the concept continued to enjoy political support within Congress.

Another issue was the lack of a focused goal for the NASP, which was doomed through attempts at turning it into all things to all people. At various points, the NASP was alternately viewed by different program partners as a hypersonic research program, a traditional X-plane, a hypersonic airliner, and a single-stage-to-orbit vehicle. These were all very different concepts, with different requirements.

There was also a fundamental difference in outlook between NASA and the NASP primary funding source, the U.S. Air Force. The Air Force needed an operational space-launch capability that would be ready as soon as possible. This implied a development program more akin to that for a new fighter or bomber rather than for a strictly research-oriented aircraft like the X-15 had been. NASA depictions of the NASP as a "flying wind tunnel" meant to "gather data," ran counter to this operational outlook. Many in the Air Force wanted to see in the NASP a "revolution, not evolution."

One reflection of these differing viewpoints could be seen in attempts to add wish-list features to the X-30. This meant having features similar to those of the operational NASP Derived Vehicle (NDV). This was a vehicle design that was more than a technological demonstrator, but less that a fully operational NASP. Some of these were simple, such as having a cargo door. Others, however, were far more complex, including the ability to go into orbit, descend into the atmosphere, and climb back into space. Another idea was a "go around" capability. If a landing attempt had to be aborted, the NDV could increase power, climb, and come around for another try. This alone would have added 50,000 to 100,000 lb to the vehicle weight and several billion dollars to project costs.

This was not the only example of internal policy contradictions. Whereas the Air Force saw the NASP in operational terms without defining the specifics, NASA focused on the research aspects. A Langley Research Center memo noted a number of different roles proposed for the NASP:

1. Single stage to orbit
2. Low-cost delivery of payloads to Earth orbit
3. Long-duration hypersonic flight (interceptor, cruise vehicle, transport)
4. Experimental hypersonic research aircraft

Of the four goals, the third and fourth were "considered sufficient" to justify the NASP. In contrast, the memo noted that many thought goals 1 and 2 "are not feasible or even a serious requirement." Yet these missions were the ones the NASP supporters used to sell the NASP to potential congressional supporters.

This did not go unnoticed by NASP personnel, and it affected the working relationship among the Air Force, NASA, and the contractors. One contractor recalled after attending a briefing, "NASA could care less if the X-30 ever flew ... as long as the data was good." An Air Force source recalled a NASA official saying, "The program is a success even if the X-30 never flies" [12, pp. 59, 235–237].

Parallels can also be drawn between the NASP program and earlier aeronautical and aerospace efforts. The twin revolutions brought by the jet engine and supersonic flight caused every aspect of aircraft design to change. One part of meeting such a challenge was the development of X-planes for use in conducting flight research. These included the X-1 series; the D-558-I and -II; the X-2, X-3, X-4, X-5, XF-92A, X-15; and the lifting bodies.

These aircraft tested different configurations and materials, and explored the pitfalls of high-speed and high-altitude flight. X-planes made it possible to investigate such stumbling blocks as stability issues with inertial coupling and pitch-up, propulsion systems, control system designs, life support, crew escape, and structural issues. High-speed and high-altitude flight put demands on aircraft structures never before seen. The results were often tragic. The de Havilland DH 110 broke up during a flight demonstration at the 1952 Farnborough Air Show in the United Kingdom. The two crewmen as well as 28 spectators on the ground were killed. The crash was traced to a wing structure that met existing design criteria, but could not withstand the stresses caused by a high-speed rolling pull-up [14].

The history of flight was a constant series of evolutionary steps made in an unexplored environment. Each new step brought new unknowns. In some cases, only minor changes were necessary. In others, all old rules had to be abandoned and new ones developed.

The same was also true with humanity's leap into space. Apollo was the ultimate example. A flight to the Moon was once regarded as impossible. In reality, the basic rocket technology needed for a flight to the Moon existed by the early 1960s. And although many unknowns remained, not the least of which was whether humans could function in space, the means to reach the Moon were in hand.

The early space theorists had developed ideas for reaching the Moon. These included a direct ascent using a huge booster, assembling the lunar craft in Earth orbit using multiple launches, and the method chosen for Apollo, designing different spacecraft for each part of the mission.

What was attempted with the NASP program was to forgo all the interim steps. It was as if the X-planes were skipped, and an attempt made to go directly from P-51 Mustangs to the space shuttle, or use the technology of Robert Goddard's first rocket to build a Saturn V.

The NASP program did accomplish the advanced research needed to transform scramjet and related technologies. Chuck McClinton noted the pre-NASP state of the art in 1984:

> From the scramjet perspective—we knew how to make an effective scramjet for operation up to Mach 7. We had cycle analysis codes, mixing "recipes," auto ignition and flameholding models, isolator models, shock-boundary layer models, boundary layer transition models, inlet starting models, multiple finite rate chemistry models, good 2-D CFD codes, reasonable 3-D CFD codes, and had just started cranking out solutions for 3-D fuel injection and mixing. We had facilities required for moderate size scramjet testing to Mach 7+. We had experimental

techniques for inlet, combustor and nozzle component testing. We also had methodologies for engine model testing to simulate airframe integration issues/effects. We also had just started looking into testing scramjet combustion in shock tunnels. ... We had limited experience with engine structural cooling—from the HRE and some small efforts funded by Langley at Aerojet. Most of this was accomplished with less than 30 man-years/year between 1965 and 1984.***

McClinton summed up the changes brought about by the NASP:

NASP's major contribution to scramjet technology was expanding the designs, analysis and experimental database and to the broad range from Mach 3 to 16 using an army including up to 10,000/year total, or at least 1000/year for scramjet technology development alone. (These numbers included government labs, university, and contractors both small and large.) NASP studied numerous design architectures, inlets concepts, airframe-integration approaches, fuel injection concepts, flame holders, etc.

NASP engineers developed improved analytical models for all parts of the scramjet from inlet to nozzle. One of the big pushes was for fully integrated analytical design methods—several were developed and utilized. CFD was widely applied to scramjet problems. Many new codes were developed, but computational technology was not much different from pre-NASP. However, serious application by an army of young engineers demonstrated the great potential of CFD to support and supplement wind tunnel testing and flight scaling wind tunnel results.

Many "unit" experimental tests were performed to verify that each code could predict the physics—like shock-boundary layer separation, shock-shock heating, glancing shock interactions, fuel penetration and mixing, recombination chemical kinetics, etc. Most of the testing used existing facilities, or copies of existing facilities. Old shock tunnels were reactivated for Mach 10–16+ scramjet testing, and a few new ones were developed. A huge database was generated for several scramjet configurations. These included inlet, combustor, isolator and nozzle component tests, and integrated flowpath tests. Engine structural, thermal management, engine control, and multidisciplinary design optimization were also addressed.***

With the end of the NASP program, the future of the scramjet and hypersonic research faced a crossroads. Would the data gained now be applied to other projects? Or, as had happened before, would the research teams be dispersed, their experience lost, and would all that had been learned have to be rediscovered yet again?

***Chuck McClinton e-mail to Curtis Peebles, "Re: NASP," Friday August 1, 2008, 8:09 P.M.

REFERENCES

[1] Fry, R. S., "A Century of Ramjet Propulsion Technology Evolution," *Journal of Propulsion and Power*, January-February 2004, p. 32.

[2] Luidens, R. W., "Arrangements of Jet Engine and Airframe for Increased Range," NACA RM E56L04, July 26, 1957, pp. 1–9.

[3] Rogers, R. C., Capriotti, C. P., and Guy, R. W., "Experimental Supersonic Combustion Research at NASA Langley," AIAA 98-2506, 1998, p. 3.

[4] Becker, J. V., and Kirkland, F. S., "Hypersonic Transports, Vehicle Technology for Civil Aviation—The Seventies and Beyond," NASA SP-292, 1971, pp. 429–431.

[5] Edwards, C. L. W., "A Forebody Design Technique for Highly Integrated Bottom-Mounted Scramjets with Application to a Hypersonic Research Airplane," NASA TN D-8369, December 1976.

[6] Henry, J. R., and Anderson, G. Y., "Design Considerations for the Airframe-Integrated Scramjet," NASA TM X-2895, December 1973, pp. 4–14.

[7] Guy, R. W., and Mackley, E. A., "Initial Wind Tunnel Tests at Mach 4 and Mach 7 of a Hydrogen Burning, Airframe Integrated Scramjet," AIAA 79-29413, 1979, pp. 2–6.

[8] Beach Jr., H. L., Mackley, E. A., and Rogers, R. C., "Use of Silane in Scramjet Research," CPIA Publication 329, Vol. 1, September 1980, pp. 640, 645.

[9] Reubush, D. E., Puster, R. L., and Kelly, H. N., "Modifications to the Langley 8-Foot High Temperature Tunnel for Hypersonic Propulsion Testing," NASA-TM-100486, June 1987, p. 1.

[10] Northam, G. B., and Anderson, G. Y., "Supersonic Combustion Ramjet Research at Langley," AIAA-86-0159, 1986, pp. 2–4, 7.

[11] Heppenheimer, T. A., *Facing the Heat Barrier: A History of Hypersonics*, NASA, Washington, DC, NASA SP-2007-4232, 2007, p. 238.

[12] Schweikart, L., *The Hypersonic Revolution: Case Studies in the History of Hypersonic Technology*, Vol. III, Air Force History and Museums Program, Washington, DC, 1998, pp. 25, 189, 190.

[13] Iliff, K. W., and Peebles, C. L., *From Runway to Orbit: Reflection of a NASA Engineer*, NASA, Washington, DC, NASA SP-2004-4109, 2004, pp. 310, 313–321, 323–325.

[14] Bullen, A., and Rivas, B., *John Derry: The Story of Britain's First Supersonic Pilot*, William Kimber & Co, London, 1982, pp. 163–166.

Chapter 5

THE BIRTH OF THE HYPER-X

> Simplicate and add lightness.
>
> *Ed Heinemann, designer at Douglas Aircraft*

With the NASP program ended, the Air Force planned to spend $20 million per year in the fiscal year 1995–2000 budgets on hypersonic research. The new Air Force effort would focus on use of scramjets fueled by hydrocarbon rather than by hydrogen, and operation in the speed range of Mach 8 to 10, instead of at near-orbital speeds. The initial engine design produced was similar to that of the NASP, despite changes in performance and fuel [1].

At Langley, post-NASP scramjet tests of the Subscale eXperimental Parametric Engine (SXPE) and the Concept Demonstration Engine (CDE) continued between 1994 and 1996. The SXPE underwent 124 tests in the Arc-Heated Scramjet Test Facility (AHSTF) tunnel. Another 27 tests of the CDE were made in the center's 8-Foot High Temperature Tunnel (8-Ft. HTT).

An investigation of the fundamental issues affecting dual-mode scramjet engine performance was also begun. The Langley Strutless Parametric Engine was selected for testing, and installed in the CHSTF. By the end of the effort, in 1996, 400 tests runs had been made using a variety of configurations, including two different sidewall compression inlets, one with both sidewall leading edges swept backward and the other with one sidewall swept back and the other forward. The tests explored inlet-combustor interaction, captured airflow profile, and test-gas contamination effects. The engine was tested both as a classic mechanically throated ramjet and as a thermally throated dual-mode scramjet. Hydrogen gas was injected from various locations using sidewall orifices. The tests were done at speeds between Mach 4 and 5.5.

All this was a pale shadow of the level of work done on the NASP by the Air Force and NASA. It seemed, once again, hypersonic research and scramjet development work done as part of NASP would be in vain. A generation of researchers, engineers, and theoreticians would disperse to other fields, and their experiences would be lost as they had been after the Aerospaceplane, HRE, and other failed hypersonic projects of the 1970s.

Clearly, what was needed was not just subscale test vehicles or simplified full-scale prototypes. Rather, a whole new concept of development and testing was needed if scramjets were to become a reality. This would come through a variety of sources—from planetary exploration, from NASA's longest-serving and most controversial administrator, and from a person whose significance in the field of hypersonics few are fully aware of.

FASTER, BETTER, CHEAPER

By the early 1990s, NASA's planetary exploration effort was in serious trouble. The 1960s had seen the initial unmanned planetary launches to Venus and Mars. These were relatively simple missions that could be launched at relatively low cost, usually with a pair of probes launched on separate rockets to allow for possible failures. Between 1964 and 1971, for example, NASA launched a total of six unmanned missions to Mars, of which data were returned on four. (The other two missions, Mariner 3 and Mariner 8, failed on launch.)

Following these probes, and as NASA Mars missions became more ambitious, costs increased, development times grew longer, and the number of flights was reduced. The Viking 1 and 2 missions, launched in 1975, each involved both an orbiter and a lander, requiring a Titan IIIE booster. Both missions were successful, but their development had taken 8 years and cost $1 billion [2].

After Viking, U.S. Mars exploration ended for 17 years, until the 1992 launch of the Mars Observer. This was a large orbiter carrying a high-resolution camera and six other experiments. Because of the spacecraft's weight, the launch vehicle was a Titan III, the largest expendable booster operational in the United States at the time. Three days prior to entering orbit around Mars, the spacecraft suffered a complete failure. The only scientific results from the Mars Observer were a few distant images of the planet [3].

The Mars Observer failure illustrated the conceptual flaws that had brought U.S. planetary missions to a halt. Ironically, the project had begun in 1981 as the first of what was intended as a series of low-cost missions to Mars. The cost was originally estimated at $250 million, but project scientists argued for more complex experiments, which the original spacecraft design could not support.

As a result, the Mars Observer grew heavier and more complex, requiring a larger and more expensive Titan III booster to carry the larger payload. Because it was now a large, costly, and one-of-a-kind vehicle, rather than part of an ongoing series, it could not fail. This meant the project had to be subjected to the paperwork of formal system-management procedures to assess risk, prolonging development time and requiring a standing army of technicians and engineers. A full 11 years passed between the project's start and its 1992 liftoff. Mission costs had risen three-fold, to $800 million [4].

Daniel S. Goldin (shown in Fig. 5.1) began arguing for a "faster, better, cheaper" philosophy for agency projects soon after becoming NASA Administrator in April 1992. He was convinced large, expensive, and complex planetary probes like the Mars Observer—which he saw as, among other things, too difficult to garner needed political support for—were creating a downward spiral for NASA's programs. The solution, he believed, was to turn the existing situation on its head. Development costs would be cut by using smaller spacecraft that didn't take a decade or more to build. Launch costs would be cut because smaller spacecraft could use smaller, cheaper Delta II boosters. Operational costs would be cut by shorter development time frames, fewer personnel, and flying the spacecraft to their destination faster.

As a result, his reasoning went, more missions could be flown, and though they would not be individually as productive as larger probes, the failure of one or more would not be as huge an impact as the loss of the Mars Observer represented. But there was a flip side—this process also meant more risk. Goldin urged engineers and scientists to "push the limits of technology even more." Potential gains, he felt, would be worth taking risks and accepting failures—and "faster, better, cheaper" would see plenty of both.

Sixteen faster, better, cheaper missions had been approved by 2000: five Mars missions, a lunar mission, three space telescopes, two comet/asteroid rendezvous, four Earth satellites, and an ion propulsion test. Of these, five missions failed in flight, and another was cancelled before launch due to cost overruns. This represented a 63 percent success rate, comparable to the 66 percent success rate for Mars missions from 1964 to 1993. In terms of their effects on the U.S. planetary program and cost of space exploration, however, the results were of great significance.

Before the era of faster, better, and cheaper, the unmanned Mars program had effectively ended. There had been many paper proposals but none had been approved beyond the Mars Observer mission. After the advent of faster, better, cheaper, Mars exploration saw a level of activity greater than that of

Fig. 5.1 NASA Administrator Daniel S. Goldin addresses a crowd in 1998.

a)   b)

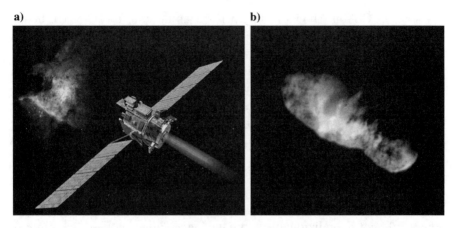

Fig. 5.2   a) Deep Space 1 probe, and b) one of its targets, the comet Borrelly.

even the 1960s and early 1970s. And this trend was not limited to Mars, but included dedicated missions to explore several asteroids and comets.

The total cost of these 16 missions was equal to that of the Cassini mission to Saturn. The Cassini project involved a Saturn orbiter and a small lander that touched down on the surface of the moon Titan. It took 8 years to develop this probe, followed by a 7-year flight to reach Saturn. Total cost for the Cassini mission was $3.3 billion [4, pp. 8, 18, 19, 49–52].

One of the faster, better, cheaper missions was Deep Space 1 (see Fig. 5.2), which tested an ion propulsion system. Xenon gas was ionized and then magnetically ejected to produce thrust. The amount of thrust was tiny, amounting to the weight of a sheet of paper on a person's hand, but the 181 lb of xenon that fueled the craft were enough to produce thrust for 20 months and increased the probe's speed by 10,000 mph. Although a conventional rocket engine could produce a similar increase in velocity far faster, a rocket needed about 10 times as much fuel as the ion propulsion system did. Deep Space 1 made a flyby of asteroid Braille, followed by another of comet Borrelly. Its primary function, however, was to test ion propulsion in flight [4, pp. 113, 114].*

The Dawn spacecraft subsequently incorporated the ion engine technology tested on Deep Space 1 in a decade-long mission. Dawn, which was launched in 2007 and carries three ion engines, is to enter orbit around the asteroid Vesta in 2011. After completing a geological survey, Dawn's ion engines will fire and it will leave orbit and travel to the dwarf planet Ceres, where it will again go into orbit, in 2015.

Hypersonics also benefited from faster, better, cheaper, though indirectly. The concept served as the basis of a hypersonic/scramjet research effort. This

---

*"Deep Space 1 Launch Press Kit," October 1998, p. 4.

effort was structured as a tightly focused technological demonstration program to be carried out in a short time frame and meet cost limitations while still providing valuable data. Like Deep Space 1, the hypersonic effort would prove an experimental technology in actual flight, to pave the way for future operational use.

LANA COUCH

The final element leading to this program was a remarkable and determined individual. Lana Couch was a young research engineer working at Langley in the summer of 1967. At the time, engineering was a man's world, and women engineers were rare. Her initial experience with hypersonics was with trying to get the HRE inlet to start. After the HRE project, she worked at the 16-ft Transonic Tunnel at Langley during the early and mid-1970s.

In the late 1970s, Couch was recruited to take over management of the 8-Ft. HTT, and she again found herself working in the hypersonic field. She continued in this role for several years before being transferred to NASA Headquarters on a developmental assignment. While at Headquarters, she showed an independent streak, advocating for adding an oxygen replenishment system to the 8-Ft. HTT to allow propulsion testing. Dave Reubush, who became the NASA Hyper-X Deputy Program Manager, recalled,

> This project was actually picked for funding instead of another facility project that Langley upper management wanted more. As a result she was "exiled" and stuck at HQ for a total of something like 11 years. During these years she had significant involvement in the very beginnings of the NASP program and stayed very close to it. Her last job at HQ was that of Director for Space in Code R[esearch].[†]

With Langley now playing a major role in the NASP program, the center's management finally allowed Couch to return there in June 1990 (see Fig. 5.3). She was named director of the NASP office and reported to the center director. Reubush joined the NASP office in March 1991 as chief engineer. He wrote later:

> As time progressed in the 1990s it became obvious to her that the NASP program was going to die before flying anything.... she did not want hypersonics to die with it and the teams and their expertise to disperse as had happened several times in the past, which then resulted in having to reinvent everything a second or third time whenever people got

---

[†]Dave Reubush e-mail to Curtis Peebles, "Re: The Lana Couch Study (DF-9)," November 24, 2008, 5:45 P.M.

Fig. 5.3   Lana Couch.

re-interested in hypersonics. She also believed that there had been a lot of technology developed during NASP and this fact needed to be made known and demonstrated. So, she bootlegged (literally took money that was at Langley for NASP and Generic Hypersonics—things were not nearly as rigid in those days as today) enough money to fund the Dual Fuel study.[‡]

The Dual Fuel study was conducted by a team from Langley and McDonnell Douglas. The Phase I concept vehicle was a manned "waverider," capable of flight at Mach 10 and designed to undertake reconnaissance missions. With this speed capability, it could take off from a base in the United States and reach a target area 8500 n miles away in less than 90 min. Between the high speeds and an operating altitude of 110,000 ft, in military applications the aircraft would have a survivability rate of 95 percent despite a total lack of stealth technology features.

The concept vehicle was about 200 ft in length, weighed less than 500,000 lb, and was fueled with either slush hydrogen or a "dual fuel" of hydrogen and hydrocarbon fuel. (This gave the vehicle its name.) The dual fuel allowed the waverider to be refueled in flight by a conventional tanker and make a subsonic return flight to the United States, which wouldn't have been possible with slush hydrogen as the fuel. The vehicle would carry a 10,000-lb payload, and take off from a runway less than 15,000 ft in length. The propulsion system was a turboramjet for low speeds (up to Mach 4 or 4.5) and a dual-mode ramjet/scramjet engine for hypersonic speeds.

The waverider configuration had several advantages that led the study team to select it. First was the shape, which had exceptional aerodynamic characteristics—in particular, high lift-to-drag ratios. A waverider was also believed to be capable of producing more uniform flowfields at the inlet than

---

[‡]Dave Reubush e-mail to Curtis Peebles, "Re: The Lana Couch Study (DF-9)," November 24, 2008, 5:45 P.M.

would other vehicle shapes. Because waveriders were primarily single-point designs, researchers believed the shape would be ideal for cruise vehicles.

However, nearly all the earlier studies of waverider configurations had focused on optimizing only aerodynamic performance. Little or no effort had been made to develop shapes that incorporated the numerous systems needed for an actual aircraft. These included landing gear, a cockpit, control surfaces, wing-leading-edge designs, and the structural and volume limitations inherent in various vehicle shapes, as well as the stability and control effects of the integrated airframe/engine configurations. The Langley/McDonnell Douglas team did address these features [5].

The vehicle was designed with payload bays in the upper fuselage, and an infrared sensor bay, an equipment bay, and an in-flight refueling receptacle for the hydrocarbon fuel in the nose. A synthetic-aperture radar antenna and equipment bay were mounted in the chine on either side of the forward fuselage. The two-wheel nose landing gear was located just forward of the low-speed engine inlet, and the six-wheel main landing gear was mounted in the chines at the mid-fuselage. The avionics bays were in the chines outboard of the two main landing gear bays. One unusual design feature of the Duel Fuel study was the location of the crew station. On the NASP, this was positioned close to the nose leading edge. On the dual-fuel cruiser, the crew station was in the left-hand chine, just forward of the main landing gear bay [6].

The operational goals of the vehicle forced configuration tradeoffs, as with all aircraft. The size of the high-speed engines required a balance between the ramjet takeover speed of Mach 4 to 4.5, and the Mach 10 cruise speed. Sizing the engine to meet the thrust requirement in the high supersonic range caused a significant drop in engine performance at the Mach 10 cruise, which reduced the vehicle's range. To correct this, the engine's size was reduced to make it more efficient at the high Mach number. This, in turn, also required the vehicle's drag at supersonic speeds be reduced.

To achieve the drag reduction, a parametric study was made using a computer program called HOLIST to determine the optimum vehicle fineness ratio, done by making an "X-Z stretch." This involved scaling up the vehicle lengthwise and in thickness by 30 percent and 60 percent, respectively, while narrowing it laterally to keep internal volume constant. The engine was kept as a constant fraction of vehicle width. The upper surface was raised relative to the chine by 50 percent and 100 percent. This was referred to as a "bump." The modified vehicle was rescaled in three dimensions to maintain the same internal volume as the reference vehicle.

The HOLIST program recalculated vehicle and performance characteristics based on geometric changes. The weight model was modified for vehicle fineness and aspect ratio changes. Engine performance, as measured by thrust per pound of airflow, remained a constant, but the captured airflow was altered to reflect the geometric changes and computational fluid dynamics (CFD) calculations of lateral air-spillage effects. Engine weight and volume were assumed to change linearly, based on the wetted area.

The amount of drag, which is the key element in improving engine efficiency, was calculated in two ways. Skin friction drag was determined by the vehicle's wetted area, and wave drag varied according to area distribution. Given the limitations of computer technology in the mid-1990s, simplifications were used to make calculations easier; the vehicle's center of gravity and control surface area, for example, both were assumed to be constant.

The measurement of vehicle performance used to evaluate the different configurations was the vehicle's range divided by its gross takeoff weight. The vehicle geometric parameter that best reflected the performance improvement was volumetric efficiency. The most volumetrically efficient design had a bump of 100 percent and a stretch of 60 percent over those of the original reference vehicle shape. Fineness, in contrast, was less reflective of efficiency. The highest fineness was characterized by maximum stretch and no bump. However, the vehicle's planform area was decreased, resulting in several performance shortcomings.

One consequence of the decreased planform area was a reduction in the amount of lift. As a result, the angle of attack in Mach 10 cruise flight necessary to maintain level flight was nearly twice that of the reference vehicle, greatly increasing drag. The low end of the performance spectrum was also adversely affected. The smaller planform area resulted in poor takeoff characteristics. The takeoff speed of this configuration was nearly 260 kt, and the runway length required even at this high speed was over 17,000 ft. The Duel Fuel study's authors concluded that the optimum vehicle shape "would reside somewhere in the middle of this design space" [7].

The Duel Fuel study was based on use of a complex propulsion system. For takeoff and acceleration to Mach-4-to-Mach-4.5 speeds, a turboramjet low-speed system was used. This consisted of four Pratt & Whitney Air (Core Enhanced) Turbo Ramjet (AceTR) engines designed for improved subsonic cruise efficiency as well as for retaining a high transonic and supersonic thrust-to-weight ratio.

The high-speed propulsion system consisted of three Pratt & Whitney dual-mode ram/scramjet engines. Although the high-speed system was sized by the thrust requirements of Mach-4-to-Mach-4.5 ramjet takeover speeds, it was optimized for Mach 10 cruise speed. The two sets of engines were mounted in an over/under dual flow path, with the ram/scramjets below the turboramjet low-speed engines.

With an airframe-integrated scramjet propulsion system, all elements of the multi-mode system had to be optimized to provide necessary thrust. The Langley/McDonnell Douglas team examined four different high-speed inlet designs for optimum performance. The inlet geometries were defined by nearly shock-on-lip at the cruise point of Mach 10, a 4-deg angle of attack, and a dynamic pressure of 1000 psf. Comparisons also examined off-design conditions, which included the ramjet take-over point.

Specific variables in the high-speed inlet designs were the number of ramps, final ramp angle, and cowl lip inclination. Of the four configurations, the preferred design had four ramps, at angles of 2.5-deg, 8-deg, 11-deg, and 13-deg, and a flat cowl leading edge. This configuration had excess thrust at Mach 10. The three other ramp configurations all had three ramps, at angles of 2.5-deg, 8-deg, and either 11-deg (two of the ramps), or 15-deg. They also all had 3-deg fixed upturned cowls, though these had several serious problems and were eliminated from consideration. Problems included decreased Mach 4 or Mach 10 performance, additional drag penalties resulting from the fixed upturned cowl leading edge, and, most serious of all, the off-design shock strengths exceeded operable limits [6, pp. 5, 6].

Another design requirement was optimized nozzle integration. This involved two issues: producing the highest specific impulse possible and the need to balance a waverider's pitching moments, which were caused by the forebody compression surface. If the nozzle did not produce a sufficient nose-down pitch moment, control surfaces would have to be used for vehicle trim. This resulted in additional trim drag, diminishing vehicle performance.

The Langley/McDonnell Douglas team made parametric analyses of increasing and decreasing the nozzle length by 10 ft and reducing its expansion ratio from the baseline 15 to 13. The different length and expansion ratios changed what is called the "chordal angle" of the nozzle, which alters gross thrust-vector magnitude and direction. Higher chordal angles generally produce more axial thrust, but less lift. Propulsion performance is improved as a result, but the reduced lift produces a lower nose-down pitch moment.

Nozzle performance was calculated using two computer programs called SCHNOZ and BLIMPK. The parameter used to evaluate results was the product of the trimmed lift-over-drag ratio and the scramjet's specific impulse. These are the driving terms in the Breguet Range Equations.

Results were mixed. The highest specific impulses were associated with the highest expansion ratios. The largest nose-down pitching moment was produced by the longest nozzle with the lowest chordal angle. With short nozzle configurations, improving the nose-down pitching moment resulted in a significantly increased range, but was less successful in improving the scramjet's specific impulse. For longer nozzles, a higher chordal angle increased both the nose-down pitching moment and specific impulse; however, a longer nozzle also increased structural weight and wetted area, without resulting in a significant increase in the vehicle's useable volume.

The final issue was the vehicle's structural design. A hot structure had been envisioned for the NASP, because it was to make a rapid ascent into orbit. Peak thermal loads would be high, but only over a brief period. For the NASP, the outer skin of the vehicle would have been a titanium matrix composite (TMC) skin, with corrugated panels attaching the outer skin to a separate layer of insulation. This, in turn, would be attached to frames that supported graphite epoxy hydrogen tanks while also isolating them from frictional heating of the outer skin. The NASP airframe structure carried the vehicle's inertial, aerodynamic, and thermal loads, while the tanks would need to withstand only the internal pressure of the hydrogen fuel and inertial loads.

The Mach 10 cruiser faced a very different operating thermal environment. Its peak heat load would be lower, but the total thermal load on the vehicle would be far higher, because it would have to withstand the aerodynamic heating of Mach 10 flight for 90 min or more.

As a result, a cold structure design was chosen for the Mach 10 cruiser. The outer layer consisted of insulating material that covered the graphite epoxy hydrogen tanks and was reinforced internally with stiffeners. Unlike with the tanks in the NASP hot structure, which were separate, the skin of the Mach 10 cruiser tanks was also the skin of the vehicle. As a result, the tanks had to carry the airframe, inertial, pressure, and thermal loads. The outer insulating layer provided protection from cruise frictional heating. This design had long been used for the integral fuel tanks on rockets. The skin of the tanks formed the skin of the rocket, with an external coating of insulation to provide thermal protection. This configuration was volumetrically efficient, which was a major design requirement in the cruiser.

To determine whether a hot or cold structure was preferable for the Mach 10 cruiser, rigorous system-level trade studies of each design had to be made. These were conducted using the McDonnell Douglas Fully Automated ThermoStructural Optimization (FATSO) process, through which the relative weight and volume penalties for specific design, volume, and hydrogen boil-off constraints were compared for identical vehicle moldlines. Separate full-vehicle computer models of the two structural concepts were also created, which were used to optimize the individual skin panel thicknesses based on the stresses, stability, manufacturing limitations, and trajectory load conditions.

The differing qualities of the titanium matrix composite and graphite epoxy influenced the required panel thicknesses. Compared with graphite epoxy, TMC had a density 2.5 times greater, but it degraded significantly at higher temperatures. The FATSO analysis indicated that many of the skin panels would require only minimal thickness, which favored the choice of cold structure airframe.

Another factor was that the heat soak for long missions added weight and volume penalties to the TMC structure, and required considerable internal insulation between the TMC skin and the internal hydrogen tank walls. The cold structure, in contrast, used external insulation for both high and low temperature control, which resulted in a more volumetrically efficient design.

Thermal issues also favored the cold structure design. Frictional heating and operational requirements included keeping the graphite epoxy airframe below 350°F during subsonic, supersonic, and hypersonic flight; preventing icing during pretakeoff fueling and subsonic flight; and minimizing boil-off and venting of the hydrogen fuel.

The cold structure was also a favorable one for maintaining the slush hydrogen temperature. Slush hydrogen did not have to be recirculated, a process that caused the slush to melt and the liquid hydrogen to boil off. It remained in the tanks for the duration of the hypersonic flight, resulting in a nearly constant operating pressure in the integral tanks. The low pressure made possible by the cold structure reduced overall structural weight. The use of slush rather than liquid hydrogen was preferable, because liquid hydrogen would boil off much faster. Based on these results, the Langley/ McDonnell Douglas team determined a cold structure was preferable for the Mach 10 cruiser [5, pp. 5, 6].

Phase II of the study focused on a dual-fuel two-stage-to-orbit space access vehicle that was the same size as and a similar weight to the cruise vehicle. The flight profile for the space access vehicle was very different, however. Rather than a climb to cruising altitude followed by a prolonged level flight to the target area, the space access vehicle would make a steep climb to a peak altitude of 280,000 ft, at a top speed of 11,120 fps. The payload bay on the top of the vehicle would open, and the second stage booster would emerge, separate, fire its rocket, and head for orbit. The space access vehicle would then descend to low altitude, rendezvous with a tanker for inflight refueling, and make a subsonic flight back to the launch site.

The final version of the dual-fuel vehicle was called the DF-9. The shape, size, and weight of the Mach 10 cruiser and the space access version were similar. The vehicles' length was 202 ft, 4 in. The Mach 10 cruiser takeoff weight was 531,986 lb; the space access vehicle was slightly lighter, at 520,769 lb [7]. (To give an idea of the relative size of the Mach 10 cruiser DF-9, a Boeing 747 airliner had a takeoff gross weight of 800,000 lb and was 225 ft, 2 in. in length. The DF-9 was both smaller and lighter than an operational airliner.) The Phase II space-access vehicle was not as well defined as the Mach 10 cruiser. Indeed, the Phase II study was postponed in order to focus on the final part of the dual-fuel study [8, 9].

Phase III focused on a subscale research vehicle that would be used to flight test an airframe-integrated scramjet. Langley researcher Dave Reubush noted,

> The purpose of the study was to show what kinds of vehicles were possible with the current state of NASP technology [paper studies] and to

design a vehicle to start demonstrating it as cheaply as possible. The full-sized Dual Fuel vehicle configuration was then used as the basis for the preliminary design of a small scale flight demonstration vehicle. . . .[§]

While the dual-fuel study was under way, Langley's Hypersonic Vehicle Office learned support had been building at NASA Headquarters for a hypersonic flight demonstration program. Reubush continued:

As an add-on to this study [Lana Couch] also wanted a flight test vehicle designed which could demonstrate scramjet operation in flight at as low a cost as possible. The phase III add-on came at an opportune time. About the time the flight test vehicle study was done Dan Goldin gave a speech at Dryden and said that he wanted to "darken the skies over Dryden with X-planes." As a result of this desire Code R ran a competition for X-plane designs.[¶]

Study leader Couch, with Charles McClinton and Vince Rausch from Langley, and John Hicks, the hypersonic lead at Dryden, put together a Phase III proposal that would qualify the concepts and data developed during the NASP program under actual flight conditions. As with the Phase I and II design studies, the research vehicle configuration had to be specific and detailed. The initial NASA studies were focused on the research vehicle's minimum size. The studies indicated the 200-ft DF-9 vehicle could be scaled down significantly, yet still demonstrate that a scramjet was capable of accelerating the test vehicle. The subscale scramjet engine would be operating far enough from the ignition and flameholding limits that hydrogen could be used as the fuel [10].

Because the scramjet and airframe were integrated, the size issue also affected the engine design. Studies of the minimum engine size were done using the three-stream finite rate chemical kinetics code (SCRAM3L) computer program. Calculations showed that the ramp, engine, and nozzle could be scaled down to less than 12 ft in length. The reduction in combustion efficiency due to the smaller size was about 5 percent. Below this size, the results became problematic. Two sizes for the vehicle were considered, 10 ft and 12 ft. The larger vehicle was selected due to a small performance gain and a minimal increase in the overall program cost, because the small increase in size had no effect on booster selection [7, 10].

A second, overarching factor also drove selection of the larger vehicle. Twelve feet was the largest size that could fit in the 8-Ft. HTT. Couch wanted to put an actual Phase III research vehicle in the tunnel and "fly" it to compare the real flight data with the tunnel data. Due to cost and timing issues, Langley

[§]Dave Reubush e-mail to Curtis Peebles, "Re: The Lana Couch Study (DF-9)," November 24, 2008, 5:45 P.M.
[¶]Dave Reubush e-mail to Curtis Peebles, "Re: Hyper-X Questions #1," June 5, 2008, 8:59 A.M.

built a full-scale flow path replica. But the driving factor was the desire to test an actual vehicle in the 8-Ft HTT.**

With the overall size decided, the next step in designing the Phase III research vehicle was to refine the scramjet's flow path to gain back the performance losses resulting from such factors as high sheer forces and overcontraction caused by increased boundary-layer-displacement thickness. The analysis was done using the SRGULL propulsion code. This computer program was developed at Langley and was designed to provide simulations of integrated scramjet loads and performance. Unlike other engine cycle codes, SRGULL was able to predict most flow phenomena, including inlet mass capture, boundary-layer sheer, heat flux and transition, inlet energy efficiency, isolator performance, combustor distortion, nozzle expansion, and divergence losses. It could also predict inlet unstarts, propulsion lift and pitching moments, thermal and pressure loads, and fuel temperatures in regeneratively cooled engines.

The primary inputs used to calculate the different flow phenomena were engine geometry, boundary-layer and initial conditions, and parameters, which corrected for known limitations in the analysis. SRGULL could predict the performance of any ramjet, scramjet, rocket-based combined cycle engine, or ducted rocket at speeds of from Mach 2 to Mach 25. Perhaps the program's most remarkable capability was that a solution could be produced on a personal computer within a few minutes. In contrast, it took weeks or months to complete an analysis with earlier computational fluid dynamics programs [7, p. 2].

The scramjet that emerged from the CFD analysis used the NASP data as its starting point, and was revised to account for scale, wall temperature effects, and other design factors. It was also optimized for vehicle acceleration and engine operability. As a result, the interior design was modified, rather than simply reduced in size from the NASP engine design.

Given the focus on limiting costs and the time required to build a research vehicle, the scramjet was built using off-the-shelf materials, fabrication techniques, and components wherever possible. Although the scramjet was designed for flight tests, its construction was akin to that of the designs used in wind-tunnel scramjets. The scramjet's short burn time required only a limited active cooling system. Thus the more complex, heavy, and costly system an operational scramjet would require was not needed.

Following the NASA studies, McDonnell Douglas undertook conceptual design of the Phase III research vehicle between February and May 1995. Scramjet wind-tunnel tests began in early 1996 to verify the engine design and its flight test controls, develop aerodynamic databases for development of control laws and trajectory, and support flight research activities.

**Hyper-X Correspondence File, Dryden Flight Research Center History Office, and interviews with two individuals.

While this effort was under way, McDonnell Douglas produced a preliminary government-candidate design for the vehicle between March and October 1996. The subscale vehicle's shape was based on the Phase I DF-9, but it would be reduced to the 12-ft length for launch on an existing (but modified) ground or air-launched booster. By using the DF-9 configuration as the starting point for the research vehicle, the existing NASP database could be used. The design would rapidly converge into a controllable subscale test vehicle with the added advantage of low trim drag [10].

BOOSTER SELECTION: AERODYNAMIC CONTROL AND POLITICAL ISSUES

With the research vehicle size and shape decided, and initial wind-tunnel tests beginning with the basic scramjet design, there remained the question of the booster rocket to be used. As in the X-7 ramjet test flights, the research vehicle would need to be accelerated to the planned test speed before the scramjet was ignited [11].

At a very early stage in Phase III planning, Couch's team was looking at various possibilities for the booster with the idea of doing the project as quickly and cheaply as possible. Their plan was to ground-launch the research vehicles, much like what was proposed for the HYFLITE and HySTP concepts. Couch found that several Castor IV solid fuel boosters were available at a very low price. These were considered for use on the low-Mach-number flights. The Castors lacked the capability to reach Mach 10, however. For this profile, a Taurus booster would be required (see Fig. 5.4). The proposed launch site was Wallops Island (see Fig. 5.5), with the flights being made over the Atlantic Ocean, so Langley personnel went to Wallops to discuss ground launch options. Wallops Island was associated with Langley and had been used for the early hypersonic rocket launches. In this scenario, Phase III flight tests would be solely a Langley project.

In contrast, Dryden personnel advocated an air-launched profile. The center had experience gained in initial Pegasus flights, with air-launched research vehicles going back to the late 1940s, and the B-52B was available as the launch aircraft (see Fig. 5.6).

How the air-launched profile was selected is a matter of controversy. Not long after the initial deliberations over the booster issue, according to one account, the Phase III launch question became a political issue. A congressman reportedly called NASA Administrator Dan Goldin regarding the lack of any Dryden Flight Research Center involvement with the Phase III effort. A separate account by a former Dryden engineer is that a senior Dryden official went to NASA headquarters and personally lobbied Goldin for the air-launched option. Soon after, Langley director Paul Holloway stepped down, and Couch was moved out of her job in October 1996 and named the deputy

Fig. 5.4   Taurus booster.

of the new Aerospace Transport Technology Office. The Dryden official was also reassigned to another position. A third account stated only that the use of Dryden's B-52B and the Pegasus booster as the launch method was decreed from "on high."[††]

The effort to build a scramjet-powered research vehicle based on the DF-9 concept vehicle was in a perilous state. Paul Holloway had left Langley, Lana Couch was transferred, and the relationship between Langley and Dryden was tense. The program needed a new leader. Vince Rausch stepped into this role (see Fig. 5.7). He made the decision to give up a job at NASA Headquarters, move to Langley, and walk into the middle of the inter-center dispute. Rausch had to get both centers working together and, at the same time, get a multi-year, multi-million-dollar program under way to achieve something that had never before been done.

[††]Hyper-X Correspondence File, Dryden Flight Research Center History Office.

Fig. 5.5   Rocket launch facility at Wallops Island with a Scout rocket being prepared on the pad.

Fig. 5.6   Pegasus and the B-52B.

Fig. 5.7 Vincent L. Rausch became the Hyper-X project manager.

Langley engineers did not have a chance to undertake trade studies of the Castor IV and Taurus boosters. Some Langley personnel believed that although the ground-launched profile would have increased complexity, this would have been offset by possible lower booster costs for earlier flights. Circumstances caused the Langley engineers to harbor considerable animosity toward certain Dryden personnel.

An Orbital Sciences Corporation team did, however, perform trade studies of air- and ground-launched booster options during 1996. The early studies used a low-fidelity 3 degrees of freedom (3 DOF) analysis, which incorporated structural limits of angle of attack and dynamic pressure, mass and aerodynamic properties, control system limitations for the thrust-vectoring controls of the rocket exhaust and/or the fin actuation system for aerodynamic controls, and acceleration limits for the boosters.

The 3 DOF analyses also assumed a 2200-lb research vehicle and a 500-lb Ballast Avionics Module (BAM) to attach it to the booster. Another assumption was that research vehicle separation would occur within the atmosphere rather than in the vacuum of space. This meant the booster would have to be aerodynamically stable all the way through separation. Three different configurations of ground-launched boosters were examined at three planned test speeds, Mach 5, 7, and 10. The first would demonstrate dual-mode operation as both a ramjet and scramjet. The Mach 7 test would determine whether the scramjet could produce enough thrust to accelerate the vehicle despite the high drag. The Mach 10 profile would test the vehicle and engine at the design cruise speed.

An Orion 50S GL rocket, used as the second stage of the Taurus booster, was the first proposed as a candidate booster. This rocket was itself a modified Orion 50S rocket that served as the air-launched Pegasus booster's first stage. The Orion 50S GL was capable of powering Mach 5, 7, and 10 flight profiles. All that would be required was adjusting the amount of ballast the booster would carry. For a Mach 5 burnout speed, 11,000 lb would be required; for Mach 7, 5350 lb; and a Mach 10 final speed required just 500 lb. (The ballast also served as a margin for weight increases of the research vehicle and the BAM.)

However, the requirement for aerodynamic control of the booster through separation meant the Orion 50S booster's thrust-vectoring control (TVC) was not sufficient. The rocket would have burned out before separation began. Without thrust, the TVC could not stabilize the booster during the sequence. Fins and ailerons would have to be added to the booster, requiring a new control system development effort. This would increase project costs and technical risk.

The remaining two booster options had more serious shortcomings. The ground-launched Castor IVB rocket had been flight proven as part of the Maxus Sounding Rocket, which was a plus. (And it had been the booster originally considered for the Phase III Mach 5 and 7 flights.) Additionally, the rocket already had a TVC and four fins, so the rocket would require only an upgrade rather than an entirely new control system. The Castor IVB had a fatal performance flaw, however. A 5800-lb ballast load was required for the Mach 5 flight, and 1400 lb were needed for the Mach 7 mission. But even with zero ballast, the rocket's maximum speed with the research vehicle and the BAM was only Mach 8, well below the planned Mach 10 speed. A second, different booster would be needed if the Mach 10 flight was to be accomplished.

The final ground-launched booster candidate was a GEM-VN rocket. This was in use with the Delta II rocket as a strap-on booster. As with the ground-launched Orion 50S GL, the GEM-VN would require a TVC with fins and ailerons, with all the complications and risks this involved. As with the other candidates, ballast would be used to adjust the burnout speed. Mach 5 required 10,000 lb of added weight, and Mach 7 required 4000 lb. But like the ground-launched Castor IVB, the GEM-VN was incapable of reaching a Mach 10 separation speed. Even under the best conditions, Mach 9.8 was the highest speed it could attain.

Perhaps significantly, the Taurus booster was not considered as a potential ground launch vehicle. (Couch and the others had planned to use the Castor IV on the Mach 5 and 7 launches, and the Taurus for the Mach 10 profile.) The Orbital trade study also stipulated that *one* booster would have to meet all three Mach numbers. Because the Castor IVB could not reach Mach 10, it was eliminated from consideration. This "one rocket fits all" requirement also effectively eliminated all ground-launched options.

The final option was the B-52B-launched Orion 50S. The Pegasus first stage would be fitted with an adapter to which the research vehicle was attached. The trade studies assumed standard 40,000 ft/Mach 0.8 launch conditions. The Mach 5 profile would require 20,000 lb of ballast. For the Mach 7 missions, 10,600 lb were needed, whereas the Mach 10 profile required only 2000 lb. The trade study indicated six advantages to using the Pegasus first stage as the launch vehicle [11]:

- Reduced risk by having a single booster configuration capable of meeting all three launch profiles
- Operational advantages of using an air launch, including not being tied to existing launch pads and test ranges
- Availability of the B-52B launch aircraft, which had made the first six Pegasus launches
- Lower launch vehicle integrated heating during boost due to air launch
- Existing aerodynamic control surfaces
- A flight-proven system with successful satellite launches to its credit

In July 1996, NASA gave approval for development to begin in Phase III. The overall project was given the name of *Hyper-X*, for "Hypersonic Experiment"; the research vehicle was designated as both the X-43A and the Hyper-X Research Vehicle (HXRV). The modified Pegasus was called the Hyper-X Launch Vehicle (HXLV). The complete assembly of the HXLV, adapter, and HXRV was called the *stack*. The adapter and HXRV were called the *short stack*. The formal start of the project came in September 1996.

The Hyper-X was to provide the needed data on hypersonic scramjet operation, validate the technology and design tools inherited from the NASP program, and meet the low-cost/high-risk/tight schedule goals of the faster, better, cheaper concept. Initial plans called for a $200 million/4-year program. The primary objective specified was to advance the technological readiness levels of scramjet engines from ground testing to the actual flight environment while staying within cost and schedule limitations (something that had not been done successfully with the NASP). The test effort had three aspects [12]:

- Flight validations of design predictions
- Development and enhancement of methods and tools to improve scramjet-powered vehicle designs
- Risk-reduction methods such as preflight analyses and verification of predicted aerodynamic, propulsion, structural, and integrated scramjet operation and operability

As the Phase I and II vehicles were, the Phase III research vehicle was still a "paper airplane." It had a detailed design and its basic characteristics had been determined, but the task remained of turning the Hyper-X from a concept into real hardware, developing a workable development plan, and undertaking the related flight research program. To this extent, the Hyper-X project still differed little from the NASP.

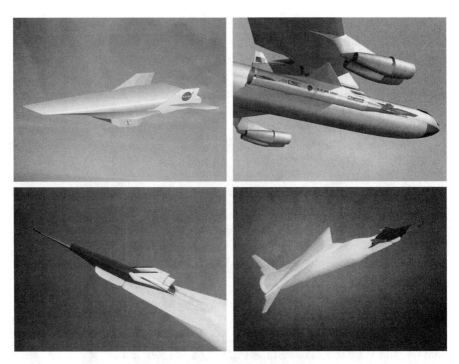

Fig. 5.8 Artist renderings of Hyper-X.

As the Hyper-X project began, several artists' conceptions of the research vehicle, stack, and launch aircraft were produced (see Fig. 5.8). It would take some 7 years for them to become a reality.

Lana Couch was what separated Hyper-X from the NASP, HRE, and Aerospaceplane programs and from the other hypersonic and scramjet proposals of the previous four decades. Through determination and ability, she had created a research plan that had the potential to succeed. Dave Reubush underscored the importance of what she had accomplished. He wrote, "Lana played a key role and has not gotten the credit she deserves."‡‡

REFERENCES

[1] Iliff, K. W., and Peebles, C. L., *From Runway to Orbit*, NASA, Washington, DC, NASA SP-2004-4109, 2004, pp. 324, 325.
[2] Ezell, E. C., and Ezell, L. N., *On Mars: Exploration of the Red Planet 1958–1978*, NASA, Washington, DC, NASA SP-4212, 1984, pp. 269, 283.
[3] Godwin, R. (ed.), *Mars the NASA Mission Reports*, Apogee, Burlington, Canada, 2000, pp. 183–200.
[4] McCurdy, H. E., *Faster, Better, Cheaper: Low-Cost Innovations in the U.S. Space Program*, The Johns Hopkins University Press, Baltimore, 2001, p. 18.

---

‡‡Dave Reubush e-mail to Curtis Peebles, "Re: The Lana Couch Study (DF-9)," November 24, 2008, 5:45 P.M.

[5] Bogar, T. J., Eiswirth, E. A., Couch, L. M., Hunt, J. L., and McClinton, C. R., "Conceptual Design of a Mach 10, Global Reach Reconnaissance Aircraft," AIAA 96-2894, July 1996, pp. 1–6.

[6] Bogar, T. J., Alberico, J. F., Johnson, D. B., Espinosa, A. M., and Lockwood, M. K., "Duel-Fuel Lifting Body Configuration Development," AIAA 96-4592, 1996, p. 2.

[7] Ferlemann, S., McClinton, C., Rock, K., and Voland, R. "Hyper-X Mach 7 Scramjet Design, Ground Test and Flight Results," AIAA 2005-3322, 13th AIAA Hypersonic Technology Conference, Capua, Italy, May 19, 2005, slide 5.

[8] Hunt, J. L., and Eiswirth, E. A., "NASA's Dual-Fuel Airbreathing Hypersonic Vehicle Study," AIAA 96-4591, 1996, pg. 3.

[9] Reubush, D. E., "Hyper-X Stage Separation—Background and Status," AIAA 99-4818, 1999, pp. 2, 3.

[10] Rausch, V. L., McClinton, C. R., and Crawford, J. L., "Hyper-X: Flight Validation of Hypersonic Air Breathing Technology," ISABE 97-7024, 1997, p. 2.

[11] Joyce, P. J., Pomroy, J. B., and Grindle, L., "The Hyper-X Launch Vehicle: Challenges and Design Considerations for Hypersonic Flight Testing," AIAA 2005-3333, pp. 2–6.

[12] Voland, R. T., Rock, K. E., Huebner, L. D., Witte, D. E., Fischer, K. E., and McClinton, C. R., "Hyper-X Engine Design and Ground Test Program," AIAA-98-1532, p. 2.

Chapter 6

FROM CONCEPTS TO A FLIGHT RESEARCH PLAN

> Second star to the right, and straight on, til morning.
>
> *J. M. Barrie, Peter Pan*

The X-43A would be both the first scramjet-powered aircraft to fly and the fastest air-breathing airplane ever built. But as with earlier experimental and research aircraft, there was no potential for large, lucrative production runs for contractors. Instead, they would have to build an aircraft to meet the extraordinary demands of a new flight environment, within the tightly scripted time frames and program costs of NASA's "faster, better, cheaper" approach. To accomplish this in the Hyper-X project, the government and its contractors would have to assemble a team, determine a research plan, make basic decisions on issues such as vehicle size and separation system design, and cope with several initial technical problems.

CHOOSING THE CONTRACTORS

The process would begin with selection of contractors. The first items that needed to be contracted for were the Hyper-X boosters. The Pegasus had already been selected as the booster that would be used, so a sole-source contract was issued to Orbital Sciences Corporation in February 1997, and it covered the modified HXLV rockets.

NASA officials decided the procurement process for the research vehicle would be a rapid one. As a result, the X-43A request for proposals (RFP) was based on the Phase III research vehicle configuration. Further, if a potential contractor accepted the Phase III design study, a detailed technical justification would not be required. Contractors with objections to the preliminary design, however, would have to give reasons for those objections and offer alternatives.

Two very different sets of contractors were in the running to build the research vehicle. The first was McDonnell Douglas and Pratt & Whitney—unsurprising, because McDonnell Douglas was a major aircraft manufacturer and had worked on the NASP, the Dual Fuel study, and the Phase III study. Pratt & Whitney was a major jet engine manufacturer that had developed the

C scramjet engine design for the NASP as well as the Air (Core Enhanced) Turbo Ramjet (AceTR) engine design and dual-mode ram/scramjet engine for the DF-9 concept.

Two of the three companies comprising the second team were also to be expected. Rockwell International's Seal Beach electronics division would be responsible for systems development. This company had built both the X-15 (as North American Aviation) and the space shuttle. The engine contractor would be General Applied Sciences Laboratory (GASL), which had been involved with scramjet development since the earliest days. The airframe and adapter contractor, however, was Micro Craft, Inc. of Tullahoma, Tennessee, which had never built an airplane of any type or size. Rather, it was a small, woman-owned company that produced wind-tunnel models.

As with other aspects of the Hyper-X project, differing accounts of the contract selection process have emerged. Dryden engineers recalled that McDonnell Douglas seemed sure to win the Hyper-X contract. But McDonnell Douglas and Micro Craft had both accepted the design study, and although there were marked differences between the technical aspects of their proposals, the contract would be awarded not on the basis of technical issues, but on cost. The McDonnell Douglas bid was much higher than that of the much smaller Micro Craft [1, 2].*

Another factor in the selection was Lowell Keel, the proposed Micro Craft program manager. Most Langley members of the source selection board knew Keel, and he was held in high regard. The Langley team felt that, with Lowell in charge, Micro Craft did have a good chance of being successful.

The NASA selection committee awarded the contract to build the X-43As to the Micro Craft team in March 1997. They and the other team members would now have to build the fastest airplane in the world.

With the contractors selected, one of the first tasks was transforming the divergent groups into a team. The Hyper-X was less straightforward than a typical flight research effort due to the pioneering nature of the project, the autonomous nature of the flight profile, the unknowns of its flight envelope, and the complexity of the task. As a result, the project was divided between Langley and Dryden.

The NASA Langley Research Center (LaRC) had overall responsibility for Hyper-X. Its specific areas were the following[†]:

• Program management
• Technology development management
• Research vehicle and scramjet development

*Paul Reukauf history interview, February 17, 2005, Dryden History Office, pp. 11–13.
†Vincent L. Rausch e-mail to Curtis Peebles, "RE: Hyper-X Questions," May 13, 2008, 9:58 A.M.

- Aeronautical, booster/research vehicle separation, and scramjet wind-tunnel testing
- X-43A aeronautical database development
- Structural design and analysis
- Thermal design and analysis
- Flight controls development
- Problem solving
- Flight data analysis

The Dryden Flight Research Center had very different duties and areas of responsibility. These included the following[‡]:

- Project management for the three missions
- Project oversight
- Software validation
- System validation and verification
- Development of aircraft-in-the-loop simulation
- Flight safety/readiness review
- Flight host (control room facilities, network, B-52B launch plane, and F-18 chase planes)
- Conducting flight missions
- Conducting emergency mission scenarios/training
- Development of test procedures
- Development of flight cards

THE TWO CULTURES

Beyond the responsibilities each bore, two different cultures needed to be melded. Langley and Dryden each had very different research activities and goals, histories, and philosophies. Langley had been the first NACA center established, and its work combined theoretical analysis, wind-tunnel testing, and flight research. It was also located near Washington, D.C., which was both the center of political power and the location of NACA/NASA headquarters.

Langley researchers had proposed development of a supersonic research aircraft during World War II. This was not, however, the rocket-powered/air-launched X-1 the Army Air Forces would build. Instead, the preference at Langley at that time was for a turbojet-powered aircraft that would fly at speeds between Mach 0.8 and Mach 1. John Stack, a senior Langley engineer, argued that the basic purpose of building a research aircraft was to gain in actual flight the compressibility data existing wind tunnels could not provide. Stack's ideas fundamentally influenced development of the Navy's

---

[‡]Dave Lux, private correspondence, January 12, 2006, Dryden Flight Research Center.

D-558-I Skystreak aircraft, which was specifically designed to meet NACA requirements.

Langley was involved in early hypersonic and scramjet research, and later moved into space-related projects. The Mercury program had been managed at Langley, as had programs involving the Echo communications satellite, the Scout rocket, the Lunar Orbiter, and the Viking probes to Mars. Langley had also supplied start-up personnel for all the other NACA/NASA centers with the exception of Marshall (established by the U.S. Army) and Goddard.

Langley could undertake end-to-end research; results of theoretical calculations could be confirmed in the center's wind tunnels and flight tested using its aircraft. It had the atmosphere of a research university. Although wind-tunnel research is the activity for which it is best known, a wide range of activities was undertaken at Langley, including research into crashworthiness, energy-efficient aircraft, vertical take-off and landing aircraft, computational fluid dynamic computer simulations, and materials studies. The center also branched out into non-aviation-related research—the Langley atmospheric sciences group was the first to prove the existence of an Arctic ozone hole. The size, layout, and scope of the Langley facility reflect this diversity of activities.

The culture at Dryden was shaped by role and location equally as much as these two factors influenced Langley. Testing of the rocket-powered X-1 required the expanse of Rogers Dry Lake, located at Muroc Army Air Field, for glide landings (see Fig. 6.1). Langley had only a standard runway, far too short for the rocket-powered research aircraft to make a dead-stick landing. The initial group of five Langley engineers arrived at Muroc Army Air Field (the present-day Edwards) on September 30, 1946. By December, to provide support for the Army Air Force's X-1 flights the detachment had grown to 13 people amid the wilds of the Mojave Desert. The head of the NACA Muroc Flight Test Unit was Walter C. Williams. His instructions were "to make all necessary contacts and decisions for the NACA ... at Muroc" [3].

This statement anticipated the independent culture of what became Dryden. The facility was isolated in the desert, far away from NACA Headquarters and Langley, both in number of miles and frequency of contact. The buildings and housing initially used by the unit were decaying wartime structures. Some of the engineers sent from Langley adapted to life in the wide-open spaces and the harsh extremes of the desert, but others did not. Although the center was established as part of Langley, it became a separate NACA center in 1954. These factors fueled a frontier mindset that, in turn, fostered a colorful collective personality among those who came to work there. In the early 1960s, for example, center personnel built the M2-F1 lifting body using center funds and without telling anyone at NASA Headquarters.

Even the different sizes and physical layouts of Langley and Dryden shaped the two cultures. One symbol of the differences between the two is the role of

a)

b)

Fig. 6.1   Dryden, a) main entrance in 1991, with the X-1E research aircraft in the foreground, and b) looking north, nearly the entire facility is visible.

hangars as architectural features. The Langley main building is out of sight of the main hangar, ramp area, and runways, which are a considerable distance away. Dryden's two main hangars, Buildings 4801 and 4802, are attached to either side of the main building. It's possible to walk out of one hangar, down the central hallway of the main building, and enter the other. The back of the building overlooks the ramp, and beyond is Rogers Dry Lake. The hangar is the iconic symbol of Dryden, whereas the wind tunnel and its supporting facilities are the symbol and substance of Langley (see Fig. 6.2).

Another set of differences entails the sheer size of the Langley facility, and the varied research activities undertaken there. It is impossible for one person to know the names and faces of all the Langley personnel, or to know all the research projects under way at a given time. Because the activities are so diverse, a person unconnected with a research flight would probably not be aware it was occurring.

In contrast, it's possible for an employee to be aware of virtually every person and project at Dryden. The entire Dryden facility is visible from nearly any area on the site. The internal Dryden television system has a channel listing upcoming flights, as well as several channels showing live air-to-air and ground video whenever a research flight is under way. These show what control room personnel see, as it happens. The facilities at Dryden, and its culture, reflect its central purpose of flight research, both for engineering and science.

BUILDING THE TEAM

The initial group of Langley Hyper-X personnel arrived at Dryden in September 1996. One of them, Lawrence D. Huebner, recalled: "It was very important that a core set of folks from Langley went to Dryden early in the project to work a lot of the preliminary work and, more important, to begin developing relationships."[§]

This was a critical issue. The era when a single center could undertake a major project on its own was gone. The Hyper-X project involved Langley and Dryden; a project like the shuttle involved all the NASA centers in one role or another. To a large measure, the success or failure of a project depended on the centers' and the contractors' ability to coalesce as a unified team.

Building such relationships was less of an engineering issue than one of personalities. Griff Corpening, who later served as Dryden chief engineer for the first two X-43A launches, recalled that there were "a lot of challenges in Dryden and Langley working together on this ... we had some extremely strong personalities early on in the projects; John Hicks [hypersonic lead] at Dryden and Lana Couch [creator of the Hyper-X] at Langley."[¶]

[§]Lawrence D. Huebner e-mail to Curtis Peebles "Re: Hyper-X Questions #1," June 4, 2008, 7:08 P.M.
[¶]Griff Corpening history interview, December 20, 2004, Dryden History Office, p. 32.

a)

b)

Fig. 6.2 **Langley, a) the vacuum spheres of the gas dynamics laboratory, and b) the 8-ft high-speed tunnel, the test section and control console were located in the domed area.**

At the start, division of authority and duties between Langley and Dryden was a major issue. Dave Reubush noted that "there was significant conflict about this at the beginning. Each had its own idea of how the program should be run.... However, as time passed and the two centers started working together a unified team developed."** Huebner added, "Once I was at Dryden we were all one team with one goal."**

Although adapting to the desert had been a problem in the early days at Muroc, it was apparently not a significant issue by the closing years of the 20th century. "For me personally," Huebner said, "I didn't really have any adaptation issues, aside from the commute being just a little longer than at Langley (10 minutes max vs. 45 minutes minimum)." This was an opinion shared by Reubush.††

Conflicts did arise on a personal level, over such issues as appointments and carpools, different working hours, and the time differences between the east and west coasts. These took time to sort out. Although an effective team emerged, there still remained a major issue that would spark disagreement throughout the project. Reubush noted: "I believe the biggest area of friction was the Dryden desire to take the vehicles apart and test the hell out of them (particularly the first one). There was a concern by the LaRC [Langley] folks that the vehicles would be broken/worn out before they could fly."††

The Langley personnel's viewpoint was shaped by experiences with the HiMAT (Highly Maneuverable Aircraft Technology) project. This had been a remotely piloted research vehicle flown at Dryden to test concepts for an advanced fighter. The HiMAT had numerous problems, and the two unmanned vehicles made only 26 flights between mid-1979 and January 1983. In addition to problems related to its poor reliability and the limitations of 1970s computer technology, HiMAT was also a tightly packed vehicle. Replacing a single failed component could take days and involved disassembling the vehicle, removing the defective part, reassembling the vehicle, then running tests to ensure not only that the new component worked, but also that nothing had been broken during the repair. Like the HiMAT vehicle's, the X-43's interior was tightly packed with electronics, tanks, actuators, and the multiple systems required to operate an aircraft.

Reubush recalled that his former branch chief

> ... was deeply involved [with HiMAT] and he stayed upset about what happened for many years. I also remember that in one of the early kick-off meetings at Dryden the retired Deputy Director of Dryden, Ted Ayers, was hired by Lowell Keel to consult and he mentioned in his

---

**Lawrence D. Huebner e-mail to Curtis Peebles "Re: Hyper-X Questions #1," June 4, 2008, 7:08 P.M.
††Dave Reubush e-mail to Curtis Peebles, "Re: Hyper-X Questions 1," June 5, 2008, 4:59 P.M.

talk not to over-test such that you got failures because it had happened in the past.‡‡

Reubush summed up the Langley attitude on preflight testing this way:

> The general issue was of trying to maintain the right balance between testing that needed to be done and over doing it such that you increased the chance of failure. It was felt that there were some (definitely not all) at Dryden who were more interested in playing with the vehicle and putting it through all they could rather than getting on with it and flying.‡‡

Dryden's practices reflected experience going back to the initial X-1 flights in 1947. Once a new research aircraft was delivered, its individual systems each were checked to ensure that they operated properly. With this completed, integrated system tests were conducted next to look for interactions among systems that might mean the loss of a research vehicle.

Some of the problems uncovered in such testing can be strange. During a combined systems test of the M2-F2 lifting body, for instance, NASA research pilot Milt Thompson pushed the cockpit microphone key, and the aircraft's pitch stability augmentation system disconnected. Once system testing was complete, captive-carry tests were conducted with the lifting body using the launch aircraft. These tests qualified the systems of both the research vehicle and the launch aircraft under such actual flight conditions as the cold, thin air of high altitudes, which could cause systems to fail or to operate erratically.

Although this extensive testing helps ensure that the vehicle will operate correctly on its first flight, it takes time. The M2-F2 was delivered on June 15, 1965, but did not make its first flight until July 12, 1966, nearly 13 months later [4].

Beyond internal NASA relations, there were also issues involving the relationships between NASA and the different contractors. Paul Reukauf, Hyper-X deputy project manager at Dryden, later recalled having mixed feelings, initially, about the selection of Micro Craft as contractor for the X-43A vehicle.

> We were all kind of, I guess, a little wide-eyed about the whole thing. Well, gee, how's this going to work ... this small company that's never really built a real flying airplane before? It actually turned out to be a great relationship. They understood their limitations. They were very willing ... to work with the government. They let the government do the things that the government was really good at, and, they did the things they were really good at and it turned out to be very synergistic.§§

---

‡‡Dave Reubush e-mail to Curtis Peebles, "Re: The Issue of Testing," December 11, 2008, 2:37 P.M.
§§Paul Reukauf history interview, February 17, 2005, Dryden History Office.

This relationship with Micro Craft and the other contractors was grounded in the different capabilities of the participants. NASA managers did not arbitrarily divide up tasks or order a contractor to use a particular procedure. Instead, when an issue arose, a meeting was held with the parties involved, and they determined together who was best suited to carry out the task. In some cases, only NASA had the facilities required. In others, a contractor had the expertise, or simply could do the work more cheaply. This was, to say the least, a departure from the stereotypical image of government/contractor relations in which the government tries to dictate every detail while contractors have only their own interests at heart.

Reukauf noted, "Each party, if you would, was allowed to bring their expertise to the table and whoever seemed to have the most expertise in an area, led in that area."¶ An important aspect was the one-of-a-kind nature of the effort. Every project component had to work properly if the mission was to succeed. The key was not to fail.

There were cases when NASA managers disagreed with a procedure a contractor was using. Over several decades NACA/NASA engineers had established standardized procedures for installing fasteners, safety wires, and electrical wiring inside research aircraft. These procedures were based on hard-won experience.

Meetings were held to resolve technical issues. The NASA representative would give reasons why a different procedure should be followed, go into the reasons why he or she wanted it done that way, and provide examples of how NASA had learned what it knew about use of such methods. Reukauf later recalled that in every case, contractors accepted the NASA recommendations. With this also came a basic sense of fairness, as well as fiduciary duty and fiscal prudence. If NASA asked a contractor to make a change to a design or procedure, NASA had to absorb the additional cost.***

"For the most part work with the contractors was true teamwork," Reubush noted. "In the vast majority of cases the work got done by the teams with great cooperation between government and contractors. Nobody cared what badge you wore. The small number of problems was where there were minor personality conflicts which might have taken some time to work out. (Equally true just within the government.)"†††

Joel Sitz, the third and final Hyper-X project manager at Dryden, described the process by which the team was built, and the beneficial effects of disagreements. The process took place on several levels. One was with the *integrated product teams (IPTs)*, small NASA/contractor groups assembled to examine

¶Paul Reukauf history interview, February 17, 2005, Dryden History Office.

***Paul Reukauf history interview, February 17, 2005, Dryden History Office, pp. 13–15, 35, 36.

†††Dave Reubush e-mail to Curtis Peebles, "Re: NASA/Contractor Relations," February 3, 2009, 7:35 P.M.

specific issues. Over time the multiple IPTs would merge into a larger group, which would merge with other groups, until a single team was formed.

Another level resulted from the merging of the technology groups and the flight groups. This took some time to accomplish, because these groups had very different cultures. This process took place from the top down rather than from the bottom up, as it had with the IPTs. Sitz described the process as involving a "healthy conflict" between the two cultures. Technological requirements flowed into flight test. Such an integration effort, by its very nature, brought the two cultures into conflict. These conflicts included debate over such issues as the use of flush air data systems and whether a blowdown test of the scramjet engine was needed. Whereas Langley wanted a 12-foot vehicle so it could be tested in the 8-Ft. HTT, some Dryden engineers wanted a longer vehicle.

In the process, as Sitz said, conflict is to be expected, and dealt with. This merging of the technological and flight cultures, he noted, was where management earns its money. From the process, a flight research program emerges.

Looking back on the experience, Sitz said he was struck by the very sharp contrast between the functionality of the team at the beginning of the Hyper-X project and at its close. The representatives from two very different centers had been transformed into a close-knit and highly dedicated team.[‡‡‡]

HYPER-X STAGE SEPARATION

The most difficult technical challenge facing the new Hyper-X team was how to separate the X-43A Hyper-X Research Vehicle (HXRV) from the Hyper-X Launch Vehicle (HXLV). The separation would take place at a dynamic pressure of about 1000 psf at Mach 7 and Mach 10 and an altitude of about 100,000 ft. From the beginning of the project, personnel had understood that separation presented the project's most difficult technical challenge. The irregular shape of the HXRV and adapter combo, and the potentially unpredictable flow fields this could create, meant that an HXRV collision or loss of control was a real possibility.

The initial separation concept was developed in the Phase III study. The HXRV was attached to the adapter with a pair of explosive bolts. At the center of the HXRV's aft bulkhead and the forward end of the adapter were three ejection rails. These held the research vehicle in position during the boost phase, and guided it as the separation occurred. The HXRV was separated from the adapter by two pistons pushing on the base of the research vehicle. Their force was directed through the X-43A's center of gravity. The pistons were to be propelled by a pyrotechnic three-cartridge breech that produced high-pressure gas. The separation would take 0.6 s from the start of the maneuver to the point at which the research vehicle cleared the forward end of the adapter.

[‡‡‡]Joel Sitz interview notes, March 24, 2009, Dryden Flight Research Center.

Among unknowns related to the separation was the effect of the shift aft in the booster's center of gravity that occurred when the HXRV was released. It was expected that the booster would pitch up as a result. To counter this possibility, several options were considered. All involved creating a counterforce that kept the booster stable despite the loss of mass. One approach was to vent some of the gas from the cartridge out the top of the adapter. Other options were a flap on top of the BAM or a rocket motor at the tail of the HXLV, either of which could be activated to prevent the pitch up from occurring.

After Micro Craft received the contract to build the X-43A airframes in March 1997, a joint NASA/contractor IPT was formed to review the separation issue and make suggestions for potential changes. Mary K. Lockwood of Langley was named the separation IPT lead; the rest of the team was made up of discipline-specific engineers.

When Lockwood quit as the separation IPT lead in June 1997, Dave Reubush was asked to take over the job. He remained the government separation IPT lead through the rest of the program. When Lana Couch was moved to the new Aerospace Transportation Technology Office in October 1996, Reubush was moved with her. He later guessed it was "guilt by association," because he had been her deputy in the NASP office. Wayne Blocker was the separation lead on the contractor side. He and Reubush were co-leads of the separation IPT. Reubush later noted that Blocker deserved at least as much credit as he did for the success of the separation event. "The whole separation development process was one of the true triumphs of government/contractor cooperation," he later wrote.§§§

The IPT contacted various organizations, in the white and black worlds alike that had experience in high-speed stage separations. Preliminary discussions were held with Sandia National Laboratories during late May 1997. Advice was also sought from the Army's Redstone Arsenal. Reubush recalled that even the Russians were queried. The IPT members found that none of the organizations had experience in high-Mach-number/high-dynamic-pressure separations of nonaxisymmetric vehicles. The general consensus was that the 0.6-s separation time was too long.

Sandia did offer three alternative separation proposals. First, engineers recommended that the X-43A make an exo-atmospheric separation from the HXLV. Sandia engineers designed warhead reentry vehicles that were separated from rocket boosters in space. This avoided the problems of separating at a high Mach number and high dynamic pressure. Although this solution was attractive, Hyper-X engineers realized that in eliminating one problem, another cropped up. The X-43A would have to make a reentry, and on the

---

§§§Dave Reubush e-mail to Curtis Peebles, "Re: The Lana Couch study (DF-9)," November 24, 2008, 5:45 P.M..

Mach 10 flight the resulting heat load would exceed the capabilities of the vehicle's thermal protection system.

Another Sandia proposal was to add a fairing to the front of the HXLV, giving it a shape like that of the standard Pegasus nose fairing. The X-43A would be mounted within the underside of the fairing with only the scramjet and wingtips exposed. After burnout, the X-43A would be ejected downward, like a bomb being dropped from an aircraft. However, this had a different shortcoming. As the X-43A moved downward, it would pass through the bow shock of the fairing. Hyper-X engineers were concerned that this "upset" would be too much for the X-43A's control system to withstand.

Sandia's third suggestion was to split the adapter with explosive charges, and eject the two halves sideways to shorten the time during which the wings would overlap the adapter structure. The Sandia engineers added that whatever separation method was used, it should not include rails. Their experience had been that no matter how well a rail system worked in ground tests, it would bind up in flight.

Despite Sandia's advice, an integrated rail/ejector system was first proposed by the IPT. Two rods would be attached to the X-43A's aft bulkhead. These would push the research vehicle clear of the adapter. Explosive bolts would fire to release the vehicle for the free flight. To avoid use of a nonsymmetrical configuration, a pair of two-step separation designs was offered. The X-43A would be mounted on a fairing attached to the front of the BAM, which would separate as a single unit. The fairing would then be separated using drag brakes or an airbag. The most unusual concept had the X-43A attached upside down to a wedge-shaped adapter at the front of the BAM. At separation, explosive bolts would fire and the HXLV would pitch up to clear the research vehicle, the scramjet would ignite, and the X-43A would fly the test mission [1, pp. 2, 3, 6–8, and 4, p. 3].¶¶¶

Griff Corpening found the process frustrating. At each meeting, engineers and managers being briefed would offer their own proposals for a separation system, and IPT members would have to analyze this input. This process was repeated again and again at each briefing. Corpening noted later that the IPT seemed unable to move the process forward to develop a workable design.****

The Hyper-X Manufacturing Readiness Review (MRR) was held at Dryden in June 1997. Part of this effort involved finalizing a separation configuration. The IPT engineers had concluded Sandia's split adapter seemed the best option, and could serve as the basis of the separation concept. But the original idea was modified. Although the Sandia concept had the two adapter halves separating, the MRR design had the adapter built as a clamshell split down

---

¶¶¶*Aerospace Projects Highlights*, May 20, 1997, p. 3.
****Griff Corpening history interview, December 20, 2004, Dryden History Office, p. 13.

the middle with the two pieces hinged at their aft edges. The hinges were added to avoid the possibility of the adapter halves striking the research vehicle as they separated. By keeping them attached to the HXLV, they could also serve as speed brakes. Their drag would slow the HXLV, and open the distance between it and the X-43A faster.

The proposed sequence began with the firing of the pistons. Once the research vehicle had moved 2 in., either a pyrotechnic piston or springs would separate the two halves. As Sandia had recommended, the use of rails was ultimately rejected. The new design sparked debate at the MRR. Engineers expressed concern that even with the clamshell design, there was still a significant period of time when the wings overlapped the adapter halves. If the X-43A were to roll during this period, there was a risk a wing could strike one of the clamshells, resulting in loss of the flight.

During the subsequent discussion, a suggestion was made to use a one-piece adapter that was hinged at the back and would swing down. This "drop jaw" design, as it was soon dubbed, would get the adapter out of the way quickly, and minimize the risk of recontact between the HXLV and the X-43A. This was selected as the baseline design for the adapter [1, p. 3].††††

Even this, however, was not the final step. Lowell Keel, the project manager for Micro Craft (which later became part of ATK GASL), concluded at an early stage in the separation debate that the best solution was the simplest one possible. He believed that it was better to keep the drop jaw fixed. This avoided the possibility that as the adapter rotated, shocks would form in front of the drop jaw. The shocks would pressurize the nozzle area, and cause a nose-down moment on the research vehicle. This, in turn, would lead to a loss of control. To settle the issue an interim solution would be adopted. The adapter would be built with the drop jaw configuration. Wind-tunnel testing would be conducted to determine whether the drop jaw would be moved or remain fixed during separation [1, p. 4].‡‡‡‡

STACK FREQUENCY

The separation question was not the only one involving the HXLV that had to be dealt with as the design effort began. Forces acting on a rocket in flight cause it to flex. This must be allowed for in design of the rocket's guidance system. If it is not, the guidance system will detect the flexing and interpret it as changes in the booster's angle of attack. The guidance system will then try to make unnecessary corrections, leading to a potential loss of control.

This issue appeared even before the contract for the HXLV was issued to Orbital Sciences. The company's structural modeling of the initial configuration

††††Dave Reubush e-mail to Curtis Peebles, "Re: Hyper-X Questions 1," June 5, 2008, 4:59 P.M.
‡‡‡‡Griff Corpening history interview, December 20, 2004, Dryden History Office, p. 14.

indicated the stack's first longitudinal bending mode did not meet the HXLV design requirement that the bending modes be in excess of 10 Hz. The Pegasus's guidance system was designed to ignore vibrations with a frequency above 7.5 Hz, because these would be caused by the vehicle bending. The HXLV used the same guidance system, so it faced identical "stack frequency" limitations.

The X-43A's large mass, and its location at the front of the booster, meant the stack had to be as rigid as possible. The weak points in the design were the adapter-to-X-43A joint and the structure of the adapter and X-43A, which were a major hindrance to progress on the project. The X-43A and adapter teams had to return to the conceptual structural design phase and undertake a redesign for increased rigidity.

The lightweight materials originally planned for use in the main keel of the X-43A as well as in the entire adapter structure had to be replaced with steel. The result, not surprisingly, meant a significant weight increase. The original mass-properties of the X-43A and the adapter were about 3300 lb. After the change to different materials, the X-43A's weight alone had reached some 3000 lb. With the adapter this became 5130 lb for the Mach 7 flight. The HXLV still had sufficient performance margins to compensate for this amount of weight increase and reach Mach 7. The situation was different for the Mach 10 profile, however, because the design weight margin was now gone. For the Mach 10 flight to be successful, the BAM and adapter weights would have to be reduced. This was done by eliminating unneeded subsystems from the adapter and fabricating some components from aluminum. The BAM for the Mach 10 flight was to be constructed entirely from aluminum but would maintain the necessary stiffness [2, pp. 5, 6].§§§§ (The selection of the Pegasus, with its higher weight margin, rather than the ground-launched boosters was fortunate. The Castor IVB would have been eliminated from the Mach 7 flight because it had the same stack stiffness requirement, but lacked the margin to cope with the added weight.)

These structural changes did not resolve the stack-frequency issue. The analysis of the initial modifications in December 1997 indicated the design still did not have the required rigidity. By October 1998, further review indicated that the first flexing frequency of the stack in the pitch axis was 7.75 Hz, and in the yaw axis it was 9.6 Hz. These were lower than earlier estimates and both were below the design requirement of 10 Hz. Modifications to the HXLV flight control system would have to be made.¶¶¶¶

Orbital Sciences, Langley, Dryden, and Micro Craft engineers discussed potential solutions. Their consensus for the best fix was a software modification. The final stiffness data would not be available until March 1999. In light

---

§§§§Griff Corpening history interview, December 20, 2004, Dryden History Office, pp. 9, 10, 33, 34, 44.
¶¶¶¶*Aerospace Projects Highlights*, December 19, 1997; October 16, 1998.

of the potential delay, Orbital Sciences personnel decided to start design work on an alternative hardware-based fix. This involved moving the control system feedback sensors aft.*****

The structural tests of the adapter and stack were made in Dryden's loads lab in March 1999. They included structural flutter, mode, bending, and interactive testing. The data analyses showed the stack stiffness was actually higher than calculated, and the frequency issue was finally resolved.†††††

VEHICLE RECOVERY

Not all decisions facing the Hyper-X project managers were as significant as the separation issue or stack stiffness. Among lesser challenges was recovery of the X-43A once the flight had been completed. Recovering the X-7 ramjet test vehicle had been a design consideration. As a result, the X-7 was fitted with a nose spike and recovery parachute. This allowed the ramjet and vehicle to be retrieved and physically examined, to determine the effects of high-speed heating on their structures. Such an examination would provide information beyond that gained through telemetry data, and be an independent check of temperatures and their effects on thermal protection and other materials. Any anomalous events or readings could be clarified.

In the original NASA RFP, the research vehicle was considered expendable. After the scramjet test and maneuvers to collect aerodynamic data during descent, the X-43A would hit the Pacific Ocean and sink. But the RFP did not specifically rule out a recovery. If a contractor proposed a recovery method, NASA would consider it. However, one major problem with any attempt to recover the research vehicle was its small size. Some of the rocket-launched reentry vehicles had carried parachutes and flotation devices. The X-43A did not have adequate internal room to carry such equipment.

Both McDonnell Douglas and Micro Craft suggested recovery methods for the planned Mach 5 flight. The McDonnell Douglas concept called for the vehicle to be flown to a recovery zone where the water was shallow. The X-43A would make a controlled water landing and sink to the ocean floor. The Navy had trained seals to recover unarmed torpedoes used during training exercises. The marine mammals swam down to the torpedo and attached a line to it. A recovery vessel then reeled it aboard. The Navy had considerable success with this procedure, and it seemed a feasible means of recovering the X-43A.

Micro Craft engineers also considered the prospects for vehicle recovery, offering a more complex and unusual proposal. Concerned about damage to the scramjet during touchdown, which was virtually guaranteed during high-speed

*****Aerospace Projects Highlights, November 6, 1998; November 13, 1998; November 20, 1998.
†††††Aerospace Projects Highlights, March 19, 1999; March 26, 1999.

impact into the ocean, they proposed skid landing gear. Unlike traditional landing gear, it would not be located in the underside of the vehicle, but on its upper surface. This allowed the use of very short skids, compared to what would be needed if the landing gear were on the underside. The X-43A would glide toward the runway on San Nicolas Island, off the California coast, the landing skids would extend, and the vehicle would roll inverted before touching down.

Turning this into a reality posed many difficulties. A good low-speed database on the research vehicle's flight characteristics did not exist, nor did a database for the vehicle flying upside down at subsonic speeds. A very precise navigation system would be needed for the attempted landing. NASA engineers were not sure they could control the X-43A well enough to even get it to San Nicolas Island, much less attempt a runway landing.

The U.S. Navy, which operated facilities on San Nicolas Island, posed additional complications. NASA engineers met with Navy representatives to discuss using their facilities. The Navy was willing to allow the X-43A to land on the island, though only if NASA could guarantee that the research vehicle would land on the island and not in the ocean surrounding it, which was part of a marine mammal sanctuary. The Navy no longer conducted test operations in the waters around the island, because anything impacting the water disturbed the whales, sea lions, and elephant seals living there. The prospect of anything—such as a research vehicle—hitting and killing any form of marine life had formidable political and legal ramifications.

This controversy went on for about 6 months, against a background of other, more significant technical issues and design problems. Paul Reukauf recalled later that none of the NASA engineers felt there was much to be gained with landing the X-43A. Many felt that attempting to recover the X-43A was turning into a separate research effort altogether. The driving force behind the recovery issue was Boeing, which had, by this time, merged with Rockwell International, becoming one of the Hyper-X subcontractors in the process. Boeing had an interest in landing one of the X-43As, and continued to apply pressure for this objective.

Reukauf recalled: "We finally said, no. Landing is going to cost us way too much money and it's going to detract from the primary experiment. Just make sure we get the data and we will not land the vehicle. We essentially had to say landing is off the table; we're not going to land."[#####] Despite the decision, there continued to be low-level interest in recovering the vehicle as late as spring 1999. Rather than San Nicolas Island, the sounding rocket launch site at the Poker Flats Research Range in Alaska was suggested as a potential landing site. Pursuing this, however, would require that project personnel, the B-52, HXLV, X-43A, and support facilities all be moved to

---

[#####]Paul Reukauf history interview, February 17, 2005, Dryden History Office, pp. 29–33.

Alaska. The difficulty, cost, and effect on the program schedule were too great to justify the data to be collected.§§§§§

RESEARCH GOALS AND FLIGHT PLANS

The initial Hyper-X plan called for four flights at three different speeds—Mach 5, Mach 7, and Mach 10. The Mach 5 flight would test the ability of the scramjet to operate as a "dual mode" engine—as both a conventional ramjet with subsonic combustion and a scramjet with supersonic combustion. The Mach 7 profile would test the vehicle's ability to accelerate under the thrust of the scramjet. The Mach 10 profile would demonstrate that the vehicle could cruise at the DF-9's top design speed.

Different combinations of the three speeds were proposed. The original plan was for a Mach 7 and a Mach 5 flight, followed by a pair of Mach 10 launches. A revised plan called for two Mach 7 flights, one Mach 5 flight, and one Mach 10 flight. All four flights, despite the different speeds, had nearly identical separation conditions: 1000 psf dynamic pressure and a flight path angle of +2 deg.

These plans represented differing assessments of the risks, benefits, and usefulness of the data to be obtained at the three Mach numbers. The initial plan was influenced by the fact that most of the earlier wind-tunnel testing had been done at Mach 7. As a result, a substantial amount of data at that speed existed, and this speed was seen as the least difficult of the profiles. As for the Mach 5 mission, the ability of a dual-mode scramjet to make the transition from subsonic to supersonic combustion was critical to routine hypersonic flight.

By contrast, however, the Mach 10 flight constituted unexplored territory. Shock tubes could simulate Mach 10 flight conditions for a few milliseconds on each run, but each test provided just a single data point. The 10 s of burn time for a Mach 10 flight would surpass the total amount of existing shock tube data by orders of magnitude. Thus, Mach 10 was seen as the big unknown, and where there was the most to be learned. Two Mach 10 missions were planned because they were also seen as the riskiest flights and there was a possibility that the first could end in failure. Planning a second Mach 10 attempt reduced the risk of the entire enterprise ending in total failure.

As the Hyper-X project advanced, program managers made a reassessment of the flight plan. The critical issue was qualifying the scramjet design tools and techniques. This required collection of as much flight data at Mach 7 as possible, to allow the flight results to be compared with ground data. Although plans continued to call for four flights, the new schedule was for two Mach 7 flights, one Mach 5, and one Mach 10. Attempts would still be made to show

§§§§§*Aerospace Projects Highlights*, April 23, 1999.

that the vehicle could cruise at Mach 10, but the emphasis was on the Mach 7 flights [2, pp. 5, 6].¶¶¶¶¶

Just as the stack stiffness issue arose after the Hyper-X project was approved, so did another issue with the Mach 5 and 7 flights. Originally, the standard B-52 launch conditions of 40,000 ft altitude and Mach 0.8 speed were assumed. The burnout speed of the Pegasus/HXLV booster launched under the standard conditions was Mach 10. For the Mach 5 and 7 flights, added ballast would be used to decrease burnout speed.

Further analysis indicated that the 10,000 to 20,000 lb of ballast this would require would raise the total stack weight above the 41,400-lb capability of the B-52B with the X-15 pylon and the Pegasus adapter. Additionally, a coast period after burnout, to slow the stack before separation, would create severe thermal loads on the HXLV and X-43A. The design teams looked at three different ways of keeping the total vehicle weight and center of gravity within the B-52 pylon limitations. Although the pylon weight limit was later increased, adding ballast was not a viable option.

The first option was to offload some of the HXLV solid propellant. This would reduce both the burnout speed and the amount of ballast required. Different amounts of propellant would have to be offloaded for the Mach 5 and 7 flights. The second possibility was to perform energy management maneuvers during the ascent (rather than after burnout) to reduce the burnout velocity. The third was to perform the launch at a lower altitude and speed. The goal was to select a single option that could be used for both the Mach 5 and 7 flights. Each had different advantages and disadvantages.

The second option was quickly eliminated, due to structural and control system limitations of the baseline Pegasus and a desire to avoid making significant changes to the booster. The third option was not feasible for the Mach 5 flight. The launch altitude would have to be less than 10,000 ft to result in a Mach 5 burnout speed without energy management maneuvers. This left only propellant offloading as a solution. The booster for the Mach 5 flight had not been fabricated, so a reduced propellant load procedure could be developed.

Engineers with ATK Thiokol, which built the Pegasus stage, soon developed a simple method. In the standard Orion 50S motor a layer of insulation material separated the casing and the propellant. Increasing the thickness of the insulation layer would permit removal of 7000 lb of propellant. This method had several advantages: standard propellant casting tools could be used, the solid fuel would not have to be machined out of the finished booster, and the HXLV ballistics at ignition and for most of the burn followed baseline characteristics. Thiokol engineers demonstrated the fabrication technique during late 1997 to NASA engineers' satisfaction. The final assessment

---

¶¶¶¶¶Paul Reukauf history interview, February 17, 2005, Dryden History Office, p. 28.

confirmed that the Mach 5 propellant offloading technique was feasible and involved only a moderate level of risk.

More options were available for the Mach 7 flights. Engineers determined that both propellant offloading and a lower/slower launch profile were viable. The same use of a thicker layer of insulation would reduce the propellant load on the two Mach 7 HXLVs by about 3000 lb. This would allow the launches to be made at the standard launch altitude of 40,000 ft and Mach 0.8 speed. At the same time, the Pegasus flight models indicated a launch was feasible at a lower altitude/speed. The launch altitude would be about 20,000 ft at a speed of Mach 0.5. Though this would result in a burnout speed of Mach 7, it required the HXLV to fly at conditions very different from those of the standard Pegasus profile. The thicker air at the lower launch altitude would increase dynamic pressure on the vehicle. The slower launch speed meant the HXLV would have to fly longer at transonic speeds [2, pp. 5, 6].

There was also another factor. Difficulties in transforming plans and concepts into hardware, such as the separation issue, had raised project costs beyond the level predicted. Cuts had to be made. The easiest one was to cancel the Mach 5 flight. A flight demonstration of a dual mode ramjet/scramjet was critical to showing the feasibility of a vehicle like the DF-9, but considerable work had been done in ground testing of dual-mode engines. This was of a lower priority, therefore, than the Mach 7 and 10 flight demonstrations. The Mach 5 flight was formally eliminated from mission plans in December 1997 [2, p. 6].******

For Mach 7 flights, there were the options of using either propellant offloading or a lower/slower launch profile to achieve planned burnout speed. Both were judged technically feasible, so the question became one of risk assessment. Reukauf recalled later in the program: "We all said, no, no, no. We [don't] want to off-load a booster. That would be like having two simultaneous research experiments. We would have a Hyper-X and we would have an off-loaded Pegasus booster, which had never been flown that way before." And so the project leadership chose the lower altitude/slower launch profile.†††††† The concepts had now become a flight research plan; the next step was to take the flight research plan and transform it into hardware, and begin the march to flight.

FIRST STEPS ON A LONG JOURNEY

The effort had taken some 15 months, from the Langley personnel's arrival at Dryden in September 1996 to the cancellation of the Mach 5 flight in December of the following year. Management and engineering personnel from two very different NASA centers had become a team working toward a

******Paul Reukauf history interview, February 17, 2005, Dryden History Office, pp. 28–33.
††††††Paul Reukauf history interview, February 17, 2005, Dryden History Office, p. 28.

single goal. Although disagreements, such as those over preflight testing, would continue, what could have become a major problem had been resolved at an early stage.

Although the basic design features of the Hyper-X stack had been finalized, the central issue of separation remained. The drop jaw configuration was the chosen design; flawed ideas, such as the use of rails, were eliminated. Yet additional study and testing using wind tunnels, CFD simulations, and actual hardware still needed to be completed. The stack-frequency issue was typical of problems that appear during a project's early design phases. Initial studies often miss potential problems, and as the project moves into the design phase these may become apparent. A considerable amount of work was needed to ensure that vehicle flexing would not lead to loss of control. Part of this effort was to develop a means of predicting vibration frequencies. Unlike the separation issue, the stack-frequency issue had been closed.

Vehicle recovery was a seemingly minor issue, but potentially this could have led to major difficulties. The project was focused on testing a scramjet in actual flight, to prove or disprove ground-test data and design tools. Recovering the X-43A after flight would have added cost, complexity, weight, and unknowns. Having the capability to examine the vehicle after flight was judged not worth the disruption to the project. Had the Hyper-X project been shifted away from its tightly focused goals, it would have lost direction, costs would have greatly increased, and it also would have run the risk of losing important congressional support.

The Hyper-X research plan was refined based on an analysis of which specific data were needed, and which were less important. It began as a broad-based effort, ranging from exploring the transition from ramjet to scramjet operation with the Mach 5 flight to confirming wind-tunnel data at Mach 7, and exploring the unknown at Mach 10. The initial focus was on the unknowns of Mach 10, the risks this flight regimen posed, and the need for two Mach 10 flights to allow for the potential of failure.

As the Hyper-X project advanced, engineers and managers realized that proving the scramjet design tools was more important than exploring the unknowns of Mach 10. The plan shifted to a pair of Mach 7 launches, so as to prove both the design tools and the ability of the scramjet to demonstrate positive acceleration. Only a single Mach 10 flight would be made, to test the engine's maximum cruise performance.

To stay within the cost constraints of better, faster, cheaper, and as a matter of priorities, the Mach 5 flight was eventually dropped. Demonstrating that an engine could shift from ramjet to scramjet mode was critical to building operational vehicles like the DF-9 or the ultimate goal of a single-stage-to-orbit shuttle. But those plans were a long way from realization. The Hyper-X project was focused on flight-testing a scramjet. More complicated research goals, such as those of a dual-mode scramjet, would have to wait for later projects. This represented a reaffirmation of the clarity of purpose program managers had established, preventing them from being lured into the one-vehicle-that-would-be-all-things trap.

The question of propellant off-loading versus adoption of a new launch profile entailed "comfort level"; universally, engineers believed off-loading to be too risky. This view was supported by computer modeling indicating that the HXLV would be successful with use of the new launch profile.

The Pegasus was chosen as the HXLV because a single configuration could meet all the research goals, and because of the advantages inherent in an air launch. In addition, by the mid-1990s the Pegasus was a well-proven booster. But the HXLV was no longer a Pegasus. The HXLV was becoming an altogether new design; it would be launched using a new, untested profile. The X-43A mounted on its nose would create aerodynamic effects emanating from its wings, fins, and sharp-edge-created vortices, which could interact with the HXLV's control fins. These factors had to be taken into account to ensure that the booster would operate properly. The underlying assumption that the HXLV retained the operational heritage of the Pegasus was not challenged, a judgment engineers would come to regret.

REFERENCES

[1] Reubush, D. E., "Hyper-X Stage Separation—Background and Status," AIAA 99-4818, 1999, p. 2.

[2] Joyce, P. J., Pomroy, J., and Grindle, L., The Hyper-X Launch Vehicle: Challenges and Design Considerations for Hypersonic Flight Testing," AIAA 2005-3333, 2005, pp. 2, 6.

[3] Hansen, J. R., *Engineer in Charge: A History of the Langley Aeronautical Laboratory*, NASA, Washington, D.C., NASA SP-4305, 1987, pp. 272, 274, 288, 297.

[4] Thompson, M. O., and Peebles, C. *Flying Without Wings Before the Space Shuttle: Testing NASA's Wingless Aircraft*, Crécy, Manchester, UK, 1999, pp. 102–105, 114.

Chapter 7

HYPER-X SCRAMJET DEVELOPMENT

In thrust we trust.

Traditional aviation quote

Risk reduction was a major element of the Hyper-X technology development program. This involved preflight analysis and experimental verification of the X-43A's aerodynamics, scramjet engine, thermal-structural design, operability, boundary-layer transition analysis and control, flight control law development, flight model development, and the integrated performance of the vehicle and engine systems. Wind-tunnel and computational fluid dynamics testing of the scramjet engine was an element of this effort.

As in previous scramjet research and development, wind tunnels were part of the vehicle/engine design process, used to predict the engine flight performance and to aid in development of control laws. Once the X-43A was flown, its data would be compared with wind-tunnel data and predictions. Successful predictions, made through wind-tunnel testing, of how the flight engine behaved would validate as reliable design tools the techniques used. Any inaccurate predictions would reveal areas in which changes would be required [1].

INITIAL DFX AND HXEM ENGINE WIND-TUNNEL TESTING

The DFX (dual-fuel experimental) parametric scramjet was developed in 1996, using the NASP database. It was a long, but thin, section of the underside of the X-43A, including the inlet ramp immediately in front of the engine, the engine itself, and a section of the nozzle. The DFX had the full height and length of the Hyper-X internal flow path, as well as the correct forebody, cowl, and sidewall-leading-edge radii. The design was not simply a miniaturized version of the NASP flow path, however. The forebody and afterbody were shortened and, to accommodate test-tunnel limitations, the width of the DFX engine was reduced by 44 percent (see Fig. 7.1). (Use of a partial-length forebody is a normal practice in scramjet wind-tunnel testing.) Because the engine was a quasi-two-dimensional design, even with the partial width it still offered a good simulation of the actual engine. The only

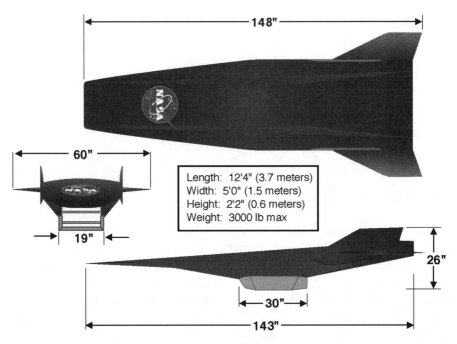

Fig. 7.1 A three-view drawing of the X-43A showing the dimensions of the research vehicle and the integration of the airframe and scramjet.

significant issue was that corner effects were more significant for the reduced-width design.

Due to its smaller size, the DFX flow path had to be modified for scale, wall temperature, and other differences. The inlet contraction, fuel injector design, and combustor length were changed, rather than being scaled down. These changes were based on the NASP data. Emphasis in constructing the vehicle was on operability and vehicle acceleration rather than on producing an actual flight-weight design with active cooling. The DFX engine had a copper heat-sink design, using standard materials and fabrication techniques developed for earlier wind-tunnel engines. Its test time would be limited to about 30 s at reduced dynamic pressure in order to avoid overheating. As a result, it could not be used at the full dynamic pressure of 1000 psf in the Langley 8-Foot High Temperature Tunnel (8-Ft. HTT).

The DFX engine was designed as a test article for flow path modifications. To do this, the initial concept for the DFX inlet used a design where the entire cowl rotated about the cowl leading edge, reducing the inlet's internal contraction. Though this allowed for easy flow-path modifications, it did not allow for verification of the Hyper-X scramjet design, which had only a forward cowl flap [2, 3].

The DFX engine was tested in Langley's Arc-Heated Scramjet Test Facility (AHSTF) between late 1996 and late 1997. Use of this facility offered the advantages of low-cost test operations combined with the capability to produce high-enthalpy test gas simulating speeds of Mach 4.7 to 8. The air heated with the electrical arc system was mixed with ambient air to produce the desired test-gas enthalpy. The test gas was contaminated with nitric oxide as a result. As with other tests made in high-temperature wind tunnels, this had to be accounted for in test results. The AHSTF also lacked the power to provide the full dynamic pressure of 1000 psf that the X-43A would experience. It was limited to 500 psf at Mach 7, and about 800 psf at Mach 5.

The initial AHSTF data were used in making minor changes to the DFX flow path. When the final flow path was determined, additional tests were made to identify any "facility effects," such as the nitric oxide contamination, that would alter the engine's performance or operability [1, p. 3, and 4, pp. 4, 11].

Subsequent DFX testing verified the Mach 7 engine performance over a range of Mach numbers and angle-of-attack simulations around the nominal values, but at the reduced dynamic pressure. Over 250 tests of the DFX engine were made, using four different configurations. The data from the tests confirmed the predicted engine forces and moments, and determined inlet and combustor component performance characteristics, ignition requirements, flameholding limits, and combustor-inlet interaction limits. The DFX scramjet demonstrated excellent performance and operability and validated the Hyper-X design methods. Finally, the data were used to determine the engine flow path and the database required for vehicle and scramjet control laws [1, pp. 4, 11]. (Mach 5 testing of the DFX in the AHSTF was planned to take place during mid-1998. This was to be followed by testing in the Combustion-Heated Scramjet Test Facility [CHSTF] in late 1998 and early 1999. This plan was canceled following the elimination of the Mach 5 flight.)

Following the DFX engine testing, the AHSTF was used for tests of the Hyper-X Engine Model (HXEM). This was also a reduced-width engine, one that simulated the center of the flight engine. The engine's inlet flow path width was reduced from the 16.78 in. of the flight engine to 6.6 in. The HXEM inlet, isolator, combustor, and internal nozzle were all two-dimensional representations of the Mach 7 flight engine. In contrast, the HXEM's fuel injectors were identical to those in the actual engine. The HXEM could be fitted with a truncated forebody to make it small enough for testing in the AHSTF. As with the DFX tests, the AHSTF was used because of its low operational costs. The testing allowed a direct comparison between the DFX and the HXEM data from the same test facility. As a result of the tests, minor changes were made to the engine design. Additionally, the data from AHSTF testing of the HXEM could then be compared to data from the subsequent tests in Langley's 8-Ft. HTT, to identify differences resulting from use of two different tunnels.

8-FOOT HIGH TEMPERATURE TUNNEL TESTING

Following the DFX and HXEM tests, a series of engine component, flow path, and control system verification tests were performed. A variety of test engines were used for these, and the tests were run in several wind tunnels. Such an extensive test program was undertaken to determine the major differences between the AHSTF database (nitric oxide contamination, reduced dynamic pressure, partial width, and truncated forebody and afterbody) and the X-43A flight database (no contamination; full dynamic pressure; full width, forebody, and afterbody).

The bulk of the Mach 7 testing was done in the 8-Ft. HTT, which could accommodate partial-width engines as well as a complete flight engine and simulated fuselage. Unlike the AHSTF, the 8-Ft. HTT could achieve the 1000 psf dynamic pressure the X-43A would experience. The high-enthalpy test gas was produced by burning methane and air at high pressure. The test gas then expanded through an 8-ft-diameter hypersonic nozzle. (This gave the facility its name.) The combustion produced about 18 percent water vapor and 9 percent carbon dioxide as contaminants, preventing an exact replication of actual flight conditions. The Mach number, static pressure, and static temperature produced by the tunnel were within 4 percent of those expected for the X-43A flight.

By using different nozzles, speeds of Mach 4, 5, and 7 could be produced. The "test cabin" was 12.5 ft long, which proved a large enough space for the entire X-43A flow path to be tested. To protect the engine models from dynamic loads produced during tunnel startup and shutdown, they were mounted upside down on a hydraulic lift. Once flow stabilized, the models were raised into the test stream, a process that took about 1.5 s. At the end of the test run, the models were retracted and the airflow stopped. About 1000 pressure measurements and 500 channels of strain gage measurements could be taken in the 8-Ft. HTT. The engine could be mounted at a 2-deg angle of attack, which was the nominal flight value, or at off-nominal 0-deg or 4-deg angles.

The 8-Ft. HTT had been modified in 1993 to support scramjet combustion research. Due to combustion, the test gas had very low oxygen content, so liquid oxygen was injected to raise the oxygen concentration to that of air. Tanks with hydrogen gas fuel, the silane/hydrogen igniter, and nitrogen were also added. The plumbing configuration, igniter/fuel mixing manifold, and pressure instrumentation system matched those of the X-43A systems.

A 20/80 silane/hydrogen mix was used to ignite the hydrogen gas. The silane/hydrogen was delivered to the fuel control system via double-walled, vacuum-jacketed, steel-braided flex lines at 1200-psi pressure—approximately the pressure downstream of the X-43A's igniter high-pressure regulator. The gaseous hydrogen fuel was delivered through similar flex lines, but at a lower pressure of 1150 psi, similar to the pressure downstream from the X-43A's high-pressure regulator.

Because the silane/hydrogen mixture and the hydrogen gas fuel both would ignite on contact with air, a nitrogen purge system was required. The nitrogen gas was at a pressure of 1200 psi and was injected into the fuel and igniter lines before and after a test run, preventing ignition through removal of any oxygen. The nitrogen also purged the scramjet model's interior cavities. The nitrogen displaced the air in the cavities, creating a nonflammable atmosphere should a hydrogen leak occur. The nitrogen also cooled internal components, such as the fuel control valves, cowl, actuator motor, and electronically scanned pressure (ESP) modules. Finally, the nitrogen, at a reduced pressure of about 750 psi, also supplied pneumatic actuation to the fuel/purge system isolation valves.

The 8-Ft. HTT also provided cooling water to the HXEM and the Hyper-X Flight Engine (HXFE). Both of the engines had water-cooled sidewall and cowl leading edges. As in the other cases, these components were identical to those on the actual X-43A. The water, at a pressure of 900 psi, flowed through the sidewalls and leading edges, then out three holes on the engines' external surface. The water was carried away in the airflow. The flow rates were the same as those expected to occur in flight. The water cooling system was active during the run, which lasted less than 30 s.

All the functions of the engine were controlled by the Propulsion Subsystem Control (PSC) computer. The PSC computer opened the cowl door and operated the fuel control valves. These were either preprogrammed to use a timed fuel sequence or used a closed-loop engine pressure feedback fuel control mode. The PSC also synchronized the data collection system, engine operation, and tunnel events [4].

For the HXEM tests in the 8-Ft. HTT (see Fig. 7.2), the Full Flowpath Simulator (FFS) was attached to the engine. This was a partial simulation of the X-43A's forebody flow path, including the leading-edge radius, flight boundary-layer trips, and a forebody surface coated with steel, copper, or alumina-enhanced thermal barrier (AETB-12) tiles.* A full boundary-layer ingestion trip or a partial boundary-layer trip would be fitted. This allowed measurement of how the resulting differences in airflow affected the truncated forebody engine. Two different forebody configurations were tested with the FFS. The first simulated flight-like, fully developed boundary-layer ingestion by the engine. The second configuration lowered the FFS forebody by 0.75 in., diverting the boundary layer to simulate its expected behavior when the engine with a truncated forebody was tested in the AHSTF.

Another modification to the actual X-43A forebody design was due to the HXEM's partial-width design. Because of this, though the forward part of the forebody had chines and boundary-layer trips identical to those on the X-43A, the FFS also had flow fences on about half of the forebody. These were not

---

*The FFS would not fit in the smaller AHSTF, requiring a shortened forebody.

Fig. 7.2   The HXEM being prepared for a test in the Langley 8-Ft. High Temperature Tunnel.

on the X-43A forebody. For wind-tunnel testing, they minimized three-dimensional airflow into the HXEM inlet. A similar measure was taken to deal with the afterbody flow. Because the engine had only partial width, the three-dimensional afterbody flow could not be simulated. To simplify the problem, a constant-angle nozzle was used to simulate the afterbody.

The quality of data needed for the engine tests, specifically direct, high-fidelity measurements of how the internal flow path axial forces changed with the fuel-flow levels, required a different instrumentation design. The force-measurement system was built into the FFS rather than as part of the tunnel's model support. The HXEM was mounted on the FFS by four metal fixtures and was isolated from the FFS by a gap, which was sealed by a heat-resistant fabric.

The 8-Ft. HTT's primary role was in validating the HXEM flow path and operations. This was accomplished by comparisons with data from several other tunnels using the truncated flow path and the full-width engine. This test also served to qualify the flight engine control software, programming that was critical to the success of the flight test. The software opened the cowl flap at the start of the engine burn, and controlled the flow rates of both the hydrogen gas fuel and the silane igniter during the startup sequence.

A more demanding task for the control software was fuel-flow-rate control. Once ignition had occurred, the program had to automatically compensate for variations caused by errors in the boost trajectory, to keep the fuel-flow rate within correct limits. At the same time, the software also monitored internal flow path pressures, adjusting the fuel-flow rate accordingly. Should the program make an error and cause too much fuel to be injected, an inlet unstart would occur, causing the engine to stop producing thrust. Finally, the software closed the cowl at the end of the test [1, p. 6, and 4, p. 7].

Testing of the HXEM in the 8-Ft. HTT was done in three phases. The first used a lower dynamic pressure of 650 psf and the diverted-boundary-layer FFS, for comparison with the AHSTF data, and was done at a dynamic pressure of 500 psf. This identified any facility effects created as a result of using two different tunnels. The second phase also used the diverted boundary layer, but at the X-43A flight dynamic pressure of 1000 psf. This identified any pressure-related differences with the data from the first 8-Ft. HTT tests. The third phase used the FFS configured for the full boundary-layer ingestion. This phase used two different dynamic pressure values—650 psf and 1000 psf— and was intended to identify model and three-dimensional effects.

Three successful unfueled runs were made with the HXEM in the 8-Ft. HTT. These were designed to measure the engine-mass-capture and inflow conditions as accurately as possible. Original plans called for making four unfueled runs. These would collect inlet flow-field data at 650 and 1000 psf dynamic pressures, and with the ingested and diverted boundary layers. However, the ingested-boundary-layer test at the 1000-psf flight pressure suffered an instrument failure.

Thirteen successful fueled runs were made with the HXEM. A run began when the engine was raised to the test height in the airflow stream, the nitrogen gas purge was completed, and the PSC was activated. The first step was opening the cowl door, a process that took less than 0.5 s. With this completed, the cowl-open tare data were acquired. The cowl door was opened at varying speeds to study the flow-field characteristics and the effects of the speeds on inlet-startup behavior. The silane/hydrogen igniter gas was injected, followed by the hydrogen fuel. The hydrogen gas flow rate was increased, with flow-rate plateaus to permit collection of steady-state engine data for post-test analysis. In addition to the preplanned fuel-flow rates, some tests were conducted using the closed-loop feedback in the PSC. This varied the fuel-flow rate based on real-time measurements of engine-pressure data, serving as both a test of the system and a means of developing and refining engine control laws. If an unstart occurred, a signal was sent from the PSC that caused the tunnel to carry out a "normal stop" procedure, to bring the tunnel to a safe "wind off" condition. At the completion of the test, the cowl door was either closed by the actuator or left open. The model retracted and tunnel airflow ceased.

The HXEM fueled tests addressed a wide range of issues. These included inlet starting, freestream dynamic pressure effects, forebody-length and boundary-layer effects, effects resulting from different silane levels, engine operability limits, closed-loop feedback controls, and development of engine control laws. All the tests were done at the nominal angle of attack of 2 deg, with a 0-deg sideslip. Based on the combined results of the DFX and HXEM tests, the 8-Ft. HTT tests proceeded to the final step.

HXFE/VFS TESTS

Though the HXEM and the FFS represented a simplified test engine model and flow path, the HXFE and the attached Vehicle Flowpath Simulator (VFS) constituted the actual X-43A flow path. This included vehicle geometry, which affected the flow entering the engine that provided sufficient mass capture for Mach 7 test conditions. The engine's internal nozzle geometrically translated the flow from the combustor to the external nozzle; the sidewall and cowl trailing edge produced the correct exhaust-plume expansion behind the engine.

The HXFE itself was an actual flight engine with instrumentation added. The HXFE inlet/isolator geometry was fixed once the cowl door opened. The engine included the pressure-sensing and feedback systems needed to monitor inlet/combustor interaction and prevent unstarts. The engine was attached to the VFS with six attachment lugs, the same way the flight engine was attached to the X-43A fuselage. All the surface instrumentation locations on the X-43A were included on the HXFE/VFS. The data produced in the 8-Ft. HTT could be directly compared with actual X-43A flight data. In effect, the HXFE tests were an end-to-end test of the complete engine, subsystem, and airframe under as realistic flight conditions as could be achieved. (The original plan called for testing an actual flight X-43A vehicle.)

The HXFE/VFS itself was fabricated to provide an accurate flow path, with all the geometric features necessary, but at minimal cost. The side chines and external nozzle surfaces were fabricated from copper plates. The side chines were also designed to allow access to the VSF for fluid connections, attachment points, structural components, and instrumentation while minimizing weight.

As with the HXEM, unfueled and fueled tests were made with the HXFE. Fourteen successful unfueled runs were made. Of these, six produced data on the inlet flowfield for the 0-deg, 2-deg, and 4-deg angles of attack; two dynamic pressure levels; and three boundary-layer trip options. The other eight unfueled tests examined cowl-door opening issues of actuation speed and the effects of prolonged exposure to simulated flight heating, as well as force and moment increments at different angles of attack and dynamic pressures.

The 45 successful fueled tests focused on engine performance and operability. Specific areas of interest were [4, pp. 2, 5–10]:

- Thermal effects on the boundary layer entering the engine
- Dynamic pressure and angle-of-attack effects

- Data repeatability from the individual tests
- Effects of the boundary-layer trips
- Effects of off-nominal sideslip angles of 1 and 3 deg
- Active fuel-control refinements
- Improvement of the engine's ignition and the shift to hydrogen-only operation
- Ability of the inlet and engine to restart following an unstart
- Effects of the ablative forebody thermal protection system on engine operations

Among the results was that the HXFE's structural integrity and internal surfaces were still in near-new condition following the tests. The only maintenance required was removal of a layer of silicon dioxide dust—a combustion product of the silane used for ignition—after each run. What was all the more remarkable was that the HXFE had been designed and built for a single flight, not for an extended series.

The only serious damage observed was to seal material, which limited flow between parts. The static seals were made of ceramic braided rope, and the dynamic seals between the cowl door and the engine sidewalls were made of 1/8-in.-square braided ceramic carbon-fiber rope held in place with a split Inconel tube. The rope withstood two unfueled runs and five fueled runs. However, the first time the engine suffered an unstart at flight dynamic pressure, the left seal was dislodged. Both the left and right seals were repaired so testing could continue. The repairs were not flight-worthy, but this was not viewed as a problem because the actual seals would have to withstand only a single engine burn, and the PSC logic was designed to prevent an unstart due to overfueling.

The HXFE testing of the PSC was one element of a multistage and overlapping effort undertaken for their development, verification, and validation. The first step was the PSC unit testing, done in an isolated computer simulation. This was quite limited, however; control laws were not integrated with the complete flight software, nor were they resident on the flight computer.

The major verification effort was done as part of the complete Operational Flight Program during hardware-in-loop (HIL) testing. The flight software was in the Flight Management Unit (FMU), but the FMU was not connected to the actual flight-vehicle systems. Simulations of both the flight systems and the flight conditions were generated. This allowed extensive testing of a wide range of potential flight conditions and scenarios, without wearing out the actual flight systems.

Fluid systems tests were also not run with the PSC software on the FMU. Testing was conducted during vehicle-in-the-loop (VIL) testing using the Operational Flight Program on the FMU. Rather than using the actual fluid system, a nominal simulation was done using a test bench. With this, all the

Fig. 7.3   The HXFE and VFS undergo a test run in the 8-Ft. HTT.

systems had been tested with the exception of engine scramjet operation using closed-loop control laws.

This final element was accomplished with the HXFE/VFS wind-tunnel tests (see Fig. 7.3). The X-43A flight conditions, including dynamic pressure, altitude, Mach number, angle of attack, and commanded vehicle attitude, were used to determine the igniter and fuel schedules. These were modified based on measurements of the vehicle's acceleration feedback, and the engine's flow path pressure feedback. (The latter was determined by the shock position and isolator pressure readings.) Igniter and fuel control position commands were generated based on the data and were sent to the valves. The resulting changes in vehicle and engine performance were detected and used to adjust the igniter and fuel flow rates in a feedback loop. Verifying the control loops for igniter and fuel flow system operation required that the fuel system plumbing, fittings, pipe lengths and diameters, control valves, venturi flow meters, injector/fuel mixing manifolds, fuel delivery pressures, fuel injectors, and instrumentation all be "flight-identical" components.

A complication affecting both the wind-tunnel tests and the actual Mach 7 and 10 flights was the very limited supplies of silane and hydrogen gas. This limited the burn time, and made it difficult to design an igniter and fuel sequence sufficient to meet test requirements. The steps included silane/hydrogen flow and ignition, switch to hydrogen-only fueling, and finally sufficient thrust to either accelerate the vehicle (for Mach 7 flights) or compensate for the vehicle's drag (for the Mach 10 flight). An added difficulty was the need for steady-state engine/vehicle data. Wind-tunnel testing was done on systems identical to those on the actual vehicle, which greatly reduced the risks in the actual flight [4, pp. 7–9].

Another requirement that had to be met with the control laws was prevention of an unstart during flight. An inlet unstart was caused by a disruption of the airflow, often called "chocking," which resulted in an abrupt reduction of the inlet's supersonic mass flow. This produced an increase in drag, and the amount of static and dynamic pressure within the engine increases while reducing pressure recovery and thrust. The speed range at which unstarts was a problem was between Mach 2 and Mach 12. They were not a problem until the engine switched to ramjet mode, or at speeds above Mach 12. The Mach 7 and 10 speeds at which the X-43A would fly were well within the danger area.

One cause of unstarts in scramjet design was *thermal choking*. This occurred when too much fuel was injected into the combustion chamber, creating excessive backpressure and pushing the terminal shock forward into the inlet throat. Once the terminal shock passed through the throat, it was quickly ejected through the inlet, causing an unstart. In some cases, the inlet restarted, but suffered another unstart. This cycle repeated until the amount of fuel injected was reduced.

Earlier ideas intended to prevent unstarts included reducing engine performance by increasing margins or adding bleed slots. This was undesirable, however, because maximum performance was critical to the success of a hypersonic air-breathing vehicle. At the same time, because maximum scramjet performance was at the boundary between the inlet starting and an unstart occurring, there was little or no margin to compensate for off-peak performance shortcomings. One of the methods eventually developed to prevent scramjet unstarts was the *isolator*. Located between the combustor and the inlet, this was designed to prevent the increased back pressure from moving forward and triggering the ejection of the shock out of the inlet [5].[†]

The anti-unstart control laws became active after the scramjet ignited, the X-43A had stabilized, and the fuel flow was increased. As long as there was no combustor/isolator interaction, flow path pressure remained above a preset limit, called "TripLevel 1," allowing the vehicle to reach the desired acceleration. Up to this point, fuel flow was held at a constant level; a controlled reduction began as the hydrogen gas was depleted.

[†]Ferri's proposed "thermal compression" concept, which involved adjusting the fuel flow into the scramjet burners to eliminate the need for a variable geometry inlet, would potentially have been very vulnerable to thermal choking. This would occur if the fuel flow controls became out of step with the scramjet combustor's internal pressure.

Should both combustor/isolator interactions occur and the pressure drop below the TripLevel 1 threshold, however, the fuel flow would be reduced for 0.75 s. At this point, the programming determined whether the acceleration goal had been met. If it had, the fuel flow would be held constant until the controlled reduction began.

If the planned acceleration had not been reached, the programming would determine whether the combustor/isolator interaction was exceeding a higher limit (TripLevel 2). If it had not exceeded this limit, the fuel flow would be increased to the planned level. If TripLevel 2 was exceeded, however, the hydrogen flow rate would be held constant until depletion. This programming was designed to maximize the probability that the X-43A would both reach its acceleration goal *and* avoid an unstart.

Researchers were also concerned about the performance effects of an off-nominal angle of attack or sideslip. The planned angle of attack for the Mach 7 flights was 2 deg, and the sideslip was to be 0 deg. As part of the HXFE/VFS testing in the 8-Ft. HTT, the angle of attack was altered to 0 deg and the sideslip to 4 deg. Wind-tunnel testing was also done at sideslip angles of 1 and 3 deg. The excessive sideslip angle created an expansion fan on the windward sidewall leading edge, and a shock wave on the leeward sidewall leading edge. These created an asymmetric inlet pressure field. Test data from a pair of sideslip runs indicated little, if any, loss of performance despite the disruption in the inlet airflow.

The HXFE/VFS test results for lift, drag, and pitching-moment coefficients were compared with database estimates. These covered the cowl-closed, cowl-open unfueled, and cowl-open fueled conditions. Database estimates were calculated by taking the experientially determined aerodynamic data for the closed-cowl engine, then adding the computed cowl-opening and power-on increments. The estimated and actual measured increments showed "very good agreement." This was a significant result, because it built confidence in the Hyper-X database.

Another result of the comparison was to show the effects of the pressurization of the forebody and nozzle areas. This resulted in aero-propulsive loads and changes in the pitching-moment values, both affecting the vehicle's trim. To compensate, control surfaces would have to be moved. The resulting trim drag would adversely affect net vehicle thrust. The ability to accurately measure this quantity in powered flow path analysis and testing was important to the success of the research flights.

Unlike conventional aircraft, where the airframe and engine are separate components, in air-breathing hypersonic vehicles the two are fully integrated. The vehicle's aerodynamic characteristics are dependent on the performance of the scramjet, and vice versa. Some of the factors involved the interrelated internal and external flow paths, the scramjet/airframe forces and moments, and the interaction between inlet spillage and exhaust pluming with the vehicle's lifting and control surfaces.

The external flow and exhaust pluming effects were studied using schlieren photographs, taken with the cowl closed, with the cowl open, and with the cowl open and fuel flowing. The images showed that all the shock waves on the forward fuselage were stable, including the bow shock, the weak shocks formed by the boundary-layer trips, and the compression from the two fore-body ramps. These slowed and compressed the air entering the engine to produce the necessary pressure recovery. In the cowl-closed configuration, a strong shock wave off the cowl leading edge formed above the engine. This shock also showed evidence of significant shear flows, along with a separation zone on the afterbody behind the cowl trailing edge.

With the cowl door opened, the shock produced by its leading edge changed, and the shear layer downstream of the cowl's trailing edge could be seen due to the alignment of the cooling water jets above the afterbody surface. When fuel flow began, the combustion produced increased pressure, resulting in an exhaust plume. The expansion of this plume was revealed through deposits of silicon dioxide (produced by silane combustion) on the afterbody. This fine white powder extended laterally beyond the surface of the afterbody nozzle surface. None of the powder was found on the wing surfaces, however, indicating that the exhaust plume, fueling levels, and other engine factors would not degrade control surface effectiveness. The combustion created the shear layer moving away from the afterbody surface. The engine's internal nozzle geometry produced waves downstream of the cowl trailing edge.

The cowl-open fueled tests were also used to show the results of sideslip on engine operation. The test conditions for two runs were identical—Mach 6.92, dynamic pressure of 1000 psf, and an angle of attack of 2 deg. With a 0-deg-angle sideslip, the exhaust flow was symmetrically aligned relative to the vehicle centerline. At a sideslip angle of 3 deg, however, the exhaust flow was asymmetrical. The shock and expansion waves from the cowl sidewall leading edges were apparent in the aftbody. Another effect of the sideslip was a slight increase in windward chine pressures near the vertical fins.

Not all the scramjet tests in the 8-Ft. HTT were directly related to the engine. Tests were made of a Boeing-developed lightweight ablative material called BLA-20. This material covered most of the X-43A's external surfaces. Two test panels were attached to the VFS—one immediately in front of the cowl leading edge and the other on the first half of the external nozzle. The goals of the test were to determine whether outgassing from the ablative material affected engine performance; to test the BLA-20's ability to with-stand repeated testing; and testing of pressure-measuring instruments and thermocouples. The results showed no engine problems from the outgassing, the ablative material survived the flight conditions with only modest damage, and the instrumentation operated properly.

A second test was made to determine the heating in the X-43A's wing gap. The vehicle's wings also acted as control surfaces. They moved in the same direction to raise and lower the nose, and in opposite directions for roll control. The wings were attached to spindles that were connected to electromechanical actuators inside the fuselage. Between the wing root and the flat fuselage was a small gap. Hypersonic airflow would cause heating in this area. Although various methods for preventing gap heating had been developed, no data were available for a longitudinal wing gap like that on the X-43A.

The gap heating tests involved four separate configurations—no wing, no gap, the flight gap, and twice the flight gap. The no-wing configuration represented baseline heating conditions, and the other three provided comparisons. The wings were half span, because the gap was the critical area. Thermocouples were mounted on the airframe adjoining the wing root to provide heating-time histories. Of the wing configurations, the flight gap showed only 1.2 times the heating of the no-wing data. The exception was just upstream of the spindle, where the heating was 1.3 times that of the no-wing data. For the twice-flight-gap conditions, the heating was about the same for the first half of the chord. Upstream of the spindle, this increased sharply to more than double the no-wing heating. The results verified the assumed heating profile used in the X-43A structural design [6].

A limited amount of Mach 7 testing was also done using the HYPULSE Scramjet Model (HSM) in GASL's Hypersonic Pulse Facility (HYPULSE). This was not directly connected with the Mach 7 flight, but was done both as an additional benchmark for the AHSTF and 8-Ft. HTT testing, and in preparation for Mach 10 testing. The HYPULSE was upgraded with a detonation-driven, reflected-shock tunnel operating mode, which allowed testing at Mach 7 and 10. The HYPULSE also had a test chamber that could accommodate scramjet engine models of the same scale as that of conventional blow-down tunnels like the AHSTF.

The HSM was a truncated, partial-width version of the Hyper-X forebody, flow path, and nozzle. It was also identical to the HXEM tested in the 8-Ft. HTT. The HSM forebody began at the start of the X-43A's second compression ramp. The engine was a 40-percent-scale version of the actual Hyper-X engine. Because of the partial width, a fenced forebody was added 1 in. on either side of the inlet to allow for bleed slots to remove the forebody boundary layer. A removable boundary-layer strip was also placed on the forebody, to trip the boundary layer before it entered the inlet. Both the HSM and the HXEM were designed and built concurrently, and both shared instrumentation locations to allow their data to be correlated. The HSM was designed with joints in the cowl and body-side flow surfaces so it could be easily modified for either the Mach 7 or 10 flow paths with a minimum of rework.

The leading edges were replaceable and had radii consistent with the flight hardware. The HSM flow path surfaces were fabricated from stainless steel, with aluminum sidewalls. A pair of windows enabled views of the isolator and combustor.

The initial tests used a stagnation temperature of 2100 K to produce a Mach 7.32 speed. The HSM was positioned in the HYPULSE to produce a shock matching that which formed at the second ramp of the X-43A in flight. The Mach 7 test plan included collecting flow path data at dynamic pressures of 500 psf, 1000 psf, and 2000 psf. As with other elements of the scramjet testing, this was to enable comparisons among the different wind tunnels, test conditions, test gases, contaminations, and engine-model effects.

The 500-psf HYPULSE tests provided a direct correlation with the database from the AHSTF runs. The 1000 psf allowed comparisons with the 8-Ft. HTT database and the actual flight data. The 2000-psf testing was not directly related, but instead provided data on fuel ignition and flameholding at higher combustion pressures. The HSM inlet flow was measured to determine the engine's mass capture. This was followed by unfueled tests and, finally, fueled tests using both hydrogen and silane/hydrogen mixtures. Both pure air and pure nitrogen were used as test gases in the HYPULSE.

The initial series of HSM engine wind-tunnel tests was made between November 16, 1998, and January 26, 1999. This consisted of 16 calibration runs, 3 unfueled (tare) runs, 18 fueled runs using a 5-percent silane/95-percent hydrogen mix, and 6 fueled runs using hydrogen fuel only. The tests used the M7-KL6 (Mach 7-Keel Line 6) flow path, with a stagnation temperature of 2100 K, producing an exit speed of Mach 7.32.

The HSM Mach 7 tests marked the first time the modified HYPULSE was used for engine testing. As a result, another goal of the HSM testing was developing proper techniques for its use and checking out the tunnel's data acquisition, fuel supply, and data processing systems. With the initial Mach 7 tests complete, the HYPULSE could then be used for Mach 10 testing of the Mach 7 flow path [7].

In all, 710 wind-tunnel tests, using the different engine configurations and wind tunnels, were performed as part of the Hyper-X Mach 7 development effort. The comparison between the HXEM and HSM showed variations of up to 10 percent in estimates of combustion efficiency. Determining the maximum and minimum of such engine uncertainties was a critical element in carrying out the Hyper-X project. These are used in the *Monte Carlo analysis*. In this computer simulation, the variables for the forces and moments are randomly shuffled to see how different combinations affect engine performance and to stress the engine control laws. These reveal any weaknesses that might occur in off-nominal conditions. To test under the worst possible conditions, a "stack up" is done. In this case, the variables used are all the worst-case possibilities [3, pp. 3–5].

The SCRAM3L and SRGULL computer simulations, as noted earlier, were used in the early conceptual design work for the Hyper-X scramjet engine, to determine the engine's minimum size and to refine the flow path to overcome problems resulting from viscous flow effects, such as high shear forces, over-contractions, and increased boundary-layer displacement. The role of computer simulations did not end once the engine design had been finalized.

The SRGULL program was also used in analysis of the ground-test data. One purpose of the program's use was to isolate and measure the engine and operability effects caused by the differing facilities' geometric scale, dynamic pressures, test gases and contaminations, test techniques, and facility limitations. Selected test data were analyzed using the SRGULL program to determine combustion efficiency and precisely measure the effects of the test methodology on combustor performance. The SRGULL results were also used to predict the engine's flight performance [3, pp. 2, 3].

THE MACH 10 ENGINE FLOW PATH DEVELOPMENT

The ground testing for the Mach 10 flight posed more difficult problems than that of the Mach 7 efforts. The different tunnels used for the Mach 7 engine models could simulate the complete 10-s engine burn sequence. In contrast, the HYPULSE tunnel produced Mach 10 airflow for a mere 3-ms test period. The performance of the engine would have to be determined one data point at a time. The data from the 11-s Mach 10 engine burn the X-43A would make in flight represented orders of magnitude more data than all the high-Mach-number wind-tunnel tests made during the previous four decades [8].

As with the Mach 7 wind-tunnel tests, the Mach 10 testing was intended to create a database to guide the flight engine's design and provide confidence in analytical tools and techniques. The Mach 10 flight was to test the maximum performance of the scramjet engine, specifically, the engine's ability to provide an amount of thrust equivalent to the drag on the vehicle. This would allow the vehicle to cruise at Mach 10.

The Mach 10 scramjet design could not be scaled photographically. Because the X-43A's shape was defined by that of the DF-9 vehicle, the resulting aerodynamic constraints also complicated the engine design task. But because an experimental aerodynamic database already existed, designs would be evaluated using trimmed net axial forces. Due to the large number of variables, a multi-step analytical approach was used to define the preliminary flow path and fuel injection. As with the Mach 7 flow path, the SRGULL program was used. It was supplemented with data on three-dimensional spillage, kinetic energy efficiency penalty, combustion efficiency, and base pressure. The effects of flight conditions were determined based on the flight Mach number, angle of attack, dynamic pressure, and five geometric variables.

With the preliminary database completed, it was further refined by adding factors such as the combustor entrance pressure, the angle of attack based on the forebody shock positions, combustor geometry, and the similarities with the Mach 7 flow path. This provided a preliminary engine flow path and an understanding of how the major design variables affected the overall vehicle.

The second stage in the development process entailed computer analysis of the design variables of the fuel injector and combustor. The GASP program was used to model the forebody and inlet, while the SHIP3D program was used for the fuel injectors, combustor, and nozzle modeling. As with the initial design analysis, propulsion forces and moments were combined with the X-43A's aerodynamic database. As a result, design decisions could be based on the trimmed vehicle net axial force, rather than a less useful measurement such as the scramjet's fuel-mixing efficiency.

The aerodynamic heating of the engine from Mach 10 flight, both with and without hydrogen fuel, and the ignition of the hydrogen gas were other design issues examined during this phase. Before and after the engine burn, a "fuel-off tare" would be performed. The cowl would be open, allowing supersonic airflow through the engine for several seconds, though no fuel would be injected during this time period. At the end of the fuel-off tare, hydrogen gas was injected to start the burn. Although the scramjet was designed to use a silane/hydrogen mix to ignite the fuel, in order for it to be as reliable as possible program managers wanted the capability to ignite a hydrogen-only mixture. This was addressed in the final design. The design process produced a fully functional engine flow path and fuel injection system, which could then be built and tested [9].

WIND-TUNNEL TESTING OF THE MACH 10 SCRAMJET

The HSM engine model faced the limitations inherent in simulating hypersonic flight in ground-based wind tunnels. The HYPULSE shock wave was produced by a hydrogen/oxygen/argon detonation wave generated by a cold helium gas driver at a pressure of 12,000 psi. The reflection of the shock wave produced near stagnation conditions. The facility design, however, limited the period of steady flow to several milliseconds only. The test gas velocity was about 10 ft/ms. The test chamber was 19 ft long and 7 ft high. As a result, the test engine, forebody, and nozzle model length was limited to about 7 ft. This ensured that three model-lengths of test gas would flow over the model during the 2-ms test period. To achieve this, the forebody and nozzle of the HSM had to be shortened from the 12-ft length of the actual vehicle.

Beyond the few milliseconds of test time for each run, other difficulties complicated the testing. Ideally, the wind tunnel should produce the flight flow path values for the local Mach number, flight dynamic pressure, and stagnation enthalpy, at the entrance of the flight hardware. With the

HYPULSE, the stagnation enthalpy was duplicated, and the nozzle design and the mounting angle of the HSM engine produced the local Mach number. The test condition that could not be duplicated exactly was dynamic pressure. The HYPULSE had a lower maximum pressure than that of actual Hyper-X flight conditions. To compensate, the HSM was mounted at a higher angle. This produced higher pressure levels, but reduced the Mach number.

The result was a Mach 10 stagnation enthalpy and a local Mach number at the nozzle matching the value on the second ramp of the X-43A forebody at Mach 10. Verification that the test conditions were correct was obtained using rakes to measure pressure profiles at the nozzle. These were cross-checked using CFD simulations of the flow. The CFD results produced a good match with the core flow. The GASP program predicted a thicker viscous layer, but was a good match for the core value. The VULCAN code was also used to produce flow solutions, including one at a higher-pressure condition and another at a higher speed of Mach 10.3.

Another issue was effects of the test gas and contamination. The test gas in the shock tube section had a composition of 24 percent oxygen, 75 percent nitrogen, and 1 percent argon. The mixture, rather than dry air, was used to replenish the oxygen lost due to the formation of nitrous oxide resulting from reflected shock heating. The result was a test gas composed of 21 percent oxygen and about 6 percent nitrous oxide [8, pp. 3, 4].

As noted earlier, the initial testing of the HSM was done at Mach 7, to calibrate the HYPULSE facility and compare its test data with those from the AHSTF and 8-Ft. HTT, before beginning the Mach 10 testing. This second test series began the day after completion of Mach 7 tests and was relatively brief. It ran from January 27, 1999, to February 17, 1999. As in the first series, the HSM used the M7-KL-6 flow path and the 5-percent silane mix.

Although the flight speed was Mach 10, the actual nozzle exit velocity was Mach 6.4. To simulate Mach 10 conditions, the HSM was positioned at a slightly negative angle of attack. The stagnation temperature was 4100 K, which was the nominal Mach 10 condition. The HYPULSE was limited to a simulated flight dynamic pressure of 500 psf. Aligning the HSM's forebody ramp with the tunnel flow provided a simulation of a flight at a slightly higher angle of attack. This also raised the simulated flight dynamic pressure to 635 psf.

The first Mach 10 "test entry" in the HYPULSE consisted of four calibration runs, two unfueled tare runs, five fueled runs with the 5-percent silane mixture, and, finally, a single fueled run with hydrogen. The test objectives included checkout of the facility operation at Mach 10 enthalpy and determining the practical upper limit of the HYPULSE stagnation pressure. Calibration surveys of the HSM's mass capture were done during the second set of tests. These data verified the calculated predictions from CFD programs. Combusting and mixing runs, along with the unfueled tare tests, were done to compare the HYPULSE results with those produced by the analytical methods used to design the Mach 10 engine flow path.

With these preliminary efforts under way, three separate test entries were done. These used a revised flow path designated M10-KL6, which stood for Mach 10-Keel Line 6. Test entries 3 and 4 were made between July 1, 1999, and March 8, 2000. Entry 5 ran between May 24 and October 13, 2000. By this time, final preparations were being made for the Mach 7 flight. Test entries 3 and 4 involved 5 calibration runs, 4 unfueled tare tests, 11 fueled runs using silane, and 21 fueled tests with hydrogen. Test entry 5 was more limited, with 8 calibration runs, 4 unfueled tare runs, 15 fueled runs with silane, and 10 hydrogen-fueled runs. One difference with test entries 3–5 was the silane mix; though the first two test runs used 5 percent silane, the usual mixture for the later tests was 2 percent silane/98 percent hydrogen. (Following test entry 5, a series of HSM Mach 15 tests, using the M10-KL6 flow path, was done in the HYPULSE tunnel. For test entry 6, the HYPULSE operated as a shock-expansion tunnel, rather than as a reflected shock tunnel, as had been done for the Mach 7 and 10 entries. The Mach 15 entry involved one calibration run and nine engine test runs [10].)

As part of the data analysis, CFD analysis was undertaken concurrently with the test entries. The results of the combined efforts showed the inlet shock train was not in the correct position at the inlet isolator section. The problem was traced to an inadequate duplication of the X-43A forebody shock system at the cowl leading edge. The values of the Mach number, pressure, temperature, and velocity at the cowl leading edge and in the inlet isolator were correct, but CFD analysis indicated the flow profiles were in error compared to those of the X-43A.

Replicating the X-43A's shock position predicted by the CFD analysis required that the cowl be moved back a few inches, producing a better simulation of actual conditions. The issue highlighted the need to correctly simulate the nonuniform and shock-dominated flow entering a scramjet engine during ground tests [7, p. 6, and 8, pp. 5–8].

The wind-tunnel tests and CFD analysis of the scramjet had been completed. The finished Mach 7 design had been tested in numerous wind tunnels, under a wide range of conditions and limitations. Results from the various facilities had been crosschecked to identify any data errors they might be creating due to differentials in pressure, test gas, contaminants, model effects, 3D, or other effects.

For the Mach 10 engine, only the HYPULSE facility could be used. And despite the different entries of the HSM, with both the M7-KL6 and M10-KL6 flow paths, the total time was in milliseconds.

At the beginning of the 21st century, the final test loomed for the Hyper-X project. Would not only the scramjet, but the X-43A, the HXLV, and their systems work in the unforgiving sky?

REFERENCES

[1] Voland, R. T., Rock, K. E., Huebner, L. D., Witte, D. W., Fischer, K. E., and McClinton, C. R., "Hyper-X Engine and Ground Test Program," AIAA-98-1523, pp. 2–4.

[2] McClinton, C. R., Holland, S. D., Rock, K. E., Englund, W. C., Voland, R. T., Huebner, L. T., and Rogers, R. C., "Hyper-X Wind Tunnel Program," AIAA 98-0553, p. 8.

[3] Ferlemann, S. M., McClinton, C. R., Rock, K. E., and Voland, R. T., "Hyper-X Mach 7 Scramjet Design, Ground Test and Flight Results," AIAA 2005-3322, p. 3.

[4] Huebner, L. D., Rock, K. E., Witte, D. W., Ruf, E. G., and Andrews, Jr., E. H., "Hyper-X Engine Testing in the NASA Langley 8-Foot High Temperature Tunnel," AIAA 2000-3605, pp. 3, 4, 6.

[5] Cox, C., Lewis, C., Pap, R., Glover, C., Priddy, K., Edwards, J., McCarty, D., and Warner, E., "Predictions of Unstart Phenomena in Hypersonic Aircraft," AIAA-95-6018, pp. 1–3.

[6] Huebner, L. D., Rock, K. E., Ruf, E. G., Witte, D. W., and Andrews, Jr., E. H., "Hyper-X Flight Engine Ground Testing for X-43 Flight Risk Reduction," AIAA 2001-1809, pp. 5–7, 9–13.

[7] Bakos, R. J., Tsai, C. Y., Rogers, R. C., and Shih, A. T., "The Mach 10 Component of NASA's Hyper-X Ground Test Program," 14th International Symposium on Air-Breathing Engines, Florence, Italy, September 5–10, 1999, ISABE-99-2716, p. 2.

[8] Rogers, R. C., Shih, A. T., and Hass, N. E., "Scramjet Development Tests Supporting the Mach 10 Flight of the X-43," AIAA 2005-3351, p. 2.

[9] Ferlemann, P. G., "Comparison of Hyper-X Mach 10 Scramjet Preflight Predictions and Flight Data," AIAA 2005-3352, pp. 1, 2.

[10] Rogers, R. C., Shih, A. T., Tsai, C. Y., and Foelsche, R. O., "Scramjet Tests in a Shock Tunnel at Flight Mach 7, 10, and 15 Conditions," AIAA-2001-3241, p. 9.

Chapter 8

HYPER-X STAGE SEPARATION RISK MITIGATION TESTING

Keep looking around; there's always something you've missed.*

Before the scramjet engine could be tested in flight, the X-43A research vehicle would have to be accelerated to the planned test speed. This was the task of the Hyper-X Launch Vehicle (HXLV), a modified Pegasus first stage carried aloft and launched from a B-52B mothership. Once the HXLV booster burned out, the X-43A had to successfully separate, stabilize, and begin the engine test. If the HXLV failed, or the separation ran afoul of uncertain flow conditions, the efforts of the X-43A project engineers would have been in vain.

The overriding issue in their minds was separation. There was no experience to draw upon as a place to start. No agencies either within the United States or in foreign countries could provide data. Two nonsymmetrical vehicles would be separating in a high dynamic-pressure environment, with highly dynamic flowfields around the adapter and research vehicle. As the airflow rapidly changed, there was the possibility that it could cause the X-43A to go out of control.

SEPARATION ANXIETY

After considerable debate and a wide range of bright ideas, an initial separation mechanism configuration was selected in June 1997. This used a clamshell adapter, hinged at the back edge and split by either explosive pistons or springs. The two halves would swing back, clearing the X-43A's wings and also serving as speed brakes. Concerns were raised during the manufacturing readiness review (MRR) that a roll upset of the X-43A, as the halves swung back, would cause the vehicle's wings to strike one of the clamshells, sending the research vehicle out of control.

During the subsequent discussion, an engineer suggested that rather than splitting the adapter, it could be hinged at the back and swing down out of the way as one piece. This shortened the period of adapter/wing overlap, and reduced the risk of contact. Referred to as the *drop jaw*, this was incorporated as the baseline design.

The X-43A was attached to the adapter by four explosive bolts—two on the X-43A's keel beam at the nozzle, and two at the aft base of the research vehicle. The X-43A was pushed clear of the adapter by two pyrotechnic pistons procured from surplus B-1B missile ejector racks. These have a powered stroke of 7 in., and a maximum stroke length of 9 in. A second set of pistons would also move the drop jaw during the separation.

Concerns over the separation were such that studies began even before the Hyper-X contract was issued. A test was conducted using an early research vehicle/adapter configuration in Langley's 20-in. Mach 6 and 31-in. Mach 10 tunnels. The X-43A model was mounted on a sting, which passed through the adapter. This setup allowed the relative positions of the X-43 model and adapter to be changed, simulating their movements during the separation. However, although data could be produced through the testing, their value was reduced by the sting's large size. As a result, this was only a preliminary step. Additional efforts, both wind-tunnel testing and CFD analysis, were needed.

WIND-TUNNEL TESTING

One area of early concern was the effects of the drop jaw on the X-43A's stability (see Fig. 8.1). With the drop jaw at a 90-deg angle at hypersonic speeds, a shock could form, pressurizing the nozzle area of the research vehicle and potentially causing a nose-down moment too strong for the control system to counter [1].

Initial testing of the potential drop-jaw effects was conducted at the Arnold Engineering Development Center (AEDC), using the von Karman Facility

Fig. 8.1 A wind-tunnel schlieren photo of the original drop-jaw configuration in the down position.

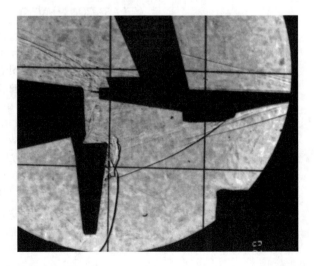

Tunnel B. This employed the Captive Trajectory System (CTS), which allowed the forces on both the booster and X-43A to be measured as their relative positions changed. The models used were 1/12 scale, and the test speed was Mach 6.

The AEDC tests had complications. Although the CTS rig greatly increased the productivity of the testing, it also imposed a minimum separation distance between the adapter and research vehicle. As a result, there were gaps in the database used to define the aerodynamic coefficients. Testing both the drop jaw and the fixed jaw also reduced the amount of data obtained at the positions where the most interference was expected during separation.

The interference effects of the mountings for the X-43A wind tunnel models posed an additional problem in both the AEDC and Langley 20-in. Mach 6 tests. A blade mounting had been used in the AEDC tunnel, versus a sting in the Langley tunnel. To understand how this altered the test data, additional tests were run in the 20-in. Mach 6 tunnel. These used a model mounted with either a blade or a sting, with either a dummy blade or a dummy sting to measure the interference effects. These data were subsequently used to correct the AEDC results [2].

Problems were also encountered with the AEDC tests made with the Langley system. A second series of tests was originally scheduled to begin in October 1997, following modifications to the tunnel. By late October, however, the first "wind-on" run had been pushed back to the first week of November. The first testing did not actually take place until November 17, 1997. Progress was slow due to continuing technical difficulties. The lateral drive motor for the CTS rig burned out. AEDC personnel found a replacement motor, installed it, and began testing. The replacement motor also failed, in a manner similar to the first. It was not until late January 1998 that AEDC testing of different drop-jaw configurations was finally completed. Despite the problems, the testing produced more data than the original research plan had called for.[†]

Determining the separation behavior was made difficult by the sheer number of factors involved. These could be organized into four groups:

- *Unsteady aerodynamic effects*: The first of these was the initial separation transient caused by airflow within the gap between the adapter and research vehicle as it was establishing a quasi-steady-state condition. The second was due to the dynamics of the separation maneuver itself.
- *X-43A aerodynamic coefficients database*: This consisted of three flow parameters, six separation parameters, and three control surface deflections.
- *Database gaps*: Due to the CTS design, there were gaps in the aerodynamic database for the initial part of the separation. Additionally, the X-43A's

---

[†]Aerospace Projects Highlights, November 14, 1997, November 17, 1997, November 28, 1997, and January 23, 1998, NASA Dryden Flight Research Center History Office.

normal and pitching moment varied depending on whether the adapter was exposed or protected from the high-speed flow.
- *Differences between wind-tunnel and flight conditions*: These included model mounting interference, freestream Mach and Reynolds number, geometry of the adapter/X-43A seal, and whether the upper surface flow was laminar or turbulent.

Dealing with these issues required more than wind-tunnel tests. The NASP program had led to development of CFD programs that could accurately simulate complex flowfields. Much of the work on the Hyper-X separation issue was done using these programs.

CFD ANALYSIS OF THE X-43A SEPARATION

One limitation of the wind-tunnel testing was that the models of the HXLV, adapter, and X-43A were held in fixed positions relative to one another. The resulting data were of steady-state conditions. In reality, the separation was a very dynamic event, taking place within a brief time period. At best, the tunnel data were an approximation. At worst, this meant problems that could lead to a failure would not be detected.

The solution represented a revolution in how engineers and researchers analyzed proposed aircraft designs. Until the end of the 1970s, computers lacked sufficient capabilities to analyze flowfields and other aerodynamic phenomena. The programs would take days or weeks to provide a solution, making the effort impractical. The development of the first Cray-1 supercomputer in 1981 meant the same problems could be solved in a few hours.

This increase in computing power meant that the flowfields around airfoils (two-dimensional) or wings (three-dimensional) could now be calculated. These were simple solutions, with factors like viscosity not included in the results. Yet they provided valuable solutions to aerodynamic problems, and, more important, showed the potential of CFD. As the 1980s progressed, some three-dimensional problems could be solved in time intervals as short as 10 s. It was now possible to write computational design-optimization programs. These could alter the wing geometry shape in order to optimize specific aerodynamic qualities. When combined with a flow solving program, the lift-to-drag ratio of the new wing geometries could also be determined, all in a matter of seconds or minutes.

The development of faster processers, linking multiple processors within a single computer, and greater memories, packaged into ever smaller computers, encouraged researchers and programmers to write more-capable software. Fluid physics being modeled went from simple panel calculations, to nonviscous Euler equations of motion, to finally achieving full viscous Navier/Stokes equations.

These advances also allowed higher-fidelity geometric modeling of complex shapes, which were surrounded with three-dimensional grids. This allowed calculations of the surface pressure on the entire vehicle, rather than the limited number of measurements made using a wind-tunnel model. Calculations of multiple body problems, internal flow calculations, and even moving grids to simulate the interactions of two or more separate bodies also were now possible.

These calculations could be performed within a few hours, a time period one engineer referred to as the "threshold of pain" for engineers waiting for the results. This balanced the ambitious goals of their programming with the risk of a computer crash [3].

Two different Euler/Navier-Stokes flow-solving programs were used to analyze the separation maneuver. The SAMcfd program was used for drop-jaw simulations prior to the AEDC wind-tunnel tests. The SAMcfd flow simulations modeled the drop jaw as it rotated downward through several different angles. When the AEDC schlieren photos of the drop-jaw model were compared with CFD results, the match was nearly identical.

Different alternatives were considered, such as delaying the movement of the drop jaw until the pistons were at half or full stroke, but debate continued over dropping the jaw or leaving it fixed until January 1999, when a group of Boeing engineers went to Langley to focus attention on the issue.

Their goal was to reach a decision on the separation scenario by February 1, 1999. Separation simulations done at Langley indicated the X-43A would lose control if the drop jaw were moved. The formal decision to keep the jaw fixed during separation was made at the end of February. Because the pistons had already been built, they were disabled on the adapters for the first two flights. The third adapter had a nonmovable drop jaw.[‡]

The solution to stabilizing the X-43A during separation proved to be relatively simple. Simulations showed that by setting the X-43A's wings to a 6-deg angle, either prior to the pistons firing or at the time of the break wire trip, the risk of recontact with the adapter was minimal. Determination of the actual risk awaited more complete CFD simulations [1, pp. 4, 9, 10].

The SAMcfd program was well suited to the stage-separation tasks, because it could manipulate the geometric positions of several objects with reference to one another with ease. It could also produce grids around the complex shapes of the adapter and X-43A. The stage-separation tests involved creating grids with 4 to 5 million cells. It was not a program that could be run on a personal computer, however. SAMcfd calculations required 5000 iterations to reach a steady state, and were run on a Cray C-90 supercomputer. The other CFD program used was the OVERFLOW, a structured, overset grid

---

[‡]Aerospace Projects Highlights, January 29, 1999, p. 4.

Fig. 8.2 A 3-percent wind-tunnel model of the Hyper-X stack undergoing a wind-tunnel test.

Navier-Stokes flow-solver program. Patches were generated that covered the surface of the adapter and X-43A. The fins' leading and trailing edges as well as the X-43A and inlet leading edges were covered with wrap-around strips.

Computational grids were constructed, and the wind-tunnel scale effects were incorporated using the proper Reynolds number. This is used to compare data from small wind-tunnel models with those from full-scale aircraft to reveal differences resulting from size and amount of air displaced. The initial grid spacing off the surfaces was adjusted to incorporate such factors as turbulent flow, compression, and temperature and Mach number.

Volume grids were extended 8 in. from the surface grids. These separate, body-fitted grids could then be moved and rotated relative to each other to simulate the movements of the adapter and X-43A. The body-fitted grids were also connected in a background grid, which provided an external reference. The entire grid structure consisted of 1.4 million points for the X-43A and the wind-tunnel blade support hardware, and another 600,000 for the adapter. The background grid added another 800,000, raising the total to 2.8 million points, in 52 component grids. Given the sheer volume of the grid, achieving the converged forces and moments required 2000 to 4000 steps on the fine grid level. The computers used for the OVERFLOW program were the Cray C-90, the NAVO Major Shared Resource Center Cray SV 1, the SGI Origin 2000, and an SGI Indigo workstation.

Just as the individual wind-tunnel results from the scramjet testing had to be compared with each other to identify facility effects, so, too, did the SAMcfd and OVERFLOW results have to be compared to those of the wind-tunnel tests. Simulations of the X-43A were done to accomplish this. The CFD results were compared to data from test T6776. This used a sting-mounted 1/8-scale X-43A model tested in the Langley 20-in. Mach 6 wind tunnel (see Fig. 8.2).

Several different wind-tunnel models were required (see Figs. 8.3 and 8.4) because of the different flight environments the stack would encounter, including a lower launch altitude for the pair of Mach 7 flights, standard launch conditions on the Mach 10 flight; the separation of the X-43A, flights of the Hyper-X Launch Vehicle with and without the X-43A, and the X-43A in free flight.

The computations assumed laminar flow and did not include sting effects. The normal force, axial force, and pitching moment coefficients were calculated using the OVERFLOW program. When the CFD calculated values were compared with the wind-tunnel data, they showed good agreement for the values of the normal force and the pitching moment. The axial force values were only a fair match over a wide range of angle-of-attack results. The level of agreement was judged adequate for predicting separation behavior.

Results from the CFD program were also compared with those of a second wind-tunnel test with turbulent boundary layers, and to another test at the

Fig. 8.3   The 3-percent-scale models of the Hyper-X stack a) the complete stack, without the X-43A and with a spoiler on the adapter; b) with the X-43A attached; and c) an adapter without the X-43A and spoiler.

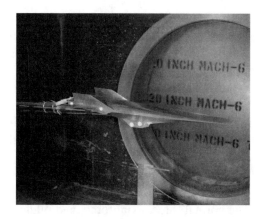

Fig. 8.4 A larger wind-tunnel
model of the X-43A in Langley's
20-in. Mach 6 tunnel.

flight Reynolds number. Both showed close agreements between the wind-tunnel and calculated values for the normal force and pitching moment. Axial force values changed as expected.

Similar comparisons were made using the SAMcfd flow solver at angles of attack of 0 deg and 2 deg. These showed agreement with the normal force and the pitching moment. The results of the comparisons between the two CFD programs gave engineers confidence that separation behavior could be predicted. The AEDC tests had provided steady-state wind-tunnel data. As a result, attention shifted to the unsteady aerodynamic effects of the X-43A separation.

The first of these was the initial transient, which occurred as separation began. As the X-43A began to move forward, a gap opened between its aft fuselage/nozzle and the adapter. The hypersonic airflow would enter the gap at the adapter lip, and vent at the sides and X-43A base. As the two vehicles moved apart, the gap would become wider and the pressure would equalize. Project engineers were concerned that a sudden pressure increase would occur as the gap opened, and this kick could cause changes in the normal and pitching force. The potential risk this posed to the stability of the X-43A depended on how fast the pressure equalized vs. the duration of the separation maneuver. With the stack at 100,000 ft and flying at Mach 7, the dynamic pressure would be 1000 psf or more.

To determine the potential consequences, two-dimensional/time-accurate simulations were performed using the centerline geometry of the X-43A/adapter, at separation distances of 0.5 in. and 2 in.. These conditions would nominally occur at 44 ms and 60 ms after the separation command was given. Because this was a two-dimensional simulation, the simulated gap flow was much more restricted than it would have been under actual conditions. There was no pressure relief along the gaps on either side. Rather, the total airflow was vented out the X-43A's base, representing a worst-case simulation.

The initial simulations were made with the SAMcfd program, and the results in both cases showed that the forces stabilized within 10 ms. A second, separate simulation was performed using the CFL3D flow solver for the 2-in. separation distance. As in the SAMcfd case, the forces acting on the X-43A became steady state within 10 ms—significantly less time than the duration of the separation. In addition, pressure relief through the gap's sides would further reduce this time period. Another, more important factor was the fact that the pistons were still pushing on the X-43A as pressure equalized. The pistons had a maximum stroke of 9 in., which was directed through the X-43A's center of gravity. This constrained the research vehicle's movements. Between the wind-tunnel data, CFD simulations, and the pistons' limitation on the X-43's motions, the initial separation transient was no longer considered an issue [2, pp. 4–6].

The second unsteady-effects issue was the dynamic separation process. This was a major challenge in CFD simulations. Other studies had been done on moving bodies at transonic speeds using a version of OVERFLOW that had 6-deg-of-freedom dynamics in response to aerodynamic forces. This meant the simulated vehicle could move up and down, left or right, could speed up or slow down, and could rotate around the yaw, pitch, and roll axes [2, p. 6].

The simulation of the mated HXLV and adapter, and X-43A separation was far more complex. This involved 6 deg of freedom for the X-43A, 6 deg of freedom for the HXLV, and 2 deg of freedom for the pistons (3 deg of freedom with the moveable drop-jaw simulations).

To produce this simulation, the SepSim program was developed by Analytical Mechanics Associates under contract to Langley. This program generated multi-body 6-deg-of-freedom equations of motion, aerodynamic forces and moments, control system characteristics, atmospheric modeling, explosive bolt and piston characteristics, vehicle dynamics, separation mechanics, and aerodynamics for the HXLV and X-43A.

Creating computer models for the SepSim program was difficult. The HXLV and X-43A aerodynamic models were drawn from the database collected in the AEDC and Langley wind-tunnel testing. This database included both the vehicles independent of one another and interference aerodynamics for the two vehicles in close proximity. The wind-tunnel data were expanded using CFD calculations. These covered the ground-to-flight scaling and the unsteady and dynamic flow occurring during the separation. Additionally, the CFD calculations allowed the data to be expanded to vehicle orientations that could not be tested in wind tunnels.

Dr. Rodney Bowersox, of the University of Alabama, analyzed the uncertainties in the aerodynamic data. From this work, the 12 aerodynamic coefficients for any relative orientation could be extracted from the database as a function of the angle of attack and sideslip of the X-43A, and the angles and distances between the X-43A and the HXLV centers of mass.

The data required for the SepSim also included the mass properties of the stack's components. The HXLV mass properties were based on Orbital Sciences Corporation's extensive Pegasus database, with allowances for the Hyper-X modifications. The mass properties of the adapter and its internal system were provided by the manufacturer, Micro Craft. The X-43A's mass properties were determined based on weight and inertial testing done at Dryden. The mass property results were cross-checked against the properties calculated by the structural analysis codes.

Though the aerodynamics of the separation was a primary concern, there was also a mechanical aspect to be addressed. The pistons used to separate the X-43A moved 9 in. in 0.1 s and in the process generated a force of 22,000 lb. This was sufficient to produce a 13-fps relative velocity between the X-43A and the HXLV, despite the research vehicle's weight of 3000 lb and the 1000-psf dynamic pressure of the hypersonic airflow. Successful separation required that the two vehicles not collide after separation, and that the X-43A be able to stabilize itself at a 2-deg angle of attack and 0-deg sideslip within 2.5 s. Should either requirement not be met, the engine test could not be completed and the mission would be a failure.

The SepSim model of the pistons' behavior was based on a number of tests made by Orbital Sciences to verify their force profiles and timing. A second issue was the possibility that the normal or side loads on the X-43A were not zero, due to yaw or pitch angles at separation. In this case, the pistons could deflect or experience higher friction than predicted.

Tests using a single piston were performed at Dryden to measure the potential results. These were done using side loads from 500 lb to 2000 lb in 500-lb increments. The tests were completed in mid-June 1999, and the results showed that such side loads, higher than those expected for the actual flight, would damage the pistons and cylinders. Even with the damage, however, they would still perform as designed and the separation would not be affected. The computer model of the piston operation was also cross-checked using the data from the original B-1B missile ejector rack qualifications. The explosive bolts holding the X-43A in place were modeled for the SepSim program. Their timing was based on extensive testing conducted by Orbital Science. These tests ranged from simple single-bolt firings on up to a full-scale ground test of the entire system.

The models of the different systems also incorporated information from a database of potential aerodynamic and hardware uncertainties and variations. These had to be assembled as part of the CFD. For the Flight Control, Guidance, and Navigation Unit (FCGNU), the flight control laws were modeled using information from Boeing, which had developed them. Honeywell, manufacturer of the FCGNU, supplied the model. Output errors were modeled, and the predictions compared to a Honeywell variation analysis to determine their accuracy.

The simulated FCGNU errors included position, velocity, and tilt-and-azimuth errors, and were assumed to be constant during the 2.5-s separation maneuver. A misalignment of the FCGNU was also simulated. The errors would result in variable feedback errors in the angle of attack, Mach number, dynamic pressure, velocity, bank angle, pitch angle, and body rates. The error states were initialized as either equations derived from the nominal boost trajectory or precalculated errors for specific boost trajectories.

Modeling the uncertainties and variations of the X-43A's control surface dynamics was a multi-step process. For the first flight, the second order actuator model with the simple freeplay element was used. (For the second and third flights, a Moog high-fidelity model was used. Moog had originally used this in the design and manufacture of the actuators. The original version was refined and qualified for flight use.) These theoretical models were expanded again from ground tests of the first X-43A's control system conducted at Dryden involving actuator and hysteresis testing.§ The final refinement was the result of the aircraft-in-the-loop ground testing of the first X-43A vehicle, also done at Dryden [4].¶

In addition, there was the issue of the atmospheric conditions the stack would experience through launch and separation. The HXLV's guidance and control software used the GRAM 95 atmospheric data during the ascent. The X-43A control system, in contrast, was programmed using the Dryden Range Reference Atmosphere model. The differences between the GRAM 95 and the Dryden Range Reference Atmosphere programs were not too significant.

These uncertainties and variations were used to create a database for Monte Carlo analysis of the separation. These were the nominal and the "3 sigma" predictions, which encompassed 99.7 percent of the predicted values. When each run was made, different values for the uncertainties and variations would be used. A month of data derived from the Vandenberg wind profiles, for example, would be picked at random as input for the Monte Carlo analysis. Though the individual variations might not have a significant effect on the separation, in combination they could determine its success or failure.

On other simulations, the maximum 3 sigma values would be used. This was called a "tolerance stack-up," and was a theoretical possibility.** In such a simulation the values all go in the wrong direction, creating maximum

---

§ *Hysteresis* refers to the delay between an action, such as a control input, and the result, in this case the response of the vehicle. A more familiar term would be *control lag*.

¶Aerospace Projects Highlights, June 4, 1999, p. 2, and June 11, 1999, p. 2.

**Tolerance stack-ups sometimes occurred during the production of aircraft. Each part had specific dimensions, and a plus or minus error. The assumption was that the random errors would cancel out; however, if the errors were primarily in one direction, the handling of the aircraft could be adversely affected.

demands on the hardware. These simulations determined whether the X-43A could maintain controlled flight despite the worst-case situation.[tt]

Just as the tests of the different scramjet models were cross-checked in different wind tunnels, so too were the SepSim program results. This validation for the first flight was done on three levels. The initial review of the input variables, implementations, and results from the simulations were reviewed by those who contributed to the program models.

The second step was modular level checks to compare the SepSim results with those from independently constructed check models. The procedure called for running the SepSim program to capture the model input/output and the Monte Carlo variables. The inputs were then run through the Simulink check model, and the results were compared. One such modular level check of the port wing commands involved the models for the actuators, aerodynamics, atmosphere, controls, inertial navigation system (INS) errors, pistons, and separation variations. When the wing deflections predicted by the two programs were plotted, the results were an exact match.

In the final step, the integrated level checks, the SepSim results were checked against results from an independent program developed at Dryden called RVSim. Additionally, the actuator model in another program, called Simulink, was used for a further check. The result was a cross-check of the simulation implementation, the equations of motion, free-flight aerodynamics, and the geometry and mass properties. A comparison of the separation results from SepSim and RVSim for five cases showed a near-perfect overlay for the 2.5-s duration of the separation sequence.

For the second and third flights, more extensive testing was done, using a somewhat different approach. A program called POST was used as the main program to check the separation maneuver. The SepDelta model, derived from SepSim, was also added to the tests, to apply forces and moments to the initial separation maneuver. (These had been lacking in the initial simulations.) The RVSim and SepSim were two separate simulations, run at two separate locations (RVSim at Dryden, SepSim at Langley). On the first flight few cross-checks were made for the two simulations. With the second and third flights, cross-checks were a major part of the test efforts. One of the purposes of the SepDelta model was to provide an end-to-end model without having to run the LVSim, SepSim, and RVSim on one machine. The SepDelta model approximated the end states to assist the RVSim. As a result, the engineers were sure that the X-43A could take any of the end states from SepSim and be successful.

The last element of the separation simulation was recontact analysis. This used the DIVISION software, which produced three-dimensional models of

---

[tt]Griff Corpening history interview, December 20, 2004, NASA Dryden Flight Research Center History Office, pp. 19, 20.

the HXLV, adapter, and X-43A. The DIVISION program would take SepSim position and orientation data for the different components, and calculate whether they interfered with each other. If none did, then 1- and 2-in. shells were "built" around the components. The DIVISION software was run again to determine whether the vehicles came within these shells, indicating the potential for a collision.

An accurate model of the Hyper-X stack—with all its systems; the variables, uncertainties, and errors in their operation; and the different combinations of Mach numbers, angles of attack, dynamic pressures, altitudes, angles between the two vehicles, separation behavior, and atmospheric and wind conditions—had been created inside the computer. The model and the associated programs had also been cross-checked for accuracy. It was now time to see whether the separation would work in cyberspace.

Monte Carlo analysis of the separation involved 2000 separate runs. Each run began with the command to fire the explosive bolts, and ended after 2.5 s. At this point, the X-43A had cleared any interference with the HXLV/adapter, had stabilized at the proper angle of attack and sideslip, and was ready to begin the cowl-open portion of the flight profile. The separation was assumed to be successful if the X-43A reached this point and was at planned flight conditions.

Of the 2000 runs, 7 failed before reaching the 2.5 s. In another 11 runs, the X-43A exceeded the ±10 deg of angle of attack or sideslip during the 2.5 s of separation. For eight cases, the X-43A was not at the correct 2-deg angle of attack and/or the 0-deg sideslip angle at the 2.5-s point. The majority of these failures were due to an assumed 0.75-deg uncertainty in the inertial navigation system (INS) position in the X-43A. These failures could be eliminated simply by making sure the INS position was accurately known.

Of the 1993 successful SepSim runs, 88.2 percent fell within correct angle-of-attack limits, and 100 percent were within sideslip limits. Data from the 8-Ft. HTT testing of the spare flight engine indicated a larger margin of error for a successful scramjet ignition than the narrow limits on angle of attack and sideslip of the SepSim. These tests showed all the SepSim runs had angle of attack and sideslip measurements within the range of successful wind-tunnel starts, which was several times larger than the more stringent requirements for the SepSim.

The 26 failures represented 1.3 percent of the total number of runs. None of the 2000 runs resulted in a collision between the adapter and the X-43A. In a single run, the two vehicles came within 1 in. of each other, and in 54 cases the margin was less than 2 in. The histories of the variables' means and standard deviations showed that the 2000 runs were statistically valid, and indicated a very high probability of a successful separation and research flight [4, pp. 3, 4, 6, 7].

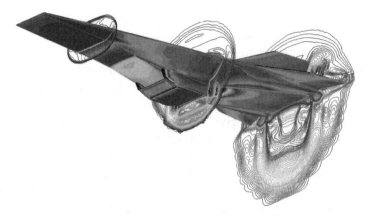

Fig. 8.5   A computer graphic of the CFD airflow around the X-43A.

The analysis of the separation maneuver involved a major effort, across many different organizations. Some of those involved were Scott Holland for the aerodynamic testing, Pieter Buning for in-house CFD, John Martin for managing the SepSim development and utilization, Dave Bose of AMA for a large part of the SepSim development, Peter Liever and Sami Habchi of CFD Research Corp. for the contracted CFD, and Walt Engelund for the aero database.[‡‡]

In the bigger picture, CFD analysis (see Fig. 8.5) provided answers to several questions about the separation maneuver, and reduced the risks involved with this part of the flight profile. This included analysis not only of the free-flight segment of the mission, but also of the scramjet operation and separation sequence.

The CFD simulations [2, p. 10]:

- Indicated the initial aerodynamic transitions of the separation were of short duration
- Showed the assumption of quasi-steady-state modeling of the separation aerodynamics was valid
- Illustrated the physics behind nonlinear variations of aerodynamic coefficients with vertical clearance
- Provided a different methodology for extending the database into regions of the nominal trajectory
- Determined the horizontal aerodynamic interference from the adapter
- Proved a methodology for gauging model support interference
- Confirmed assumptions on effects of geometric and flow differences resulting from wind tunnels and flight conditions

---

[‡‡]Dave Reubush e-mail to Curtis Peebles, "Chapters 8, 9, & 10," June 19, 2009, 2:34 P.M.

- Provided information on the usefulness and accuracy of both inviscid and viscous CFD simulations
- Established an experience base for the development of follow-on systems

HARDWARE GROUND TESTING

Though computational fluid dynamics played a central role in analyzing and predicting HXLV separation behavior, more traditional mechanical methods were also used. These addressed issues for which computer simulations could not provide answers, or determined which physical tests were needed. Methods used included tests of the explosive bolt shocks, drop-jaw tests to evaluate the adequacy of the pistons, and airframe tests to determine the flight management unit (FMU)/actuator shock level. These tests began in mid-June 1998.

One element tested the explosive bolt that joined the forward jaw and the X-43A. A portion of the structure on both sides of the joint was constructed for the test and held together by an explosive bolt. When the test was made in July 1998 the bolt fired, but the joint did not separate. Had the failure occurred in flight, the mission would have been lost.

The test was run a second time, and again the joint did not separate. Engineers consulted with several organizations that had experience with explosive-bolt operations. Sandia National Laboratories suggested a modification to the counterbores around the bolt, and Micro Craft looked at three counter-bore design options in late July 1998 to deal with the problem. This design change proved successful, but the twin failures raised doubts about the first explosive bolt's suitability. Additionally, engineers wanted the joint to be as stiff as possible. Project managers began examining potential replacement bolts. The decision was finally reached to replace the bolts originally selected.

In August, the new explosive bolts were tested in three different fixtures to determine their performance. These simulated the forward jaw to X-43A joint, the adapter bulkhead station 144 to X-43A aft bulkhead, and the jaw to aft adapter. No problems were encountered with these tests and the bolts operated flawlessly. The bolts supplied by Pacific-Scientific had an added advantage of being stronger than the bolts originally selected. This improved stack rigidity as well.

The second hardware testing was of the drop-jaw mechanism, and was conducted by Orbital Sciences engineers during October and November 1998. (At this time, the issue of the movable vs. fixed drop jaw was still unresolved.) Rather than using the actual hardware, a mass simulator was used. This duplicated the jaw's moment of inertia within 0.2 percent. The jaws and drive pistons were mounted in a framework, and the pyrotechnic cartridges were fired. Instrumentation was attached to the two pistons and the jaw hinge. A transducer measured the driving pressure of the pistons, while load cells measured the force produced by the pistons. Video and still photographs completed the documentation. The test determined that the ejection mechanism would both force the jaw to drop and cause the pistons to force the X-43A away from the HXLV at low and high pressures alike, without complications.

Another concern regarding the explosive bolts was that the shock from their detonation would be greater than the FCGNU, the actuator controllers, and the actuators themselves were designed to withstand. An airframe shock test was performed using the adapter for the first flight, and the second X-43A. This provided shock data from several airframe locations.

The shock test was followed by testing of the pistons using dummy mass simulators of the empty HXLV stage and the X-43A, using concrete slabs with the same weights as the vehicles. To allow the two mass simulators to move freely, both were supported on air pads. They were instrumented with pressure transducers on both pistons, and piezoelectric force rings on the X-43A mass simulator. An energy absorber was attached between the two mass simulators. The mass simulators would determine whether any yawing moments would be produced when the pistons fired; this would indicate the forces generated by the pistons were not equal. A cross-feed was included in their design to equalize the pressure produced by the pyrotechnics. The test was made twice, and showed no yaw movement.

These separation tests were not without a measure of excitement, however. On one, the restraining device on the 3000-lb slab simulating the X-43A was either not engaged or failed outright. As a result, it went sliding across the floor, but caused no damage.

The final test was a full-scale ground separation test using actual flight hardware. The second X-43A was ballasted to its proper flight weight and center of gravity location. The X-43A was supported by an overhead crane, and was attached with a long cable in order to minimize swinging. The research vehicle was then attached to the first adapter using flight explosive bolts. An engineering version of the FCGNU was fitted to determine whether the shocks from the explosive bolt firing and piston push would affect its operation. All of the separation instrumentation, flight pyrotechnics, and initiators were also fitted as they would be for an actual flight.

The final feature was a late addition. During the debate over the movable vs. fixed drop jaw, in late January 1999 Dryden engineers suggested the addition of a camera package to the adapter. Two black-and-white cameras would provide views from the left and right sides as the research vehicle moved away from the booster. The images were grainy, but could give an indication of the cause of a failure. Approval for installation of the cameras was given in early February. As with the other adapter/X-43A components, engineers were concerned with the effects of the explosive bolt and piston shocks on the camera systems. One complete camera system was installed to test its susceptibility to shock.

The hardware test was conducted on September 16, 1999. When the pistons were extended, instrumentation indicated that no yaw was imparted to the X-43A. The FCGNU was unaffected by either of the shocks from the explosive bolts or the pistons. All of the onboard instrumentation, including the cameras, functioned as planned [1, pp. 5, 11].§§

ASSESSING CFD SIMULATIONS

CFD simulations, such as those done for the Hyper-X separation, represented a revolution in aircraft design methods. The development of supercomputers gave researchers the ability to perform calculations that were impossible with earlier mainframe systems. The researchers who were involved stressed that the CFD simulations were not done in isolation, but were one element in a much larger set of tests, which gave an understanding of what the Hyper-X would face. David Reubush, who was in charge of stage separation, commented:¶

There were CFD calculations, both static and dynamic (also, both in-house and contract) of the separation event. The results from these calculations, along with the results of a number of wind tunnel tests, and the results of the various hardware tests fed into a Monte Carlo simulation of the separation. The simulation was exercised extensively, including a number of very stressful cases. The results of the simulation exercises gave us a high degree of confidence that the separation would work as desired.

One specific thing that the dynamic CFD gave us was it verified that the flow set-up time in the gap between the X-43 and the adapter was significantly faster than the movement of the vehicles. This allowed us to be confident that the quasi-static wind tunnel data would be applicable to the dynamic event.

Lowell Keel of ATK GASL added:***

The combination of sound design efforts, a rigorous experimental effort to validate that design, wind tunnel testing of the multiple body separation event, CFD (both steady state and dynamic) and Monte Carlo studies

---

§§Aerospace Projects Highlights, June 12, 1998; July 24, 1998; August 21, 1998, p. 2; September 4, 1998, p. 4; October 9, 1998, p. 4; October 16, 1998, p. 4; November 6, 1998, p. 4; November 13, 1998, p. 3; November 30, 1998, p. 3; January 29, 1999, p. 4; and February 5, 1999, p. 3.

¶David Reubush e-mail to Curtis Peebles, "Re: Separation CFD simulations," March 23, 2009, 3:34 P.M.

***Lowell Keel e-mail to Curtis Peebles, "RE: Separation CFD Separations," March 24, 2009, 9:49 A.M.

all contributed to our confidence and worked interactively to mature the understanding that was its basis.

Crediting any one tool out of the context of the combined set, would not be a clear representation of what the X-43 project did or how we worked.

Randy Voland, a Langley researcher, expanded on this: "... there were a large number of wind tunnel tests, flight hardware functional/developmental and qualification tests, and a very large number of separation trajectory simulations that also contributed to our confidence in the separation event. The CFD on its own was not what gave us confidence. At least that's my opinion."†††

HXLV LAUNCH PROFILE WIND-TUNNEL TESTING

Analysis of the separation involved extensive testing in different wind tunnels, with different CFD software, and ground tests with actual flight hardware, as well as cross-checking the results to ensure their accuracy. This was a reflection of the concern raised by the unknowns of the maneuver.

In contrast, analysis of the HXLV launch profile was handled much differently. Though there was not the level of concern there was with separation, the lower/slower launches for the pair of Mach 7 flights had risks. During the initial period of the rocket burn, the HXLV would be flying at transonic speeds at a fairly high angle of attack. With its odd shape and sharp edges from the nose, wings, and tail fins that could generate vortices—all of which could interact with control surfaces in unpredictable ways—the X-43A was very different from the shroud normally on the front of a Pegasus. Several wind-tunnel tests were made to measure the aerodynamic forces and moments on the stack under those conditions.

Griff Corpening, who served as Dryden's chief engineer on the first two Hyper-X flights, later recalled one of the assumptions of the project, one that reflected its better, faster, cheaper heritage, and pointed out both an advantage and a limitation of the approach. He recalled:‡‡‡

> We also had a certain amount of direction at the beginning of the project that we would treat the launch vehicle as a quasi off-the-shelf system. It [the Pegasus] was proven hardware. It was a proven launch system. We were just modifying it slightly—or not so slightly.... what we wanted to do was focus on those changes in the launch vehicle, not the things that were similar between us and Pegasus. We inherited a number of things.

---

†††Randy Voland e-mail to Curtis Peebles, "RE: Separation CFD Separations," March 23, 2009, 12:13 P.M.

‡‡‡Griff Corpening history interview, December 20, 2004, Dryden Flight Research Center History Office, pp. 4, 42, 43.

One of them we inherited was the launch vehicle wind tunnel model. We'd modified, of course, with the research vehicle on the front end. Now this model was set up where the control surfaces that rolled the vehicle—the two downward dihedral fins would go in opposite directions to roll the vehicle ... was set up to roll in five degree increments.

So we went into the wind tunnels, and we didn't want to design a new wind tunnel model; we didn't see a reason to. So we went in there, and we would test at an angle of attack with the fins neutral, with the fins at plus or minus 5 degrees, plus or minus 10 degrees. And we would do that at various angles of attack and side-slip angles and things to develop the database. And that you take the data from zero to 5 degree deflections, and you would connect those directly in a linear fashion. So if you were at—let's say 2 1/2 degrees—you'd simply take the force that you get at zero degrees and at 5 degrees, and divide it by 2. So it's a linear average. And what we were looking at was our roll authority. In other words, how quickly and with what authority the launch vehicle would roll the stack. And we had that data at zero, 5, and 10 degrees. Then we just interpolate linearly between those conditions.

The fins remained in a fixed position during each wind-tunnel run. To measure the forces on the fins at the transonic flight conditions, strain gages were used, and no excessive forces on the fins were detected that would cause a loss of flight control during the boost phase.[§§§]

In February 1999, a new trajectory analysis using new drag estimates indicated a 15 percent increase in the maximum dynamic pressure during the boost phase over the previous estimate. Dryden's Aerospace Project Highlights noted that the drag estimates raised concerns that thermal and structural analysis would be needed to determine the impact.[¶¶¶]

The HXLV tests were expanded following an incident during a Pegasus launch on March 5, 1999. The booster had experienced a sideslip excursion at transonic speeds, but recovered. The launch carried the WIRE (Wide Field Infrared Explorer) satellite, which reached orbit, but failed soon after. Additional Pegasus transonic wind-tunnel tests were recommended by the WIRE mission failure review panel. Langley also reviewed the WIRE launch incident, and as a precautionary measure additional tests of the 6-percent Hyper-X stack model were scheduled in the Langley16-ft. tunnel in the summer. These didn't constitute a complete reexamination, however. As part of the tests, a few HXLV test points were also run. The HXLV tests were under way at the end of July 1999, and were expected to be wrapped up by the end of the week. The Aerospace Project Highlights for August 13, 1999, noted: "The Hyper-X transonic wind tunnel runs (part of the Pegasus WIRE mission

---

[§§§]Walter C. Engelund interview notes, April 13, 2009.
[¶¶¶]Aerospace Projects Highlights, February 26, 1999, p. 3.

tests) were completed. A data review in progress, preliminary results show no issues with the Hyper-X stack configuration."****

This was not the only reexamination of the Hyper-X launch profile for transonic stability issues. Aerospace Project Highlights for January 21, 2000, reported, "Additional HXLV wind tunnel testing was completed at Vought at Mach 0.4, 0.5, and 0.6. Test to reduce uncertainty level at B-52 drop conditions."†††† Again, no concerns were raised about the new launch profile.

REFERENCES

[1] Reubush, D. E., "Hyper-X Stage Separation—Background and Status," AIAA 99-4818, pp. 3, 4.
[2] Buning, P. G., Wong, T.-C., Dilly, A. D., and Pao, J. L., "Prediction of Hyper-X Stage Separation Aerodynamics Using CFD," AIAA 2000-4009, pp. 1–3.
[3] Consentino, G. B., "Computational Fluid Dynamics Analysis Success Stories of X-plane Design to Flight Test," NASA/TM-2008-214636, pp. 2–4.
[4] Reubush, D. E., Martin, J. G., Robinson, J. S., Bose, D. M., and Strovers, B. K., "Hyper-X Stage Separation—Simulations Development and Results," AIAA 2001-1802, pp. 2, 3.

****Aerospace Projects Highlights, April 30, 1999, p. 2; June 4, 1999; July 30, 1999, p. 2; and August 13, 1999, p. 2; and WIRE, http://www.astronautix.com/craft/wire.htm.
††††Aerospace Projects Highlights, January 21, 2000, p. 1.

Chapter 9

X-43A Research Vehicle Design and Construction

> We are thinking of building a machine next year with [a] 500 sq. ft.
> surface.... If all goes well the next step will be to apply a motor.
>
> *Wilbur Wright, letter to George Spratt, December 29, 1902*

Micro Craft and its partners faced a daunting task. A small company that built wind-tunnel models had been chosen as the prime contractor to build the fastest aircraft ever to fly, using a new air-breathing engine that had logged no actual flight time. The vehicle would also have to operate autonomously. And finally, the electronic systems; flight controls; hydrogen, silane, and nitrogen tanks; coolant supply; batteries; instrumentation; ballast; wiring; connectors; telemetry system; and transponders all had to be packaged into a vehicle only 12 ft long.

The team that built the X-43A originally consisted of Micro Craft, North American Aviation, and GASL. During the course of development, construction, and flight tests, the U.S. aerospace industry continued its post-Cold War consolidation. North American Aviation became part of Boeing, as did McDonnell Douglas (the losing X-43A prime contractor). Micro Craft and GASL merged, becoming part of ATK, and were renamed ATK GASL.

The breakdown of the ATK GASL/Boeing team's duties was as follows. ATK GASL was to be in charge of the overall contractor team and also had responsibility for most of the work on the X-43A. This included design of the X-43A vehicle and its structural details, manufacture of the vehicle itself, the final machining of the vehicle's outer mold lines, and assembly of internal systems. The company would also be given responsibility for the adapter structural design, manufacture, internal assembly, and outer mold lines.

GASL had built the Hyper-X scramjet engine, and, following the merger, ATK GASL retained this responsibility. This included the design and manufacture of the engine itself as well as design and manufacture of the fuel and igniter systems.

Boeing absorbed the roles North American had originally been given with the Hyper-X project. These were responsibility for the vehicle's aerodynamic and structural design; thermal management; designs and specifications for several vehicle subsystems, flight control laws, vehicle control-system

Fig. 9.1   At the X-43A's nose was the tungsten ballast needed to maintain the center of gravity.

software, and vehicle management systems; and production of the vehicle's thermal tiles.

Major subcontractors on the X-43A included Honeywell for the vehicle management systems, and Moog for electromechanical actuators and control units. Carbon-carbon leading edges for the two Mach 7 vehicles (see Fig. 9.1) were designed by SAIC and manufactured by B. F. Goodrich. For the Mach 10 flight, MER Inc. would manufacture the leading edges. Subsystem contractors included EaglePicher, which supplied the vehicle's batteries; VALCOR, for the high-pressure solenoid valves and regulators; Jansen Aircraft System Controls, for the hydrogen and silane flow control valves; and L-3 (previously Aydin Vector), for the modular instrumentation system [1].

STRUCTURAL DESIGN: BUILDING A STEEL AIRPLANE

The X-43A vehicle shape was derived from the DF-9 study; however, its internal structure was influenced by the demands of the research mission. The initial structural design was based on the preliminary studies. As the detailed structural design process began, the stack-stiffness issue arose. It was soon apparent that the X-43A and the adapter would have to undergo a major redesign due to the limitations of the Pegasus guidance system. Complicating the problem was the position of the short stack on the forward end of the booster.

The research vehicle's first bending frequency would have to be about 65 Hz, with additional stiffness in the joints between the X-43A and the adapter, and the adapter and the booster, to achieve a first bending frequency above 8 Hz. A further complication was the need for the research vehicle to have nearly neutral longitudinal-stability characteristics. To move the X-43A's center of gravity far enough forward to accomplish this, a large amount of ballast would have to be added to the nose. An 865-lb wedge of a tungsten alloy called Densalloy 180 was added, which amounted to more than a quarter of the research vehicle's total weight. The ballast may have also been the largest single piece of tungsten ever produced. Such a large mass on the forward end of the research vehicle made the stiffness issue all the more severe.

To find a solution, an extensive series of finite element analyses were undertaken. The results showed the need to add considerable amounts of steel to the research vehicle and adapter structure. The skin thickness of the research vehicle would also have to be increased beyond that originally planned. Not surprisingly, this increased the vehicle weight considerably, to about 3000 lb.

Several different steel alloys were used for the internal longerons, internal bulkheads, and chine supports comprising the X-43A structure that emerged from the analysis. Aluminum was also used in areas where stiffness requirements were lower. In areas where higher temperatures would occur, such as the aft bulkhead, titanium alloys were used. The bulk of the outer skin was aluminum alloy.

The manufacturing readiness review (MRR) for the X-43A airframe was conducted on July 16, 1997, at Boeing North American's Seal Beach, California, facility. The contractor team was cleared to undertake the fuselage primary structural design and to begin materials purchase and parts manufacturing. By early September 1997, Micro Craft management reported that the company had begun machining the first part of the first X-43A vehicle [1, pp. 2, 3, 5, 7].*

A separate issue was the design review of the vertical tail and stabilizers, rudders, wings, and their secondary structures. These components faced greater manufacturing, structural, and thermal requirements. Accordingly, the MRR pertaining to them would take place at a later date. The initial design studies were done by the Hyper-X contractors. Their results were then submitted for review by Langley engineers. By mid-August 1997, the preliminary analyses of the wing thermal-structural design studies were completed by the Langley engineers and presented to contractors. Three different wing designs emerged from the review. Each was made of Haynes 230 nickel steel alloy. One used a solid piece of Haynes 230, the second featured a hollowed-out wing covered with Haynes skin, and the third was a wing with carbon-carbon leading edges for improved thermal protection. Each design model underwent three iterations, and the data were encouraging for both the solid and hollow wing designs; however, the Haynes 230 leading edges would not survive the heating. The Langley study concluded a carbon-carbon leading edge would be required.†

The wing and fin designs were unusual. Normally, hot structures such as the control surfaces would be solid or semi-solid in construction; however,

---

*Aerospace Projects Highlights, July 18, 1997, July 25, 1997, and September 5, 1997.
†Aerospace Projects Highlights, July 25, 1997, and August 15, 1997.

center-of-gravity requirements meant most of the structures had to be machined out to reduce weight. What remained was the outer edge and thin spars. The triangular spaces between these were voids. The outer skins were welded onto the internal structure; thus, they had to be designed as built-up welded structures.

This design resolved the weight issue, but raised new problems in parts manufacturing. The Haynes 230 nickel steel alloy was very hard and retained its strength at high temperatures. Both the machining and welding processes required several iterations before the wings and fins met the design requirements. The high temperatures required to weld the alloy caused distortions in the skin. These exceeded the very strict waviness specifications for the wing and fin surfaces. Boeing and ATK GASL began a welding research effort to correct the problem. After considerable effort, they were able to weld the material and meet the design specifications, but several attempts were necessary.

Langley engineers' analysis of the aerodynamic heating indicated the leading edges would become too hot to survive if they consisted of Haynes 230 only. The final design for both the Mach 7 and Mach 10 wings used carbon-carbon leading edges to withstand heating. In contrast, the Mach 7 fin leading edges were made of Haynes 230. The Mach 10 fins required the addition of carbon-carbon leading edges.

A passive approach to the X-43A airframe thermal protection was taken wherever possible. This meant the vehicle's structure was designed to be able to absorb heat without damage. In some areas, a heat sink was used to carry off the increased temperature. This approach avoided the weight and complexity of active cooling methods.

The vehicle's tungsten nose simplified the thermal protection required for much of the forward fuselage. The duration of the high-speed flight would be too short for the nose to undergo surface heating. The edges of the tungsten slab were protected from heating by carbon-carbon leading edges and chines. The bulk of the airframe aft of the tungsten nose and leading edges was covered with alumina-enhanced thermal barrier ceramic tiles. These were applied to most of the upper surface and parts of the vehicle's underside.

The tiles were sized to the heating requirements of the Mach 10 flight profile, resulting in a typical thickness of 0.5 in. In some areas, tile thickness was increased to create the proper aerodynamic shape. Normally, thin tiles posed problems when attached to flexible surfaces. Ironically, this problem was resolved by an earlier problem. To address the stack-stiffness issue, the X-43A's structure was made so rigid by the steel structure and thicker skin that flexing was effectively eliminated. For this reason, gaps—such as those between the space shuttle thermal-protection tiles—to allow skin flexing were not needed. The X-43A's tile covering was continuous.

The carbon-carbon leading edges also presented design and manufacturing difficulties. The edges were designed to take advantage of the heat-sink effect of the metal structures to which they were attached. The Mach 7 and Mach 10 leading edges and chines had differing layups, however, leaving them with different mechanical properties. These parts also were hard to manufacture due to the small radius of their edges, which was required in order to minimize drag at hypersonic speeds. The Mach 7 vehicle's leading edges had a radius of curvature measuring 0.030 in. This requirement was eased for the Mach 10 leading edges, to 0.050 in. Nor did problems end with the manufacturing process. As with fabricating the wings and fins, several attempts with the leading edges were made before success was achieved. A joint contractor/NASA team was assembled to develop procedures to produce the small-radius edges.

Though the airframe relied on passive cooling, a combination of both heat-sink and active cooling was used on the GASL scramjet engine. The engine structure was made of a high-strength heat-sink copper alloy called Glidcop. The engine was attached to a strong stainless steel back that itself was attached to the X-43A airframe. During boost and descent, a movable cowl door protected the engine's internal flow path. After separation, the door opened and the engine ignited. Once the burn was completed, the door closed during descent to the ocean. Zirconia ceramic coatings on the forward section of the engine and in various places within it added passive thermal protection.

Though this provided sufficient protection for most heating, other areas of the engine required active cooling due to the higher heating rates they would experience. Both the door's and the engine's vertical leading edges were actively cooled. Geometric constraints did not permit the recirculation of coolant, and this posed a challenge, because the coolant was dumped overboard after passing through the internal cooling passages. As a result, the amount carried had to be carefully calculated. To meet the cooling requirement, the coolant supply was divided between the adapter, which contained the glycol/water supply used during the boost, and a smaller water-only supply from a tank within the X-43A, used during the engine burn.

The airframe structural analysis review by GASL was completed on September 12, 1997. This was an important milestone, because it cleared the way for fabrication of nearly all the vehicle parts. The exception was the cowl leading edges; their fabrication was delayed until the boost-uncertainty study results were available [1, pp. 3–5, 6, 9, and 2].‡

BUILDING THE VEHICLE

With the design of the research vehicle completed, attention shifted to its construction. The critical design review for the X-43A was conducted in early February 1998, with most of the team in attendance. By early April, team

---

‡Aerospace Projects Highlights, September 22, 1997.

members were reviewing schedule options for completing fabrication of the X-43A, and for verification and validation testing with it at Micro Craft. With this testing accomplished, the vehicle could be delivered to Dryden for preflight testing. Further discussion about the assembly and testing was needed, however, and a final decision was not expected until early May.[§]

Manufacturing problems were soon to arise. Not surprisingly, the weight of the tungsten ballast created an early setback. The only practical means of producing this large mass of metal, given tungsten's high melting point, was a process called "sintering." This involves heating tungsten metal powder to a temperature below its melting point. In the process, the individual particles adhere to form a solid mass. The sintering process is commonly used in both powder metallurgy and production of ceramics [3].

Because of its great weight, several blocks of tungsten had to be sintered together in a large furnace. The process was completed and the first ballast was delivered to Micro Craft's Ontario, California, facility in late April 1998. When it was examined, 14 voids were discovered. Teledyne, the supplier, estimated fixing the voids would require about 6 weeks. This posed a potential delay in the separation shock tests then set for August. But the initial estimate proved too pessimistic, and reworking of the first tungsten ballast was completed by early June 1998. With the fixes completed, the bond joints were within the design specifications.

Sintering of the second tungsten ballast went much more smoothly. A single void was discovered close to the nose-airframe joint. The ballast was to be attached to the second airframe the following week. Micro Craft began negotiations with the supplier on a repair effort. The only concern was that any major delays would make it difficult to perform a January 1999 stack-stiffness test at Orbital Sciences.[¶]

Though the airframe's interior structure was now complete, the outer skin had yet to be manufactured. This involved a process that was exactly the opposite of the one used to produce the space shuttle, which was coated with similar thermal-protection materials. The shuttle had an aluminum airframe covered by thousands of individual tiles. Before being glued onto the fuselage, each tile was individually machined to fit a specific location on the vehicle. Use of the tiles eliminated the need for a hot structure, but required a time-consuming production process and installation of the tiles by hand. The protracted tile-installation process was a major reason for schedule delays that plagued the shuttle's development.

The X-43A tiles were attached to the vehicle's upper and lower surfaces, forming a single outer skin without gaps to allow flexing, as on the shuttle. Only after they were in place were the tiles machined to form the vehicle's

[§]Aerospace Projects Highlights, January 29, 1998, and April 10, 1998.
[¶]Aerospace Projects Highlights, May 1, 1998, June 5, 1998, and October 5, 1998.

outer mold line. This was accomplished using a three-axis numerically controlled milling machine. The entire X-43A fuselage section was installed in the machine, which shaped the tiles with a minimal number of setups. Once the surface shaping was completed, the outer mold line was measured. The final step was coating the tiles with silicon carbide, which gave them their signature flat black finish. The coating was not baked on, but rather dried at room temperature. This unorthodox procedure resulted in an exceptionally accurate outer mold line, and greatly simplified the production of the vehicle's external thermal protection [1, pp. 5, 7].

SYSTEMS INTEGRATION: GETTING THE PARTS TO FIT

As issues with the nose ballast were being worked out, an engineering review was being held at Micro Craft's Tullahoma, Tennessee, facility on June 25, 1998, regarding the X-43A's internal layout. The internal layout was soon drawing an inordinate amount of attention. The biggest concern: space—or the near-total lack thereof. The Dryden Aerospace Projects Highlights noted, "Internal volume is beginning to drive the design." Several layout changes were discussed at the meeting. These included adding fuel through the vent lines, though only if hazard analysis concluded that the procedure would be acceptable. Other changes included modifying the water tank and an external fill line, and reviewing the need for the actuator control box to be isolated from vibration.**

The airframe was rectangular, formed by two longerons and divided by several bulkheads. Its shape was that of a shallow box, with separate compartments formed by the bulkheads. The tungsten ballast was attached to the forward bulkhead. The carbon-carbon leading edges and chines were attached to the ballast. The rear bulkhead formed the aft end of the X-43A airframe. This housed the connections with the adapter for external power, and connections between the X-43A and the adapter/HXLV systems. Outboard of the longerons were the chines. Their outer edges were formed by curved bulkheads onto which the outer skin was attached. Various systems were squeezed into these areas as well [1, pp. 3, 4, 6].

A miniature unmanned research vehicle such as the X-43A offered advantages but imposed difficulties. Because total project cost was tied directly to vehicle size, the miniaturized approach was cheaper than construction of a full-size manned research vehicle. Given the unknowns of scramjets, any attempt to build a manned aircraft would inevitably mean a repeat of the unsuccessful NASP experience. Because it cost less and had no pilot, tests too risky for manned aircraft could be undertaken with a miniature research

---

**Aerospace Projects Highlights, July 2, 1998 p. 2.

vehicle. In the event of mission failure, the vehicle would be lost but no pilot was endangered.

The X-43A, however, still had to contain all the mechanical systems of a manned aircraft. These included actuators, fuel and coolant systems, power supplies, communications and data systems, and a wide range of other components. But because there was no pilot in the loop, the vehicle also needed to be capable of detecting and correcting errors autonomously—not being positioned at proper angles of attack and sideslip for an engine test, for example. Having eliminated the pilot to reduce project costs and vehicle size, a mechanical substitute now had to be created.

A further complication was the lack of redundant systems, which were not included because of the vehicle's size, the project's tightly scripted schedule, and the nature of the mission. The X-43A was to be a one-shot, throwaway vehicle. Even if enough room were available on the vehicle, redundant systems made no sense. They are costly in terms of both time and money to design, package, and test. They also posed software-complexity issues. The same was true with the HXLV; like the heritage Pegasus booster, the HXLV had only "single-string" systems. Except for the flight termination system, there was no redundancy.

The best that could be accomplished was to design fail-safe systems. The hydrogen and silane systems were designed so that, if a failure occurred, it would not develop into a hazardous situation. The pressure relief valves were located downstream of the regulators. If a regulator failed, the relief valve would open and relieve the pressure before a pipe burst. The HXLV had software and mechanical interlocks that prevented the rocket motor from igniting prematurely or pyrotechnics being triggered at the wrong time. In sum, there were safeguards to prevent explosions, fires, and incorrect operations. But if a system failed, of course, so would the mission.[††]

The X-43A vehicle systems were controlled by the flight management unit (FMU; see Fig. 9.2), which contained preprogrammed flight instructions. These were carried out based on both the mission timeline and readings from onboard instrumentation and vehicle flight data. The FMU also included the scramjet control software. One of the FMU's requirements was to detect and prevent an engine inlet unstart. The first FMU was delivered by Honeywell to Boeing in late July 1998.

The X-43A systems, including the FMU, the actuators and actuator controller, instrumentation system, and batteries, used off-the-shelf hardware wherever possible. Efforts were made to use components that had already been flight proven. These offered the advantage of low cost, because they did not have to be developed from scratch. But, again, there was another side to

---

[††]Griff Corpening history interview, NASA Flight Research Center History Office, December 20, 2004, pp. 30, 31.

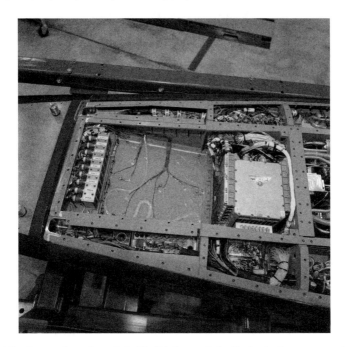

Fig. 9.2   The forward section of the X-43A housed the flush air-data sensors (small cyl-inders on the front bulkhead), the battery (not yet installed but located in the center section), and the FMU (the box at the aft end).

the low-cost coin. The conditions under which the components would operate were not those for which they had necessarily been designed. All systems had to be qualified to operate at altitudes of 100,000 ft and above. This meant they would be exposed to very low atmospheric pressure but not to an actual vac-uum. Ironically, qualifying the systems for these low-pressure conditions was more difficult than it would have been to qualify them for operation in the vacuum of space.

The major problems this created were with the electrical/electronic sub-systems. Correcting the problems required use of large, aerospace-quality connectors and fittings—something that would probably not have been an issue on a full-size vehicle. The X-43A interior, however, was limited in volume, and large connectors took up too much of the space available. Solving the packaging problem required design iterations of the system layouts and custom brackets, a limited number of connector modifications, and use of nonstandard/high-density Deutsch connectors.

A related issue was that of the data-collection system. Despite the research vehicle's diminutive size, the Hyper-X stack was heavily instrumented. Over 500 separate instruments were to record data during flight. Specific data to be collected were those measuring scramjet performance and flowpath, aerody-namic data from preplanned maneuvers taken during slowing from the test

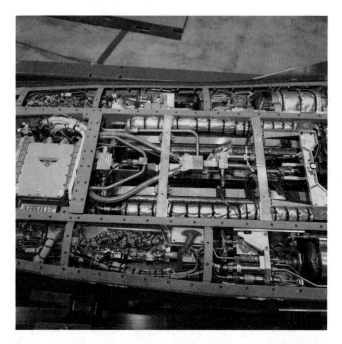

Fig. 9.3   The center of the X-43A housed the hydrogen tanks and related plumbing; the left chine held the instrumentation stack and the nitrogen purge system and related tubing, and the right chine held the silane system.

speed to Mach 1, flight trajectory data from the inertial navigation system and global positioning system satellite receivers, data measuring external and internal temperatures and pressures, and structural strain gage data. S-band transmitters and C-band transponders provided tracking and telemetry from the vehicle. Because the research vehicle would not be recovered after the flight, it was critical that data be successfully received by ground stations and tracking aircraft. The most critical readings, such as the scramjet burn data, were to be retransmitted throughout the flight.

Recording the data required extensive wire and tubing bundles to connect the instrumentation and recording devices. This required care during the buildup process and posed a major manufacturing challenge.

The tight fit of the instrumentation and data systems had other effects. One was the internal heating created by the electronics. Telemetry transmitters each produced an amount of heat equivalent to that of a light fixture. The original design assumed the X-43A's structure would act as a heat sink, providing cooling for the systems. Early tests of the telemetry system, however, showed this assumption had been incorrect, largely because the FMU had to operate during the prelaunch phase. This exceeded the airframe's heat-sink capability, requiring a modification to the nitrogen purge system (see Fig. 9.3).

Originally this was done only to prevent oxygen from entering the vehicle's interior, to prevent a fire in the event of a hydrogen or silane leak.

Modifying the nitrogen purge system so it could also provide cooling for the systems involved several changes. Additional tank capacity had to be added to the B-52 launch aircraft and the adapter. Modifications also had to be made to transfer nitrogen from the B-52 to the research vehicle before the launch, and from the adapter to the research vehicle during the boost phase. Once separation occurred, the research vehicles' internal nitrogen supply would be used.

Still another issue affecting the interior layout was that of the actuators and their power supply. The X-43A control surfaces were located at the rear of the vehicle. These were built based on off-the-shelf systems, which required limited modifications to the battery case but significantly more changes to the actuators themselves. Due to the small space available, all five actuators were packaged in a single case. Four actuators operated the control surfaces, and the fifth opened and closed the engine cowl. The latter was a particularly demanding task, because the cowl had to open and close fully, even in the event of distortion due to heating. The linkages also took up space within the airframe.

Moog, the actuator contractor, was able to package the five actuators, their electronics, the power supplies, and voltage controllers within the volume of a car's glove compartment. The company had never before packaged these devices within so small a volume. The task was made more difficult because standard-sized aircraft connectors, which were huge compared to the research vehicle's internal volume, had been used in all the power supplies and actuator system components [1, pp. 4, 6–8].‡‡ Figure 9.4 shows the aft bulkhead, which contains the external power, fluid, and gas connections. The large box in the center is the controller box for the actuators. On either side of the controller box are the actuators for the left and right wings; the actuator for the left and right rudders are ahead of those. The S- and C-band transponders are located forward of the controller box. The large cylinder in the center houses the coolant system, and the smaller tanks in the chines are the nitrogen purge tank (left) and silane system (right).

Both the 28-V and 150-V power supplies that powered the electric motors were contained within the Moog box. The close proximity of the components resulted in cross-talk between the adjoining controllers, and several design iterations were required before Moog was able to resolve the issue. Making the situation more critical was that the problem did not become apparent until late in the vehicle's checkout. With the first flight date looming, there was considerable pressure to address these problems. The solution was installation of metal sheathing in some areas and rearranging some components in

---

‡‡Aerospace Projects Highlights, July 24, 1998; Corpening history interview, pp. 26, 27.

Fig. 9.4   An aft view of the X-43A.

others. Care also had to be taken in the wiring design and placement to prevent electromagnetic interference effects.

There is a difference of opinion on how the actuator issue came to be dealt with. Griff Corpening, at Dryden, recalled that Moog had put a "full-court press" on solving the problem. Langley's Dave Reubush, on the other hand, felt that this had constituted the only significant problem encountered with the entire NASA/contractor relationship during the project. He commented:

> ... when we had problems with the actuators/controllers, we had diffi-
> culty getting Moog to put their best people on the problem for much
> longer than we would have liked. While we didn't like it, it was under-
> standable as the work for us was a very minor job compared with Moog's
> bread and butter.[§§]

The fact that the Moog actuator/controller problem was the only contractor issue to be raised by any of the project personnel was an indication of just how well the project team had meshed. Although the issue's last-minute nature probably heightened the concern, it was successfully dealt with, and played no part in the events to follow.

[§§]Corpening history interview, pp. 26, 27; Dave Reubush e-mail to Curtis Peebles, "Re: NASA/ Contractor Relations," February 3, 2009, 7:35 P.M.

The adapter structure (see Figs. 9.5 and 9.6) contained the tanks for the high-pressure nitrogen and water-glycol coolant used during the boost phase, the separation system, ballast, cameras used to record the separation, and the electrical and fluid connections. The first and second adapters were delivered with the movable jaw. The mechanism was disabled for launches, leaving the drop jaw fixed in place. The third adapter had the ramp built in one piece.

The fuel system was another area that posed difficulties from both a design and a manufacturing viewpoint. Again, the starting point was off-the-shelf technology. Hydrogen fuel and a hydrogen/silane mixture igniter gas had long been used in scramjet wind-tunnel tests. These wind tunnel facilities typically used bottles of hydrogen and the hydrogen/silane at pressures of about 1800 psi. Even though the scramjet fuel sequence would last only 10 s, the volume of hydrogen and hydrogen/silane mixture required was too great to fit within the airframe at these pressures. The solution was to increase the tank pressures. For the hydrogen/silane mixture, a pressure of 6500 psi was originally specified. However, the ATK GASL team expressed concerns this would cause the silane to condense. A rigorous analytical and experimental program was undertaken to better understand the limits of mixture stability. This effort resulted in a maximum allowable mixture pressure of 4500 psi with the added requirement of an elevated initial tank pressure. A new heated tank system was also designed to provide a sufficient mass of the igniter gas [4].

This was still a far higher pressure than had ever been used before. Due to the high pressures and pyrophoric/combustible nature of the gasses on

Fig. 9.5   A front view of the adapter showing the various ports for the connections.

Fig. 9.6   A side view of the adapter.

board, extensive fuel system integrity provisions including vibration isolation were necessary. The fuel system design also required use of small,
very-high-pressure valves. These became a separate design effort, because
high pressures were required for the valves to seat properly. Repeated testing and redesigns were needed to achieve the level of reliability required of
the valves.

Underlying all this was the issue of the fuel system's safety. The hydrogen/
silane mix would ignite on contact with air, and there was a relatively large
volume of silane aboard the research vehicle, at very high pressure. During
the checkout of the X-43A, the tanks would be empty. When high-pressure
testing of the fuel system's integrity was necessary, inert gasses would be
used to ensure that the system worked properly. At other times, however, tests
would be required using actual hydrogen and silane. Prelaunch preparations
would also require the hydrogen and silane tanks to be filled. Both activities
would expose checkout crews to the risks of fires and explosions.¶

To meet the safety requirements of working around silane, several venting
and purging loops had to be added to the fueling system. Vent stacks were
attached to the research vehicle during ground tests, and these carried the
combustible gasses away from service personnel. Mechanical joints in the
fuel system underwent a design review in mid-June 1998 in efforts to identify
potential problems [1, p. 8].***

Anthony Castrogiovanni of ATK GASL later recalled the first "real fuel"
run of the scramjet:

When it came time to run the first "real fuel" blowdown tests at ATK
GASL in Ronkonkoma, NY there was a lot of discussion and deliberation

¶When silane burns, it produces very fine silica deposits. Ironically, these deposits are useful in locating leaks in the fuel system piping.
***Aerospace Projects Highlights, July 24, 1998.

regarding how we would vent the hydrogen/silane igniter and hydrogen fuel into the atmosphere considering the propensity to burn with the pyrophoric igniter gas. We designed and built flare/vent stacks that extended approximately 16 feet above the roofline of the main GASL building which is in an industrial development adjacent to Islip MacArthur airport on Long Island. After a long day of carefully filling the tanks to the full pressure, the test was finally conducted at about 11 pm. A few of us watched from outside to watch the flame while we remained in communication with the engineers operating the system inside the control room by a radio. The test sequence was a complete flight profile including an initial hydrogen/silane igniter segment followed by the primary hydrogen flow which was at a fairly high flow rate for about 10 seconds. The flame extended over thirty feet above the vent stack and lit up the sky with a roar. All of us simultaneously reacted with cheers for the successful test. After the adrenaline wore off, we thought about what that must have looked like from the perspective of an approaching aircraft— it was one of those experiences you never forget.[†††]

Sometimes, external issues affected final assembly of the first X-43A. One example of this was the problem with the explosive bolts. During tests of the different bolts, vehicle assembly was halted. Only after selection of the Pacific-Scientific bolts could work restart. The bolt issue combined with delays arising from changes to wiring and plumbing design details resulted in a 7-week delay in transferring the vehicle to Tullahoma so integration work could begin. It was not until early December 1998 that the first research vehicle was finally at Tullahoma.[‡‡‡]

SYSTEMS CONFIGURATION, VEHICLE PERSONALITY, AND CHOOSING YOUR PROBLEMS

The X-43A's delivery to Dryden on October 11, 1999, marked the culmination of four decades of research. The airframe was 144 in. long and had a body/wing span of 60 in. From the top of its fins to the base of the scramjet engine measured 26 in. The engine was 19 in. wide and 30 in. long.

Even with careful layout of the instrumentation, the subsystem design, and packaging, there were numerous accessibility problems. Systems were stacked one on top of the other. Yohan Lin, a Dryden engineer who participated in the checkout of all three of the X-43As, later recalled:

> ... there [was] also tubing around the tanks, underneath the tanks, and valves that were inverted because of volume constraints; [the valves] had to be put upside down so that in order to service it or repair it, you have

---

[†††]Anthony Castrogiovanni added this comment to a draft of this chapter.
[‡‡‡]Aerospace Projects Highlights, September 4, 1998, and December 7, 1998, p. 3.

to tear it all apart and then take it out of the vehicles. That requires moving other tubing that's around. So, it was a big mess.§§§

The risk was that in attempting to fix a leak, or just by removing and reassembling the parts, more leaks could be created. The procedure used during checkout was to leave the upper skin of the research vehicle off at all times. Test cables were connected to the on-board components. To protect the chines from being damaged by the cables, plastic foam sheeting was laid under the cables. Clear plastic covered the open top of the X-43A to prevent dust and debris from falling into its interior. Not until the Hyper-X stack was finally attached to the B-52 was the upper skin attached, the holes in the surface fitted with fasteners and ceramic plugs, and the surface finished.

Lin also noted that each of the X-43As had its own "personality."¶¶¶ "The first vehicle was basically our baby," he said later [1, p. 8].

> It was a learning experience for us. We learned a lot. We ... had to swap out and replace a lot of different components and spares.... Ship 2 went very smoothly, no problems. By that time we pretty much understood the system fairly well and, in terms of operations there were no significant problems.
> ... Vehicle 3 was by far the most difficult vehicle to test. It did not want to fly.... every step of the way there would be issues or things we had to address.... some were small, some were identifiable, others just kind of came up and went away on its own.§§§

Despite all the difficulties created by the X-43A's limited volume, the decision to keep the research vehicle's length at 12 ft contributed to the Hyper-X's ultimate success. Had the length been increased by even a foot and a half, the initial problems might have been solved much more easily but still others would have resulted.

The first of these would have related to risk reduction. At 12 ft, the vehicle fit into the 8-Ft. HTT at Langley. This allowed testing of the exact hardware and software configuration to be flown, at the Mach 7 speeds, enthalpies, and dynamic pressures the vehicle would experience. Use of a larger vehicle would have eliminated this advantage, because the wind-tunnel data would no longer have been representative of the actual flight vehicle.

A potentially more serious issue was the disruptions to the program an increase in the research vehicle size might have meant. The early stages of a new project are always difficult, and this was borne out in the Hyper-X

---

§§§Yohan Lin history interview, NASA Dryden Flight Research Center History Office, August 12, 2005, pp. 27, 28, 32, 33.

¶¶¶The X-43A research vehicles, like earlier vehicles, were all hand built, which ensures each vehicle has its own personality—quirks that distinguish one copy from another.

project. The stack-stiffness issue required a considerable amount of work, including redesign of the X-43A and adapter, and this led to an increase in vehicle weight. The separation issue was studied extensively, yet unknowns remained. Initial project costs were more than had been projected. As a result, the number of flights was cut from four to three, with the dual-mode Mach 5 flight being dropped.

With the size of the vehicle fixed, engineers then had to fit the systems within it. That required time and effort, but the alternative posed a greater risk. By fixing the research vehicle length at 12 ft, an endless series of design iterations was avoided. If changes to the size had been allowed, the aerodynamic and structural design would have had to undergo a new iteration. Griff Corpening recalled later,

> ... now you're a year into your program, and you still don't have a design because you keep iterating; it's called requirement creep.... even with the time and the money and the effort to make the 12-foot vehicle work, it was still simpler to do than to have all those version 1, version 2, version 20 vehicles, during the very early stages.****

Corpening also noted that the issue of separation was a factor in determining vehicle size. Sandia engineers had recommended that separation take place in space. In that scenario the X-43A would reenter the atmosphere and make the engine test, having avoided the unknowns of a high-Mach-number/high-dynamic-pressure separation. When this profile was examined, engineers found that additional thermal protection would be required. He recalled,

> ... because we were constrained by 12 feet on the outside, the additional [thermal protection] had to go on the inside and take volume away from the inside of the vehicle. We had already established there was no volume left to take away from if we were going to meet our objectives.****

As a result, the reentry profile was rejected.

Corpening observed that although many ideas sounded good on the surface, in actuality the result was just trading one set of problems for another. He added, "I have no doubt in my mind that had we had the volume and gone exo-atmospheric, we would have had a whole set of different challenges and problems to deal with that would have been equally as challenging as separating those two vehicles within the atmosphere."****

The matter of size and the tradeoffs that accompanied the choice underlined the contrast between the tightly focused Hyper-X project and the ill-fated NASP. By limiting Hyper-X project goals, many of the problems that beset the NASP were avoided. Had this approach not been taken, the Hyper-X effort could easily have met a similar fate as the NASP.

****Corpening history interview, pp. 32–34.

Fig. 9.7 The rocket motor.

Although checking out the X-43A represented the major efforts in preparing for flight, the HXLV also underwent its own tests. These are documented in Figs. 9.7 through 9.11. The rocket motor, shown in Fig. 9.7, and the other components were delivered to Dryden and assembled in a specially modified building. Due to the different flight profiles and loads on the HXLV as compared to those of a Pegasus satellite launch, modifications were made to the

Fig. 9.8 The fillet being mounted onto the rocket.

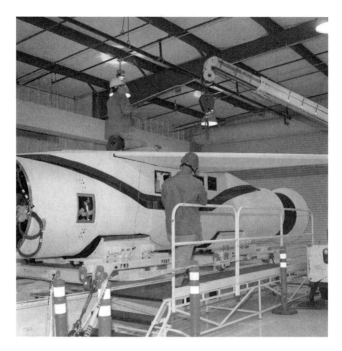

Fig. 9.9   The wing being attached.

Fig. 9.10   The three tail fins being attached to the booster.

Fig. 9.11   The HXLV on its trailer.

fillet's heat protection, nose and midsection structure, and attachment to the rocket motor (see Fig. 9.8). Once this was completed, the wing was attached, as shown in Fig. 9.9. Because of stagnation-point and shock-on-shock heating it would encounter, the wing also required additional thermal protection on its leading edges and surface. The three tail fins (see Fig. 9.10) both stabilized and guided the HXLV. Like the wing, these also required modifications to the standard thermal protection. This was difficult to achieve, due to limited space between the fin root and the rocket motor. The fins' leading edges were also given added heat protection, causing a change in their radius.

The completed HXLV is shown in Fig. 9.11. The trailer was used during both the assembly and checkout, as well as to move the stack out to the B-52B mothership and attach it to the pylon.

In the larger picture, when exploring the unknown, problems cannot be avoided. One can, however, to a certain extent, choose the problems to be faced.

But there still will be problems nonetheless.

REFERENCES

[1] Harsha, P. C., Keel, L. C., Castrogiovanni, A., and Sherrill, R. T., "X-43A Vehicle Design and Manufacture," AIAA 2005-3334, p. 2.
[2] Marshall, L. A., Bahm, C., Corpening, G. P., and Sherrill, R., "Overview and Lessons Learned of the X-43A Mach 10 Flight," AIAA 2005-3336, p. 4.

[3] "Sintering," *Answers.com*, http://www.answers.com/topic/sintering [retrieved October 25, 2010].

[4] White, R. E., Calleja, E. A., and Castrogiovanni, A., "Blowdown Characterization of a High Pressure Silane-Hydrogen Mixture for the Hyper-X Research Vehicle," presented at the 36th JANNAFCS/APS/PSHS Joint Meeting, Cocoa Beach, Florida, October 18–22, 1999.

Chapter 10

GETTING READY: TESTING THE HYPER-X

Flying is not Nintendo. You don't push a button and start over.

A space vehicle undergoing checkout in a clean room is a familiar image after five decades of spaceflight. Less familiar is the preparation of research aircraft before their flights. This involves tests on everything from individual components to the entire vehicle. The vehicle then "flies" simulated missions with inputs from ground-support equipment. Once these preparations are completed, multiple reviews are held to determine whether the first flight is ready to proceed. All these efforts are the results of long and often bitter experience with research into the unknown.

The test and safety procedures used with the jet aircraft and rocket planes of the late 1940s and early 1950s differed little from those used to test biplanes two and three decades before. Engineers planning those flight research programs also assessed their safety. Adding to the risks were the new aircraft configurations being tested—delta wings, swept wings, straight low-aspect-ratio wings, flying wings, and variable-geometry wings that could change their sweep in flight. These new aircraft designs often suffered from poor stability and control characteristics or vicious stall behavior, making them difficult and dangerous to fly. Though such risks could be expected and allowed for with new aircraft, unknown aerodynamic phenomena, such as inertial coupling and pitch-up, were also being encountered.

With the development of the X-15, the old procedures were no longer satisfactory. A formal review process was developed to ensure that the risks of a flight were understood and procedures established to deal with them should they occur. Flight rules were developed for what to do in the event of specific malfunctions. Should an X-15 suffer an engine failure, for example, the pilot would land on the closest of several preselected dry lakebeds along the flight route. These rules would not be violated or changed during a flight.

During the flight, engineers monitored data transmitted from the X-15 in real time. If they detected a problem, a warning would be radioed to the pilot. As speed and altitude build-up testing of the X-15 was begun, these same procedures were being adopted by the Mercury program. The result was the Mission Control procedures that are used today.

TESTING, TESTING

Though prelaunch testing of the Hyper-X vehicles was undertaken as a standard part of research aircraft preparation, it also reflected these aircrafts' unique nature. On the surface, the Hyper-X project was similar to those of the remotely piloted research vehicles that had been flown at Dryden since the late 1960s. What made the X-43A different from earlier research aircraft were its autonomous flight profile and lack of a manual backup system. In the event of a malfunction, the only action ground controllers could take with the Hyper-X stack was to transmit the destruct command. The X-43A's autonomous nature made it more akin to the early rocket-launched hypersonic reentry vehicles than a traditional research airplane [1].*

For this reason, risk reduction was critical to the success of the Hyper-X project. Like the ICBM-launched test vehicles, the Hyper-X was not a long-term project as the X-15 had been, but would involve only three launches. To ensure a high probability of success, a great deal of work was done on the vehicle's aerodynamic and thermal properties, boundary-layer transition, flight control law development, and flight simulation modeling. The CFD studies used several independent programs, and each set of results was then cross-checked to detect any possible errors. Multiple levels of scrutiny increased confidence in a successful flight.

With the completion and delivery of the flight hardware to Dryden, testing focused on the mechanical systems and software programming. Building 4847 at Dryden was modified for assembly and testing of the Hyper-X stack. Modifications included adding humidifiers, a high-voltage power supply, and anti-static floor paint. The integration of the HXLV rocket motor and wing was scheduled to begin in July 1999, with the launch set for 2000. Orbital Sciences personnel began arriving at Dryden on July 21, 1999, and ground support equipment began arriving July 23. The rocket motor was scheduled for delivery on July 27, and ordinance was due July 29. By mid-August, integration of the HXLV rocket motor and wing was under way. The work was completed in early September.†

On October 11, 1999, as the first HXLV was being delivered and assembled, the first X-43A arrived at Dryden. Once the receiving inspection was complete, the X-43A underwent structural testing at Dryden's Flight Loads Research Facility during November. Plans for structural tests and initial systems testing were already under way before the vehicle arrived. The adapter

---

*Paul Reukauf history interview, Dryden Flight Research Center History Office, pp. 10, 11.

†Aerospace Project Highlights, May 7, 1999, p. 1; July 2, 1999, p. 1; July 23, 1999, pp. 2, 3; August 13, 1999, p. 2; September 10, 1999, p. 2.

Fig. 10.1   An X-43A undergoing preflight testing.

was originally due to arrive at the end of September 1999; however, the complete adapter and aft engine skirt were not in Building 4847 until March 31, 2000. No problems were identified in the receiving inspection of the adapter and skirt. Orbital Sciences personnel were at Dryden to assist with the integration of the adapter and the rocket motor [2].[‡]

In anticipation of the hardware's arrival, Dryden and contractor engineers had to devise a wide range of tests to be performed on the hardware and software. There was no precedent in this regard for the X-43A, so the tests had to be developed as the vehicle was being designed. There was no finished set of hardware that could be used as the basis for planning a test program.

As their starting point for developing the test plan, engineers used the hardware's basic design, and the fact the vehicle had to be physically supported in certain configurations. Devising the test procedures had to address how to access components and wiring that were difficult to reach and how to test specific functions, as well as the design of any specialized test equipment this would require.[§] Figure 10.1 shows an X-43A with the top panels removed

---

[‡]Aerospace Projects Highlights, April 9, 1999, p. 1; August 23, 1999, p. 2; August 27, 1999, p. 2; September 3, 1999, p. 2.

[§]Yohan Lin history interview, August 12, 2005, NASA Dryden Flight Research Center history office, pp. 2–6.

and cables connecting the ground test equipment to the onboard systems. The test equipment generated signals simulating those the onboard systems would receive in flight. How the systems responded to the signals indicated any hardware or software problems.

In testing any new and one-of-a-kind vehicle, problems were to be expected. The Hyper-X's performance, its flight environment, the tight packaging of the systems, and the exotic technologies involved made these difficulties even greater. The initial inspection of the X-43A revealed several minor issues, most of which were soon resolved. However, failure of the bearing on the right wing was discovered and required correction. Additionally, efforts to plug internal leaks were under way. This was important, because during the flight to the launch point the vehicle's interior was purged with nitrogen gas to prevent a fire should a leak of hydrogen or silane occur. Other tests required that fluid and gas be supplied to the onboard systems (Fig. 10.2). These were supplied from external tanks via connectors to the adapter. The tests identified problems such as leaks or damaged tanks.

The next problem appeared during antenna-pattern testing. The test was made using the second X-43A and completed on January 28, 2000. Engineers' analysis of the data indicated cross-coupling between the antennas. The data review began, with an initial report due the following week, but by mid-February the antenna issue and the inability to maintain positive internal purge pressure had become the project's highest priorities.

Fig. 10.2   Testing of fluid and gas supplies.

Though these represented significant difficulties, progress continued elsewhere. The failure modes and effects testing had been successfully completed. This involved inducing both nominal and off-nominal conditions in the vehicle systems. The objective was to make sure the vehicle behaved as expected, in both normal and abnormal situations, and to catch and correct any errors that might have slipped through in the design phase. Minor issues were discovered and corrected.

Whenever a potential problem or a failure was discovered, the first step in fixing it was writing a discrepancy report (DR) and sending it to the configuration control board, which held weekly meetings. An engineer was assigned to develop a fix, and a configuration change request (CCR) was written. The modification was completed and the test repeated. If the fix was successful, a system test report (STR) was written, describing the problem, the fix that solved it, how the fix was tested, and why engineers believed the problem was now corrected. The STR, like all other records, became part of the Hyper-X configuration control documentation. Work with the first X-43A research vehicle resulted in 119 DRs, 384 CCRs, and over 100 STRs.

In other cases, such as when a design specification or performance requirement could not be met, a waiver was requested by project management. The process began with an assessment of the consequences of not correcting the problem. In some cases, the effort needed to meet the specification was seen as unjustified. In others, engineers might have felt original specifications could not be met and that the situation would have to be accepted. Finally, the analysis could indicate that failing to meet the specification would not increase the risk of failure. In these cases, a waiver was issued.

The goal was to document every step made in every configuration change the vehicle underwent, the detailed process through which the change was made, and the means used to qualify the change. In the event of a mission failure, the documentation would allow a Mishap Investigation Board (MIB) to reconstruct the chain of events leading to the failure. This reflects standard systems engineering procedures.[1]

In mid-February, the wing bearing issue was resolved, with the modification to the X-43A expected to be completed later that month. The right-hand wing assembly was completed in the third week of February. Additionally, the left-hand assembly underwent modification, with the work due to be completed the week after that. The work was finished on schedule, closing the issue. With the wings reinstalled, surface calibrations and frequency sweeps were conducted. In the latter, the wings were moved at different rates

[1]Yohan Lin history interview, August 12, 2005, NASA Dryden Flight Research Center history office, pp. 30, 31; Linda Soden history interview, August 25, 2005, NASA Dryden Flight Research Center history office, pp. 44, 46, 56, 58; Aerospace Projects Highlights, January 28, 2000, p. 1; February 14, 2000, p. 2.

Fig. 10.3 The second X-43A undergoing antenna pattern testing in the Benefield Anechoic Facility during January 2000.

to test the actuators and control systems. Guidance-system checks were also in progress and would be completed the same week.

An analysis of the antenna problem was under way at the same time (see Fig. 10.3). This looked at removing the aft C-band antenna on the X-43A to avoid data transmission while the research vehicle was attached to the HXLV. A separate C-band antenna was mounted on the HXLV for tracking and range safety requirements. Alternatives for the S-band system, to eliminate cross-coupling of antenna patterns, were also being considered. By the beginning of March 2000, a preliminary design that would correct the problem had been approved by project management. The C- and S-band pattern testing was scheduled for late March**

At this point, "hardware-in-the-loop" testing could begin. This involved feeding simulated flight data from a simulation bench to the flight control computer. The flight computer responded to the simulated data and generated commands as though the vehicle were actually flying. It verified the propulsion and the flight control algorithms loaded onto the flight control computer, that propulsion control laws were operating correctly and with the proper timing, that the flight control computer commands to open valves were

**Aerospace Projects Highlights, February 14, 2000, p. 2; February 25, 2000, p. 1; March 3, 2000, p. 1; March 17, 2000, p. 2.

correct, and that the flight control laws sent the servo commands at the right time and with the proper magnitude.

Several hardware and software problems with the X-43A and adapter were identified and corrected in the course of testing. These included controls and guidance software errors, which were being fixed in mid-March. The tests also uncovered several timing errors in the simulation software. As a result, further testing was delayed by about a week. The first Mach 7 flight was now tentatively scheduled for mid-August 2000.

Checkout delays translated into rising project costs, of course. To make up for lost time, in March 2000 project managers began planning multiple shifts, which would allow parallel testing. Boeing engineers would undertake verification testing, while NASA personnel would continue hardware-in-the-loop tests. As March drew to a close, nominal hardware-in-the-loop testing of the X-43A was completed. The off-nominal testing was to begin, and was scheduled to last about 2 weeks. At Boeing's Long Beach facility, changes to the software were being implemented. Antenna isolation testing was under way on March 24 to evaluate potential cross-talk between the C-band and S-band antennas. Testing had revealed mechanical interferences between the scramjet engine sidewall and the cowl-door actuator, and the X-43A wing and vertical tail. The cowl-door issue was fixed and successfully retested, and a fix was identified for the wing and tail interference issue. Finally, on March 31, a milestone was reached with the arrival of the X-43A adapter at Dryden. All the elements needed to complete the first stack were in place.[††]

April 2000 saw an increase in test activity. The adapter underwent weight and balance checks during the first week of the month, and the fuel system was assembled and prepared for bench leak checks. Successful calibration checks of the X-43A control surfaces were completed. Tests of the X-43A vehicle management system and propulsion system control also were successful, as were hardware-in-the-loop test cases of the guidance and flight control nominal trajectory.

A nominal trajectory aircraft-in-loop test was also completed. This involved the flight control computer and the simulated flight data, which provided the forces and moments sensor data to the vehicle. But the actuation system was now connected and moved according to the commands sent by the flight control system. The aircraft-in-loop test showed the flight control system was operating correctly.[‡‡]

By mid-April, low-pressure fluid system checks were under way. This was a relative term, because the three "low-pressure" systems operated at about 2000 psi. The silane and hydrogen systems held the pressure level as planned.

    [††]Lin history interview, pp. 7, 8; Aerospace Projects Highlights, March 10, 2000, p. 1; March 17, 2000, p. 2; March 24, 2000, p. 1; March 31, 2000, pp. 1, 2.
    [‡‡]Aerospace Projects Highlights, April 7, 2000, p. 2; Lin history interview, pp. 7, 8.

Fig. 10.4   The scramjet undergoing systems and fluid testing.

In contrast, a nitrogen solenoid valve had an internal leak. The valve was replaced and tests continued.

Engine-only pressure-port leak checks were also under way (see Fig. 10.4). With completion of these, the engine would be electrically and pneumatically connected for end-to-end pressure-port checks. Once these were successfully completed, the scramjet would be permanently attached to the X-43A. The hardware testing was also tied in with computer simulations. Nominal and off-nominal hardware-in-the-loop trajectory tests with the X-43A were made in mid-April to investigate a reproducibility issue with the simulation. Several items were identified as the cause of the simulation errors, but reviews of the controls and propulsion systems test runs indicated that no changes to the research vehicle's software were required.

Langley, Orbital Sciences, Micro Craft, and Boeing North American representatives were at Dryden in mid-April to begin Flight Readiness Review (FRR) planning. Detailed agendas for each discipline were developed and reviewed by the team. The Dryden independent software review team was also looking at requirements for verification and validation traceability. Though the FRR meeting was an important step, it also highlighted the delays that beset the Hyper-X project. This meeting had originally been scheduled for the end of February.[§§]

[§§]Aerospace Projects Highlights, April 14, 2000, p. 2.

The second half of April 2000 saw significant progress in checkout and testing. The adapter inspection was completed, the scramjet was connected electrically and pneumatically, end-to-end checks had begun, the fuel service cart used for the high-pressure testing had been delivered, and final modifications were under way. Testing included completion of the aircraft-in-loop control surface calibration checks and analyses of the results.

Hardware- and aircraft-in-loop testing of the vehicle management system, the propulsion control system, and the guidance navigation and control nominal testing were completed with no major issues. The hydrogen and igniter systems both checked out successfully and the fluid system underwent bench leak and function checks.

Not all of the results were so positive, however. Aircraft-in-loop checks of the built-in tests (BIT) were also run, with results indicating that software changes would be required. The X-43A's hydrogen tank was tested and found to be defective. A replacement tank arrived on April 20, but then the nitrogen solenoid valves proved defective. Change-out of the failed parts was soon under way.

By the end of April 2000, aircraft-in-loop testing of the X-43A and adapter had been completed with no significant issues. The Aerospace Projects Highlights noted this as "a major milestone," and with good reason, considering the difficulties encountered to this point. Fluid system bench and leak testing neared completion. The solenoid valve seating at low pressure was the dominant issue; a change in the servicing sequence appeared to be a solution. Processing of the second HXLV would soon begin, with completion of the work scheduled for the end of June.[¶]

May 2000 saw completion of scramjet instrumentation checkout and final installation of the instruments on the first X-43A. Then began the final internal vehicle purge testing. This involved filling the interior of the research vehicle with nitrogen gas at pressures slightly higher than those of the atmosphere, to force the internal air out and replace it with nitrogen. With a B-52 source pressure of 500 psi, the X-43A's internal pressure could be maintained at 0.5 psi above outside atmospheric pressure, preventing outside air from leaking into the research vehicle. Over 30 additional propulsion system control hardware-in-the-loop tests were run.

Langley and Dryden engineers were also completing a detailed review of the component qualification testing that remained. One element of this was replacement of a GPS relay for satellite navigation data. This would require some additional testing, but was not expected to significantly affect the schedule. A detailed work-schedule meeting was also held, and analysis of the plan made.***

[¶]Aerospace Projects Highlights, April 21, 2000, pp. 1, 2; April 28, 2000, pp. 1, 2.
***Aerospace Projects Highlights, May 5, 2000, p. 1; May 12, 2000, pp. 1, 2.

As a result of the schedule review, the planned launch date for the first Mach 7 flight was pushed back, from mid-August to late September 2000. The new date was predicated on a significant limitation, however. The Aerospace Project Report noted: "This assumes no major problems found during final full vehicle tests and stack integration. Team is working hard to meet this schedule."†††

For the moment, preparations were going well. Checks of the fuel system high-pressure bench leak had been completed, the heater blankets were installed, the final fuel-system inspection was completed, and final installation of the fuel system was under way. Additionally, instrumentation checks on the pressure ports and the hydrogen sensor wiring were completed, and the S-band transmitter cooling modification and wiring was installed.†††

In mid-May, the leak and functional-fluid system testing began. This was completed successfully by the end of the month, at full pressure and with inert gasses. The X-43A emergency vent and purge test was completed, also with inert gasses. Early June 2000 saw testing of the X-43's oxygen and hydrogen sensors as well as of internal temperature and purge cooling. NASA/ contractor personnel completed these successfully, with the nitrogen purge maintaining the vehicle's internal temperature at or below planned values. Testing of the research vehicle's mass properties was also completed, with results matching predictions; no roll inertia issues remained.‡‡‡

PROBLEMS

By early July 2000, however, other problems surfaced. Tests of the adapter found that one of the glycol/water tanks, used to cool the scramjet during ascent, had a damaged bladder. Nitrogen gas pressurizing the tanks squeezed the bladder, pushing the coolant mixture out of the tank. The design prevented bubbles of nitrogen from mixing with the coolant and then flowing into the pipe. This would disrupt the smooth flow, resulting in inadequate cooling of the scramjet. The damage occurred during checkout at Boeing's Seal Beach facility, where a valve had allowed the nitrogen gas to be expelled faster than the water after the testing. Both tanks were replaced.

It was also at this time that the actuator issue appeared. Moog notified project managers that the actuator qualification test schedule would be delayed by 1 month. The reason: spikes in the data output caused by electromechanical interference and cross-talk generated by power sources. But by the end of August, progress was being made with the Moog controller qualification, and a flight unit was scheduled for delivery on August 8, 2000.

---

†††Aerospace Projects Highlights, May 19, 2000, pp. 1, 2.
‡‡‡Aerospace Projects Highlights, June 2, 2000, p. 2; June 9, 2000, p. 2; June 16, 2000 pp. 1, 2; June 30, 2000, pp. 1, 2.

During early July, Boeing, GASL, and Langley engineers were at Dryden to conduct a "blowdown" test of the scramjet. This was an engine test using inert gasses rather than flammable hydrogen and silane gas, conducted with the purge/cooling systems operating. The first run was scheduled for Saturday, July 8, with a second planned for the following Monday. The first inert gas blowdown went well, "with few anomalies noted." The most serious one, the weekly report noted, "was the discovery that the coolant valve was installed backwards."§§§

Although the initial blowdown test was successful, a series of failures soon followed. The coolant tanks in the X-43A and the adapter were opened and inspected. Both tanks had holes in the bladders, and there was no indication of the cause. The vendor began tests to determine the cause of the failure. Until it could be understood, subsequent blowdown testing was delayed.

By late July, project engineers were testing various options to prevent the bladders from being damaged during ground testing. To prevent recurring damage, changes in the cooling system design would be required. In the course of the testing, a review of all relief valves was made to ensure they were the correct size to cope with possible failure scenarios and installed in the correct direction.

The ongoing problems of the late summer drove project managers and engineers to consider ways of getting checkout plans back on schedule. Vince Rausch, the Hyper-X program manager, came to Dryden in early August to review both the status of the first flight and the overall state of the effort. Once again, checkout problems meant a delay for the first flight, this time pushing it to December 2000.

Among changes made that altered the preflight test schedule were modifications to the X-43A's purge system, to prevent damage, and changes to the fuel system testing. The X-43A fluid system leak and function test, and the second inert gas blowdown test, were planned for the second half of August. This would be the final blowdown test, and would be followed by a test using the hydrogen and hydrogen/silane ignition mixture, rather than inert gasses. The cause of the damaged adapter coolant tank bladders was eventually traced to excessive pressure during the tests. The solution was to limit the test pressure to 50 psi.§§§

Continuing delays in scheduling the first flight led to a teleconference between Langley and Dryden center directors. From the project's outset, Langley engineers had been concerned about what they saw as Dryden's excessive testing of flight hardware. With the first flight now delayed 2 years beyond the original schedule, and costs higher than planned, these early concerns were exacerbated. Dave Reubush recalled that "during that period . . . I

---

§§§Aerospace Projects Highlights, July 7, 2000, pp. 1, 2; July 14, 2000, p. 1; August 11, 2000, p. 1; August 25, 2000, p. 2.

was aware that, now that we had real hardware at DFRC to work with, things were not going as fast as everyone had anticipated so the schedule kept slipping. There were a lot of discussions that took place on what to test and how much to test . . . ."¶¶¶

Paul Reukauf, the Hyper-X deputy project manager at Dryden, recalled the situation as it stood in 2000.

> We were supposed to fly in 1998 and in 2000 . . . we still had not flown. And so, we got into a mode of our management asking us when was the soonest we could fly. We essentially got into the situation where we would project, given the best possible scenario, [and] with no further problems, we could fly on a certain date. The date I remember the most was in the middle of December 2000. We had been slipping as problems arose. We were going to fly in July of 2000 and then it was going to be September of 2000, and when we did another estimate it was December of 2000. But, our management seemed to be happy and they kept asking us for the best case.
>
> It turned out that the Director of Langley, Jerry Creedon and our Director [Kevin Petersen] were in a meeting at some point where Jerry Creedon took our director to task. He said: You know this is ridiculous. You guys keep telling us you're going to fly on this date, then on this date, and then on this date and I want to know a real date when you're going to fly.
>
> So, all of a sudden, our management came back and said: We want to know a real date when you're going to fly, not the best possible date. We said, "you were asking us for the best possible dates." They said, "we don't want that anymore. We want you to be able to guarantee us a flight date." And so, we did exactly what anybody else would do. We looked at the work that was left and we essentially doubled the time we thought it would take to do it and we came up with a date, like May or June of the next year [2001]. We thought that will shock them and we went to them and they said: "okay" and we said "okay."****

Dryden engineers were directed to find ways to make the first flight happen sooner. Analysis indicated that extending shifts and working on weekends would cut 15 days from the schedule. Extended shifts were planned leading up to testing of the first X-43A with actual gas in mid-September. Discussions were also held with project team members to plan extended shifts for integration work with the X-43A and HXLV.

Leak and functional testing of the X-43A/adapter "short stack" had been completed by the start of September. The short stack was scheduled for a move to Building 4847 so instrumentation checkouts with the HXLV could begin. As a result, an intensive bottom-up planning effort was under way in

---

¶¶¶Dave Reubush e-mail to Curtis Peebles, "Re: 2000 Hyper-X schedule revision," April 22, 2009.
****Paul Reukauf history interview, Dryden Flight Research Center History Office, pp. 9, 40, 41.

mid-October of all activities required for the first Mach 7 flight. This involved creating a "realistic schedule"—one that entailed more than 900 separate elements.[††††] By the latter part of October 2000, validation of the new schedule was complete and it was reviewed first with Dryden senior management and then by the Hyper-X program office. This was followed by a second examination, by a four-man independent review team, all members of which were external to Dryden. When the review was complete, the independent review team concluded that the new schedule had "adequate reserves to be considered reasonable."[‡‡‡‡]

A new schedule notwithstanding, the checkout process continued to see successes and failures. In late September 2000, successful aircraft-in-loop tests of the loaded control deflection were completed. These tests simulated the flight aerodynamic loads on the control surfaces to determine how the actuator linkages and electronics would operate. This was a critical issue, because loading on the control surfaces would cause them to bend slightly, and the actual control deflection might be greater than commanded. The loads also caused the linkages and gears of the actuation system to respond differently than in the absence of the loads.

The HXLV cooling system testing was also completed, but results indicated minor modifications would still be required to reach the proper nitrogen flow rates from the onboard B-52 supply. A more significant issue was discovery of a leak in the hydrogen/silane system, requiring that the system be taken out and reinstalled before a leak and function test could begin. While this test was under way, Dryden engineers were expressing concern about the motorized control valve in the system. Despite these problems, the preparations were on schedule for a test with actual gas the following week.

Because of the combustible nature of hydrogen and hydrogen/silane ignition mixture, the test would take place on a taxiway (see Fig. 10.5). Technicians would make a practice run of procedures to be used on Monday, setup would be on Wednesday, and the test on Thursday. The test was uneventful, given the combustible gasses, but several components had problems that required repair. These were addressed by early November 2000.[‡‡‡‡]

By mid-November, the third inert blowdown test was completed. This was the final major test, and with its completion the final assembly of the X-43A, adapter, and HXLV could began. The electrical and mechanical mating, as well as functional testing of the X-43A and adapter, had been successfully completed by December 1, 2000. Preparations were also under way for the X-43A/adapter integrated coolant/purge system test. Once this was successfully completed, the short stack would be ready for integration with the

---

[††††]Aerospace Projects Highlights, August 25, 2000, p. 2, September 1, 2000, p. 1, October 13, 2000, p. 1.

[‡‡‡‡]Aerospace Projects Highlights, September 22, 2000, p. 1; October 6, 2000, p. 1; October 30, 2000, pp. 1, 2; November 3, 2000, p. 2; Lin history interview, pp. 13, 14.

a)

b)

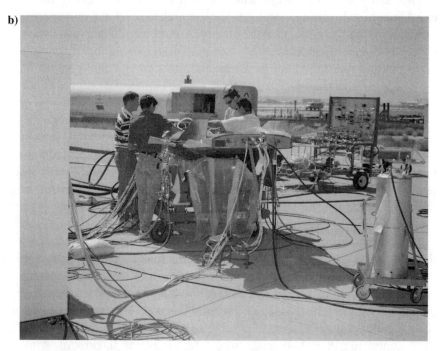

Fig. 10.5 The real-gas tests, using hydrogen gas and the hydrogen/silane ignition mixture, were done on a taxiway.

c)

d)

Fig. 10.5   The real-gas tests, using hydrogen gas and the hydrogen/silane ignition mixture, were done on a taxiway (continued*).***

HXLV. Additionally, troubleshooting and replacement of the side antenna/ amplifier in the X-43A telemetry system, which had created noise in the data, thus degrading its quality, was completed. By December 18, 2000, the electrical mating of the vehicle with the HXLV had been successfully completed.§§§§

FINAL STEPS TO FIRST FLIGHT

With the New Year and new millennium, the Hyper-X project began the final steps to flight. Beyond the continuing tests and checkout, these steps also involved formal reviews of the vehicle's readiness to fly, control room training, and a captive-carry flight of the complete stack to test its systems in flight conditions.

By late January 2001, work was in progress on the Dryden Independent Review (DIR), scheduled for February 13 and 14. This process had several stages and involved about 120 people. The first stage consisted of review by the DIR committee, which included both Dryden and Langley personnel. The Dryden review process was, in turn, reviewed by another body, and still another review team, from the X-plane office at NASA Headquarters, was in attendance as well.¶¶¶¶

Members of the DIR made a total of 80 requests for information (RFIs) from project personnel, which were given to the appropriate team members for resolution. Of the 80 RFIs made, by early April, 58 had been answered. Of these, 47 were considered closed by the DIR. The remaining 11 required further action. A week later, 70 had been answered, 57 of which were considered closed, with the DIR requesting further analysis on 13 RFIs. By mid-April, 64 of the 80 RFIs were closed.*****

While the remaining RFIs were being addressed, final preparations continued. A B-52B rehearsal flight took place over the Point Mugu range on January 25, 2001, as a training exercise to qualify timelines and procedures. The B-52 was flown by Dana Purifoy and Roger Smith, while the two-seat F-18 chase plane was piloted by Gordon Fullerton with the photographer Lori Losey in the rear seat.

The HXLV, adapter, and X-43A were also fully assembled by the end of January. In the rush to see things through and adhere to the schedule, it was noted in the weekly Aerospace Projects Highlights that: "as far as we know no one was sealed inside the adapter."†††††† A month later, at the end of

---

§§§§Aerospace Projects Highlights, December 1, 2000, p. 1; December 18, 2000.

¶¶¶¶Aerospace Projects Highlights, January 17, 2001, p. 1; January 22, 2001, p. 1; February 20, 2001.

*****Aerospace Projects Highlights, February 22, 2001, p. 2; April 2, 2001, p. 1; April 9, 2001, p. 2; April 16, 2001.

††††††Aerospace Projects Highlights, January 29, 2001, p. 2; February 26, 2001, p. 1; March 19, 2001, p. 1; January 29, 2001, p. 2; Gordon Fullerton history interview, NASA Dryden Flight Research Center History Office, pp. 10, 11.

February, the HXLV was electrically connected with the B-52. End-to-end function checks were successfully accomplished with the X-43A's monitoring station and the launch panel aboard the B-52. The X-43A and adapter were also serviced in preparation for the combined system test (CST), scheduled for March 15, 2001. This involved attaching the completed stack to the B-52 and checking out the systems, for which the B-52 was put on a parking apron away from structures. At the completion of the CST, the B-52 made a taxi test. Gordon Fullerton, a former shuttle astronaut and a Dryden research pilot, recalled:

> ... we taxied but I think we went ... down the runway at a hundred knots mainly to bounce the rocket around on the hooks that hold it and the people in the control room are watching the strain gauges on the hooks. ...
>
> So, we're giving them [the ground controllers] a look at real life loads other than just sitting still on the ramp. And, it gives us a feeling for all the interfaces. The crew is on board so they can check that the paths of data coming from the rocket are all working and show up on the screens. And so, it's a good end-to-end test and there were some dynamics. Actually the west end of the runway is less than perfectly smooth so you get some jouncing around when you go down there.

No major issues arose during the CST, and its successful completion cleared the way for a captive-carry flight of the Hyper-X stack by the B-52 in mid-April.[†††††]

Being held simultaneously with ground tests were training exercises for ground controllers (see Fig. 10.6). They were engineers, both NASA and contractors, working on Hyper-X systems. They knew the systems' behavior and the warning signs of trouble. In their role as ground controllers, they had to learn the checklists, procedures, and specialized language they were to use. Standardized phrases were used for clarity. "In work," meant a controller was looking at a problem. If a serious problem appeared, the controller called "abort." It was a precise, disciplined, and direct form of communication, designed for a situation where fractions of a second count. Although the problems controllers faced in these tests were simulated, the practice runs also simulated the potential stress of genuine emergencies.

A training exercise was held in the mission control room in January 2001. The focus of the training was on the captive-carry phase of the mission. Controllers' displays showed real-time data generated by the Hyper-X simulation. The failure scenarios were preplanned, but were unknown to the controllers in advance.

The result was unusual—the controllers knew that the flight was simulated, the data they were seeing on their displays was not real, and the B-52 was not in flight. Yet, their physical reactions—increased pulse rate and respiration,

Fig. 10.6 Ground controllers undergoing training exercises that included simulated malfunctions and emergencies.

and experiencing the sensation that time seemed to slow or accelerate—were real. The training exercise tested the team's responses to the failures and ensured they could recognize a specific problem and respond with proper emergency actions. The controllers also participated in the CST on March 15, 2001, and control room training exercises were conducted on April 6 and April 26. These involved scenarios that included several emergency procedures as well as an abort.‡‡‡‡‡

As summer approached, the final steps were nearing completion. The HXLV and X-43A were mated and successful functional tests were completed. Work on the X-43A's thermal protection system was completed for the captive-carry flight. All the RFIs had been submitted to the DIR, with 64 of the 80 requests now considered closed by the committee. The next step was a meeting of the airworthiness and flight safety review board (AFSRB), held on April 17, 2001. At the meeting, board members raised a number of issues including an analysis by the AFSRB of the telemetry link, a DIR review of contractor discrepancy reports, and an AFSRB review of the emergency procedures, "go/no-go" items, and mission rules. Despite issues raised on these topics, provisional clearance was given for the captive-carry flight,

---

‡‡‡‡‡Aerospace Projects Highlights, January 5, 2001, p. 1; April 9, 2001, p. 1; and April 30, 2001, p. 1.

though the flight was contingent on the issues first being successfully resolved.§§§§§

The AFSRB follow-up meeting was held on April 24, and approval was given for the captive-carry flight. The pace of activity now began to accelerate. A technical briefing was held on April 25, and the Captive Carry Flight Request was approved by both the Langley and Dryden center directors. The Captive Carry Crew Brief came on April 27, 2001, with the flight scheduled for the next day.

As with similar missions going back to the X-1/B-29 captive-carry flights in 1946, the flight would test the responses of onboard systems to prolonged exposure to low air temperatures. Subsequent X-planes, the lifting bodies, and the other air-launched research vehicles underwent captive-carry flights before making initial free flights. The Hyper-X captive-carry flight was successfully completed over the Point Mugu test range on April 28, with only a few "squawks" noted. Fullerton and Purifoy flew the B-52B, with Richard Ewers and Losey in the photo chase F-18. Control room personnel monitored data from the stack and its onboard systems.

Orbital Sciences personnel made an initial review of HXLV data during the first week of May that indicated all the booster systems operated properly. A review of the X-43A and adapter data was scheduled for the following week. As this was under way, final prelaunch activities went forward. The adapter underwent a successful leak and function check, and the ordnance installation for the X-43A/adapter coolant system was completed. A battery sensor that had failed during the captive-carry flight had been replaced and successfully tested. Work on the X-43A's thermal protection system was also nearing completion. Dryden and Langley Public Affairs offices began making plans for the Mach 7 flight. The two centers planned to hold a live news conference on NASA TV about an hour and a half after the flight.

Perhaps the most important milestone came at the hands of the DIR. All the RFIs submitted by the DIR had been answered. With its actions complete, the board sent a memo to the Dryden Airworthiness Board saying its portion of the review was finished. A tech brief for the first Mach 7 flight followed on May 24. The final DIR briefing and a mini-tech brief were held before the flight. Close-out work was completed on the stack.

The launch was scheduled for Saturday, June 2, 2001. The flight was set for a weekend to accommodate scheduling of the Point Mugu range and to prevent conflicts with trans-Pacific airliners flying into Los Angeles International airport, because the B-52's and X-43A's flight paths took them directly into those of inbound transpacific flights.¶¶¶¶¶

---

§§§§§Aerospace Projects Highlights, April 16, 2001, p. 1; April 23, 2001, p. 2.

¶¶¶¶¶Aerospace Projects Highlights, May 7, 2001, p. 1; May 14, 2001, pp. 1, 2; May 29, 2001, pp. 1, 2; June 4, 2001, pp. 1, 2.

Looking back on the atmosphere just before the first flight, Griff Corpening later recalled:

> At our flight-readiness review ... every one of the presenters' final slide was essentially "we are go for flight." And we had been working I think, at that point, four years or so. We had done everything we could think of to understand and mitigate the risk within the constraints of the program schedule and budget. I did my own polling. I would walk around and ask people if they were ready, and every person on the project was "go" for flight. We were confident we had done what we could think of to mitigate and understand the risk.
>
> We also knew that there were probably "unknown unknowns" out there. But we could not think of anything left to better investigate or uncover those unknown unknowns. We were ready. We were going into the control room. We were confident that we had done due diligence. And in an emotional sense or a human sense, at least I did not have any doubt in my mind that we should launch and find out what we didn't know.

He also noted that the development and testing was a team effort, saying:

> And everyone took their job so seriously. They realized that they had an important part to play, and they had to play it correctly. Down to—and very much so—the mechanics and avionics technicians. If one wire isn't crimped right, the whole vehicle comes down. If one system isn't put in correctly, the vehicle comes down. So it's an interesting situation: The responsibility for the success of the mission weighs equally with every person involved in the project. And everybody took that responsibility very seriously and maintained focus and dedication.******

Yohan Lin, a Dryden engineer who developed some of the test procedures, later said: "The first vehicle ... was a learning experience for us. We learned a lot. We, in going through the growing pains, had to swap out and replace a lot of different components and spares because; again that's part of the learning process."††††††
The learning was over. Now came the test.

JUNE 2, 2001

The color video from the B-52 launch plane was displayed on a big screen in the Dryden cafeteria; the camera looked at the stack, with the X-43A as the main focus. The black and silver X-43A rode the nose of the white booster. Below was the blue Pacific and scattered low clouds. A large group of reporters

******Griff Corpening history interview, December 20, 2004, NASA Dryden History Office, pp. 21, 28, 29.
††††††Lin history interview, p. 27.

had assembled to watch as the countdown neared its end. The audience watched the HXLV fin sweeps, one of the last activities before launch. The final 2 minutes drew to a close. The video switched to one of the two-seat F-18 chase planes, showing the entire B-52 and stack from off the right wing. After some 5 years of work, the call "Launch, Launch, Launch" was finally made.

The pylon hooks released, and the stack fell away. The 5 seconds before booster ignition seemed to drag on. Finally, the orange flame erupted from the HXLV's nozzle. The rocket began to accelerate and climb. Because of the lower/slower launch profile, it was at transonic speed. Then, a white piece of debris fell from the HXLV. The rocket went out of control, and spiraled down toward the low cloud deck.

The first Hyper-X flight had lasted 13.5 seconds, and then the mission was over. It had failed. The investigation now began.

REFERENCES

[1] Voland, R. T., Rock, K. E., Huebner, L. D., Witte, D. W., Fisher, K. E., and McClinton, C. R., "Hyper-X Engine Design and Ground Test Program," AIAA-98-1532, pp. 2, 3.
[2] Harsha, P. T., Keel, L., Castrogiovanni, A., and Sherrill, R. T., "X-43A Vehicle Design and Manufacture," AIAA 2005-3334, p. 2.

Chapter 11

"Lock the Doors . . ."

> If you are looking for perfect safety, you will do well to sit on a fence and watch the
> birds; but if you really wish to learn, you must mount a machine and become
> acquainted with its tricks by actual trial.
>
> *Wilbur Wright, The Western Society of Engineers in Chicago, September 18, 1901*

When ground controllers walked into the Dryden mission control room on June 2, 2001, they were told not to bring their personal effects. Should a mishap occur, they would have to get up and leave. Their notebooks, checklist cards, flight plans, the strip charts and data, computers, Palm Pilots, and other personal items would be left behind. As one individual noted, if a pen was on the table, it remained on the table. The lockdown was designed to preserve any and all evidence following an accident.*

Before the launch, Hyper-X personnel had felt pride at what they had accomplished. They were also anxious; there was frustration over the long delays, and a lot of excitement, in a cascade of conflicting, simultaneous emotions. Finally, they had the sense that it was time to fly. Separation was seen as the big risk, and the push over (when the flight path began to level out) represented an uncertain flight regime. But the launch itself was not among the concerns.

As the Hyper-X spiraled down (see Fig. 11.1), Griff Corpening recalled feeling stunned that the failure had occurred so soon—even before the stack had reached supersonic speed. The personnel left the control room and met as a group in a large conference room in Building 4800. Dryden Center Director Kevin Peterson stood in front of the group. Corpening recalled Petersen saying, "You know, we knew it was high risk. We obviously ran into some problems here. We've got to sort them out and push through. We've got two other vehicles to fly."†

After this larger meeting broke up, a smaller group assembled that included Mike Pudoka, Orbital Sciences' chief engineer; Phil Joyce, the Orbital project manager; and several other Orbital Sciences engineers. On the NASA side

---

*Linda Soden history interview, August 25, 2005, NASA Dryden Flight Research Center History Office, pp. 18, 19.

†Griff Corpening history interview, December 20, 2004, NASA Dryden Flight Research Center History Office, pp. 38, 39.

Fig. 11.1   The Hyper-X stack falls toward the Pacific Ocean after the control fins break off.

were Griff Corpening, Brad Neal, Joel Sitz, and Vince Rausch. The Orbital personnel took the loss personally. Phil Joyce immediately accepted responsibility. "We let you guys down," he said, "Our job was to get you to separation, and we didn't get there. It's our responsibility to do that, we take that responsibility."‡

When the initial discussions were through, most of the engineers then went to the postflight party at a nearby bar called Wing and a Prayer. Canceling the party had been considered, but the group decided to go ahead with it. Corpening thought this was "a real good idea." He recalled: "It was therapeutic to get together with everybody." From the failure came a resolve to succeed. Corpening added, "I never had any hesitation or doubt in my mind that if my management allowed me to, I would stay with this through the second flight and do everything I could to make sure we understood what happened and fix it and move on."‡

MEETING THE MIB

This resolve to triumph over the failure would be sorely needed, because the Hyper-X's fate was now in the hands of the Mishap Investigation Board (MIB), and no one on the board was directly connected to the program. Paul Reukauf recalled how jarring this shift was for participants:

It's a very difficult time because number one, you're in a situation where [the Hyper-X personnel] no longer get to direct what's happening. And,

---

‡Soden history interview, p. 13; Griff Corpening history interview, December 20, 2004, NASA Dryden Flight Research Center History Office, pp. 38, 39.

number two, you're doing lots of work going down avenues which you know are not direct contributors to the accident. But you have to do it. So it seems like you're doing a lot of extra work.[§]

Reukauf added that Joel Sitz, Dryden project manager for the X-43A (Fig. 11.2), played a major role during this time.

He was able to keep the morale high enough to get through the year and a half of investigation plus waiting for the investigation results.... I give him a lot of credit for that. I don't think many people could have done it. I think a lot of personnel would have drifted away and done other things without his leadership.[§]

Jeremiah F. Creedon, NASA Associate Administrator for Aerospace Technology, gave verbal approval for the MIB to assume responsibility for the Hyper-X investigation on Tuesday, June 5, 2001. Creedon's formal letter of approval was issued on June 8. The MIB's duties were [1]:

- Obtain and analyze the evidence, facts, and opinions it considers relevant.
- Use reports, studies, findings, recommendations, and other materials by NASA and contractor personal. Conduct inquiries, hearings, tests, and other actions deemed appropriate. Take statements from witnesses.
- Impound property, equipment, and records as needed.
- Determine actual cause(s) or probable cause(s) of the X-43A mishap, and document their findings in terms of (a) the dominant root cause(s) of the mishap, (b) contributing cause(s), and (c) significant observation(s).
- Develop recommendations for preventive or other actions.

Fig. 11.2 Joel Sitz had responsibility for the overall flight test program.

[§]Paul Reukauf history interview, February 17, 2005, pp. 43, 44.

- Present a verbal report to the Associate Administrator and a final written report.
- Provide a lessons-learned summary.
- Perform other duties requested by the Associate Administrator.

The MIB had a mass of data to work with from the first Hyper-X's short flight. This included telemetry measurements taken in the Blue and Gold Rooms (Dryden's primary control rooms), the Telemetry and Radar Acquisition Processing System control room, and the Structural Analysis Facility. Additionally, there were the data from the B-52B launch aircraft, the video and still photos from the F-18 chase planes, and the images from one of the adapter cameras.

Background data included statements given by NASA and contractor engineers, controllers, and technicians; records of the HXLV, adapter, and X-43A testing and checkout; configuration control records; test results of the different components; Pegasus accident and anomaly reports from previous launches; preflight wind-tunnel test results; and other studies, tests, and analyses. All of this material had to be secured, cataloged, and analyzed.[¶]

This information was especially critical because the evidence normally associated with aircraft accident investigations was not available; the debris was out of reach, at a depth of 1200 ft, and there would be no attempt to recover it.

The MIB was composed of Chairman Robert W. Hughes (Marshall Space Flight Center), Joseph J. Lackovich, Jr. (Kennedy Space Center), Frank H. Bauer (Goddard Space Flight Center), Michael R. Hannan (Marshall Space Flight Center), Luat T. Nguyen (Langley Research Center), Victoria A. Regenie (Dryden Flight Research Center), Karen L. Spanyer (Marshall Space Flight Center), and Pamela F. Richardson (Office of Safety and Mission Assurance, NASA Headquarters). Two advisers from Dryden were part of the MIB, Chauncey Williams of the Office of Chief Counsel and Fred Johnson of the Office of Public Affairs.

The initial MIB meetings were held at Dryden June 5–23, 2001. The group met daily to review the data as they were processed and interpreted. They then relocated to the Orbital Sciences Chandler, Arizona, facility, to review records and interview personnel there. Review of the HXLV failure scenarios continued in Chandler from June 23 to August 31. The group then moved to Langley Research Center for the final part of the investigation. This involved new Langley wind-tunnel testing, and lasted from September 10 to December 7, 2001. Subsequent investigative efforts were made via teleconferences and e-mail.

From the flight data, the MIB was able to construct a timeline of the brief flight's events. It began with the release of the Hyper-X stack from the B-52's

---

¶Soden history interview, pp. 17, 18.

Fig. 11.3   The Hyper-X stack begins its flight after being dropped from the B-52B.

pylon at 0.0 s mission time (see Fig. 11.3). The HXLV autopilot was enabled at 0.38 s. The stack fell until, at 5.19 s mission time, the HXLV motor ignited. All these events occurred at planned times.

After the rocket ignited and the stack accelerated, the first malfunction occurred. Between 6.23 and 7.1 s, the gaseous nitrogen pyrotechnic valve on the adapter opened, and resulting pressures exceeded the relief valve setting. This caused an uncontrolled venting of nitrogen gas, which was logged by the MIB as an anomaly that required correction but was not related to the loss of the Hyper-X stack.

At 10.18 s mission time, the HXLV path steering guidance was enabled, and the stack was commanded to begin the pitch-up maneuver. At 10.82 s, as the HXLV pitched into the climb, telemetry indicated the beginning of a divergent roll oscillation. The rudder actuator stalled, and stopped responding to steering commands at 13.02 s.** At this point, the HXLV's angle-of-attack, dynamic pressure, and sideslip measurements all exceeded nominal values. Between 13.30 and 13.48 s, the right fin shaft strain gage, actuator motor temperature, and fin leading-edge temperature values went to their maximum. This indicated gage wiring had broken. At approximately 13.5 s, the right fin broke off from the HXLV. Between 13.8 and 15.0 s, the left fin and the rudder broke away. The stack then went out of control and began a corkscrew descent toward the Pacific Ocean. At 18.84 s, the X-43A's left wing linkage failed, causing the wing to spin on its pivot shaft. Telemetry from the HXLV failed at launch plus 20.87 s.

**The term *stall* refers to the FAS being unable to move the rudder against the aerodynamic forces acting on it. This is different than a wing's loss of lift at too low a speed or at an excessive angle of attack.

The destruct signal was transmitted at 48.31 s, blowing a hole in the motor casing. Despite the violent gyrations and the detonation of the self-destruct package, the stack was still a single unit up to this point.

At 49.31 s the X-43A broke away from the adapter. One of the adapter cameras relayed images of the research vehicle tumbling away, its left wing turned backwards from its normal position, and of a section of the sheet metal that broke free of the adapter. The X-43A's aft S-band transmitter also activated at this point. Telemetry signals continued to be received until 77.57 s, when the research vehicle hit the water, after which it sank [1, pp. 15–19].

Fig. 11.4 shows a series of images taken by one of the cameras aboard the adapter after control was lost. In the background is the B-52 smoke trail and the larger rocket exhaust. In Fig. 11.4a, the X-43A has broken free of the adapter. The left wing has also rotated 90 degrees from its normal position, indicating damage to the mechanism. In Fig. 11.4b, a metal panel has broken off the adapter, and the X-43A's left wing has rotated nearly 180 degrees. In Figs. 11.4c and 11.4d, the X-43A is visible at an increased distance from the adapter.

To analyze this mass of data, the MIB used fault-tree methodology. This was based on a series of requirements. One of these was the availability of the fault trees, which had been developed and used in the project's earlier risk assessments. Another was the complexity of the X-43A and its systems, and the multiple organizations involved in the project. Finally, MIB personnel

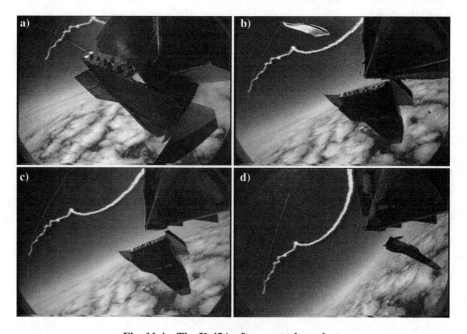

Fig. 11.4   The X-43A after control was lost.

all were familiar with fault-tree investigation procedures, so it came as a natural choice.

As the term suggests, a fault tree begins with potential failure scenarios, which are based on the assessments of a project's physical, functional, operational, and engineering characteristics. This forms the top level of the "tree." This is then broken down into a series of lower branches, each based on possible scenarios of failures that may have caused specific top-level faults. This continues through several branches, with each fault labeled as a confirmed contributor, a potential contributor, or a confirmed noncontributor.

This examination process forms the basis for determining the individual causes of the failure. Additionally, the fault tree is used to determine possible system weaknesses that did not cause the failure, but that could lead to another. No fault was added or removed from the tree until the MIB gave its approval. In the case of the X-43A fault tree, 613 specific faults were examined. The complete fault tree chart extended from ceiling to floor, and took up much of a wall in a Hyper-X meeting room that was about 30 ft long.

The top level of the Hyper-X mishap fault tree consisted of seven branches—(1.0) loss of B-52 flight safety, (2.0) loss of X-43A stack, (3.0) failure to drop/ignite, (4.0) failure to reach desired separation point, (5.0) unsuccessful HXLV/X-43A separation, (6.0) unsuccessful free flight, and (7.0) ground operations stack damage. Only the failure of the stack to reach the separation point (4.0) contributed directly to the launch failure. All the other specific faults were eliminated as confirmed noncontributors because their different sublevels had no connection with the failure, and they were closed out.

With 4.0, failure to reach desired separation point, identified as the initial branch in the fault tree of the mishap, the next set of branches was then checked. This consisted of six possible scenarios—(4.1) loss of control, (4.2) structure, (4.3) flight termination system initiation, (4.4) fire/explosion, (4.5) collision with air vehicle, and (4.6) loss of HXLV data. Again, all but one of the potential branches were eliminated, leaving only 4.1, loss of control.

The third set of branches consisted of possible failure scenarios that led to this loss of control. Five possibilities were identified—(4.1.1) external disturbances, (4.1.2) structures, (4.1.3) aerodynamic/control, (4.1.4) motor, and (4.1.5) avionics. Once again, all but one branch was eliminated, leaving 4.1.3, aerodynamic/control, which had led to the loss of HXLV control.

To this point, the sequence of the failure was straightforward. Each branch had only a single fault connected with the mishap. But it is a truism than no accident has a single cause. The aerodynamic/control fault branch had eight elements. Of these, the first three, (4.1.3.1) autopilot design, (4.1.3.2) autopilot implementation, and (4.1.3.3) structural dynamic modeling, did not contribute to the failure. Four branches—(4.1.3.4) mass properties modeling, (4.1.3.5) aerodynamic modeling, (4.1.3.6) vehicle configuration, and

(4.1.3.7) fin actuation system—all contributed to the failure in a confirmed, quantifiable manner. The final element, (4.1.3.8) aeroelastic effects, was a potential contributor [1, pp. 22, 23]. After the analysis was complete the MIB issued its report, concluding [1, p. 5]:

> The mishap occurred because the control system could not maintain the vehicle stability during transonic flight. The vehicle instability was observed as a divergent roll oscillation. An effect of the divergent roll oscillation was the stall of the rudder actuator. The stall accelerated loss of control. The rudder actuator stalled due to increased deflections that caused higher aerodynamic loading than preflight predictions. The deficient control system and under prediction of rudder actuator loads occurred due to modeling inaccuracies.
>
> Determining the cause of the X-43A mishap consisted of in-depth evaluations of the Pegasus and HXLV system and subsystem models and tools as well as extensive system level and subsystem level analyses. To support the analyses, extensive mechanical testing (fin actuation system) and wind tunnel testing (6 percent model) were required.
>
> The major contributors to the mishap were modeling inaccuracies in the fin actuation system, modeling inaccuracies in the aerodynamics and insufficient variations of modeling parameters (parametric uncertainty analysis). Pegasus heritage and HXLV specific models were found to be inaccurate.
>
> - Fin actuation system inaccuracies resulted from:
>   - Discrepancies in modeling the electronic and mechanical fin actuator system components
>   - Under prediction of the fin actuation system compliance used in the models.
> - Aerodynamic modeling inaccuracies resulted from:
>   - Error in incorporation of wind tunnel data into the math model
>   - Misinterpretation of wind tunnel results due to insufficient data
>   - Unmodeled outer mold line changes associated with the thermal protection system (TPS).
> - Insufficient variations of modeling parameters (parameter uncertainty analysis) were found in:
>   - Aerodynamics
>   - Fin Actuation System
>   - Control System
>
> Less significant contributors were errors detected in modeling mass properties. Potential contributing factors were found in the areas of dynamic aerodynamics and aeroservoelasticity.
>
> Linear stability predictions were recalculated using the corrected nominal models. Stability gain margins were computed for all axes. Aileron gain margin (roll axis) was examined in particular and showed a

sizeable reduction from the 8 dB preflight prediction. Model corrections led to a revised prediction of less than 2 dB at nominal conditions. This was well below the requirement of a 6 dB gain margin. Although this reduction was very significant and close to instability boundaries, the revised prediction was still stable. This meant that the nominal model corrections alone were insufficient to predict the vehicle loss of control and that parameter uncertainty had to be included. Accounting for parameter uncertainties in the analyses replicated the mishap. This was confirmed by nonlinear time history predictions using the 6-degree of freedom (6-DOF) flight dynamics simulation of the X-43A stack.

No single contributing factor or potential contributing factor caused this mishap. The flight mishap could only be reproduced when all of the modeling inaccuracies with uncertainty variations were incorporated in the system level linear analysis model and nonlinear simulation model.

The report noted that the divergent roll oscillation, which set the events in motion, was primarily caused by excessive control system gain. As a consequence of the roll oscillation, the rudder actuator stalled and accelerated the loss of HXLV control. Neither the roll oscillation nor the rudder actuator stall had been predicted by preflight testing.

The HXLV flight conditions were considerably more severe than those of a Pegasus booster flying a satellite launch. At the point of failure, the HXLV was at an altitude of 22,244 ft. Higher dynamic pressure at transonic speeds was a major factor in the vehicle's loss. On a Pegasus launch, the booster would be at about 40,000 ft. The lower altitude translated into a significantly higher dynamic pressure of 650 psf on the HXLV, versus about 300 psf for a Pegasus booster used in a satellite launch.

In its investigation the MIB also looked at the near-loss of the WIRE Pegasus launch on March 4, 1999. That incident began when the Pegasus was at Mach 0.9 to Mach 1.2, between 6 and 12 s after launch, and at an altitude of 40,000 ft. The booster began to experience large sideslip and bank excursions. These began in the roll axis, and quickly coupled into yaw and finally pitch axes as well. But once the Pegasus accelerated through the transonic regime, the booster recovered from the excursions, stabilized, and successfully placed the WIRE satellite into the correct orbit. Modifications were made to the Pegasus autopilot, as were improvements in aerodynamic modeling and upgrades to the fin actuation system (FAS). The WIRE launch incident was studied as part of the qualification of the HXLV, but did not result in any suggested changes. The significant difference between the WIRE launch incident and the HXLV launch was the higher dynamic pressure experienced by the Hyper-X stack [1, pp. 11, 18, 19, 24].

As part of the WIRE launch incident investigation, a limited number of additional wind-tunnel tests were made with the HXLV stack. These tests, like others made over the course of the Hyper-X development effort, did not raise any issues of HXLV transonic stability. This failure to detect the scale of the HXLV's transonic instability was due to multiple factors, particularly the limitations of the mathematical models used in predicting the stack's flight behavior, shortcomings in the wind-tunnel model and its instrument system, a failure to realize the differences between Pegasus heritage conditions and those the HXLV was required to meet, and a failure to look deeper at assumptions that had been made about the lower/slower launch profile.

Wind-tunnel tests of the Hyper-X stack had been conducted with a Pegasus model, one modified with the adapter and X-43A. The model was built with fins that were adjusted in 5-deg increments. The forces acting on the fins at transonic speeds were measured using a strain gage. The model was mounted at different angles of attack and strain gage measurements were taken. For the forces at fin positions between the 5-deg increments, simple linear interpolations were used. (The fins were kept in a fixed position during each run.)

Dave Reubush noted later: "... Orbital did not know the booster characteristics as well as they thought they did. Since we thought they knew the booster we did a minimum of wind tunnel testing for the launch to save money."[††]

When the data were reexamined after the loss of the first flight, engineers realized they had understood neither the technical shortcomings of the tests nor the issues of compliance and friction and the interconnections between the FAS and the air loads the fins experienced due to transonic conditions. Instrumentation shortcomings were also a cause. The stress gage used in the initial model was simple, and did not simulate compliance and other factors. It also lacked resolution and precision sufficient for detecting the problem. This hid the effects that had caused the fins to stall and control to be lost.

After the loss of the first Hyper-X, its wind-tunnel model was modified. The fins could now be adjusted in 2.5-deg increments. New Langley-designed strain gages were also fitted to the model. Tests in the Langley 16-Ft. Transonic Tunnel were conducted, using the new model at the 2.5-deg fin increments. When the data were analyzed, a nonlinearity was discovered between the 0-deg and 5-deg fin settings. This nonlinearity was at the fin position where divergent roll oscillations had begun on the first launch. Reubush noted, "... the data from those tests matched what actually happened."

The new wind-tunnel data also showed more rolling moment than had been predicted, although part of this may have been due to vortex interactions. With the new model and strain gages the problem became apparent. Another problem testing uncovered was that the compliance model had been inaccurate. The fin and the linkages as well as other portions of the flight article's structure were more flexible than predicted. Finally, the delay between flight control system inputs and response was longer than had been predicted. As a

---

[††]Dave Reubush e-mail to Curtis Peebles, "RE: HXLV Flight 1 Launch Profile," April 7, 2009, 11:15 A.M.

result, there was more roll authority than originally thought, the fins were moving farther than they should have, and the computer was slow to correct the booster's trajectory.[‡‡]

This was not the only testing done to fill gaps in the aerodynamic database. A new and intensive program of wind-tunnel and CFD testing was undertaken to increase understanding of the HXLV's complex aerodynamics. Other factors that affected stack stability included the geometry of the X-43A and the additional thermal protection added to the HXLV wings and fins.

The initial tests were done in the Lockheed Martin high-speed wind tunnel between April and October 2002. The Lockheed tests used a 6 percent model of the stack, at speeds from Mach 0.6 to Mach 4.6. This testing also collected fin data at high dynamic pressures. The testing showed several nonlinearities at transonic speeds, which were traced to asymmetric wing loading. Following these tests, additional wind-tunnel runs were made in December 2002 at NASA Langley's Unitary Plan Wind Tunnel. These tests were made to check critical data in the aerodynamic models that were based on analytical approximations rather than wind-tunnel measurements. The new round of tests also served to confirm previous data collected from different facilities.

Hypersonic wind-tunnel tests were also done as part of the return-to-flight (RTF) program. These used a 3 percent model of the stack at speeds from Mach 6 to Mach 10. The first series, in March 2003, was made in Langley's 20-In. Mach 6 tunnel. This was followed in August 2003 by a second series in Langley's 31-In. Mach 10 tunnel. These tests focused on discovering any shock impingement on HXLV control surfaces at hypersonic speeds [2].

To correct the problems that had destroyed the first Hyper-X, modifications were required in several areas. These posed new difficulties, one of which would threaten the very survival of the Hyper-X project.

FIXING THE ISSUES

The lower/slower launch Mach 7 profile was the basic cause of the failure of the Hyper-X launch. A way now had to be found to launch the HXLV at the standard Pegasus altitude and speed, yet with a Mach 7 burnout. Some means of using up the rocket's excess energy had to be found. The initial idea was to add more weight. Though adding ballast to alter burnout speed had been among initial design concepts for both the ground and air-launched boosters, engineers soon recognized a problem.

Although the dynamic pressure would be lower in a 40,000-ft launch, the weight of the stack would be greater. As the stack pitched up into the climb, forces on the fins would be greater, not less, because the control surfaces

---

[‡‡]Dave Reubush e-mail to Curtis Peebles, "RE: HXLV Flight 1 Launch Profile," April 7, 2009, 11:15 a.m; notes from a telephone interview with Walter C. Engelund, April 13, 2009, regarding Hyper-X wind tunnel testing; Corpening history interview, pp. 42–45.

would have to move a heavier vehicle. This ran counter to the requirement of reducing fin forces. A secondary question was whether the Pegasus adapter and the B-52B's pylon could support the added weight. A reduction in stack weight seemed advisable.[§§]

The initial and final energy states of the HXLV trajectory were fixed by launch conditions and the speed and altitude of burnout and scramjet test conditions. The booster would be launched at an altitude of 40,000 ft and a speed of Mach 0.8, and would burn out at a speed of Mach 7 at about 95,000 ft. The total amount of velocity change, Delta V_{ideal}, that the rocket could produce was fixed by the propellant. This ideal total speed was reduced by three factors—Delta V_{atm}, which was the velocity loss due to the rocket operating within the atmosphere; Delta V_{drag}, the velocity loss due to vehicle drag; and Delta V_{grav}, the velocity loss due to the change in altitude.

The terms Delta V_{atm} and Delta V_{grav} could not be changed, and they were, in any event, small compared to the values of Delta V_{ideal} and Delta V_{drag}. Increasing the Delta V_{drag} value required either significantly changing the stack mold lines, which would invalidate the vehicle's aerodynamic database, or increasing the integrated dynamic pressure profile the stack would experience during ascent. The latter option was not viable because it would require major structural and thermal protection changes, and the vehicles had already been built and delivered. Moreover, dynamic pressure greater than what the booster's control system could overcome had caused the first failure, so increasing it further was ruled out.

The only remaining option was to reduce the value of Delta V_{ideal}. But because adding ballast was not a viable option, the removal of about 3300 lb of solid propellant now seemed the only genuine option if a Mach 7 burnout speed was to be achieved from the higher launch altitude [3].

There was a great deal of concern about this. A team of government personnel was assembled under the direction of Langley's Mel Lucey, who brought in several outside experts. Some of the outsiders had experience with offloading boosters, and had not been involved in the earlier debate over a lower/slower launch vs. propellant offloading. Lowell Keel, the X-43A project manager at Micro Craft, later commented: "It is always easier to identify options and results when the specifics of the needs are better defined and motivations are high."[¶¶]

By way of explaining why the offloading option had not been pressed before, Randy Voland noted:

> At the time the decision was made to go low and slow in 1996 or 1997,
> there was not a lot of experience with "off-loading" solid rocket booster

---

§§Corpening history interview, pp. 49, 50.
¶¶Corpening history interview, pp. 49, 50; Lowell Keel e-mail to Curtis Peebles, "RE: HXLV Flight 1 Launch Profile," April 8, 2009, 1:13 p.m.

propellant. By the time we had the mishap on flight 1 and looked into the issue again several years later, a significant amount of work had been done by one or more other programs on machining out propellant from existing boosters. So the risk balance had shifted. Off loads were now a more known quantity and thus considered significantly lower risk than they had been earlier.***

The team concluded that it was possible to offload enough propellant to get a Mach 7 burnout speed from the new launch altitude. That the two remaining HXLV boosters had already been built, however, was a complication. One of the original ideas, increasing the thickness of the insulation layer between the propellant and the motor casing, was not possible, so the propellant would have to be machined from the motor. Trade studies of Mach 7 trajectories were done to determine how much propellant to remove: 3345 lb, or about 12 percent of the total propellant weight, was selected based on these studies. Removing this amount also minimized the risk of the machining process, maintained viable ignition grain geometry, and was within the Orion 50S design constraints. The propellant was machined from the area just forward of the nozzle, rather than along the motor's entire length.

This raised several issues. A large cavity was formed by removal of the propellant, and there was concern that this would prevent sufficient backpressure from building to ignite the rocket and keep it burning. A second concern was the possibility that the cavity could act like a large loudspeaker. As the propellant burned, it generated a large amount of acoustic energy, creating an acoustic resonance.

The "HXLV Lite," as the modified booster was nicknamed, and the Pegasus Orion 50S booster had distinctively different vacuum thrust profiles due to the propellant offloading. At ignition, both reached full thrust within less than a second. The Orion 50S had a slightly higher initial thrust than the Hyper-X Lite. Both boosters' thrust increased, with the Hyper-X Lite's thrust peaking at 15 s and the Orion 50S's at about 17 s after ignition.

At this point in the trajectory, the two boosters' thrust behaviors became radically different. The Orion 50S thrust began a slow decline until shortly before an elapsed time of 75 s. The thrust dropped sharply over the next several seconds, before the booster burned out after just over 80 s. In contrast, the HXLV thrust would drop sharply between 15 and 20 s elapsed time. This drop-off in thrust occurred at the point of the trajectory's maximum dynamic pressure, reducing the aerodynamic loads on the Hyper-X Lite vehicle and its control system. The booster's thrust then began a gradual increase between 20 and 80 s elapsed time, peaking slightly higher than that of the Orion 50S. After that, the Hyper-X Lite thrust would rapidly drop, with burnout at around 90 s.

***Randy Voland e-mail to Curtis Peebles, "RE: HXLV Flight 1 Launch Profile," April 7, 2009, 2:01 p.m.

Normally, so extensive a change in a rocket's thrust profile would require a static test to ensure that it would work as predicted. But only two HXLVs remained. A static test would mean the Mach 7 flight would be the only launch. As a result, an independent review panel was formed to assess the Hyper-X Lite's complete design, and the need to make a static test. Specific areas examined by the panel included ballistic modeling, combustion instability, thermal margins of the longer burn time, and igniter performance with the larger area at the nozzle.

ATK completed the structural, thermal, ignition, and ballistic design analysis, developed the machining techniques needed to remove the propellant safely, and demonstrated the technique on an inert motor. With the review panel satisfied, the second flight motor was disassembled. Machining of the motor began on February 12, 2003, and just over a week later the process had passed the halfway point. By the end of the first week in March, the 3340 lb of propellant had been removed, and the motor had successfully passed an X-ray inspection. The motor was reassembled, and arrived back at Dryden on April Fool's Day 2003. The process had been accomplished in 6 months [3, pp. 9, 10].[†††]

THE FIN ACTUATION SYSTEM MODIFICATIONS

Though the propellant offloading was completed with minimal problems, other issues also had to be dealt with as part of the return-to-flight effort. One element was improving the fidelity of the HXLV aerodynamic model. The aerodynamics of the vehicle were very complex. Particularly important were interactions of the HXLV's rudder with the vortices generated by the X-43A's wings and tails during flight at high angles of attack and sideslip. These interactions generated forces on the rudder too great for the fin actuation system (FAS) to overcome. The rudder FAS stalled due to the highly nonlinear behavior of the rudder hinge moment created by these vortices. The problem had not been detected before the first flight due to a paucity of wind-tunnel data under these conditions [3, pp. 10, 11].

To ensure that a similar error was not missed, larger aerodynamic uncertainties were included in the 6-deg-of-freedom simulation conducted as part of return-to-flight efforts. Additionally, more conservative control margins were assumed in the simulation. The simulation indicated, not surprisingly, that the dynamic pressure on the HXLV Lite would be lower than that experienced on the first launch. However, the simulation also showed there was still the possibility that the standard Pegasus FAS could experience a fin stall. Eliminating this required a significant increase in FAS torque output. No

---

[†††]Aerospace Projects Weekly Highlights, February 14, 2003, p. 4; February 21, 2003, p. 4; March 7, 2003, p. 5; April 4, 2003, p. 5.

existing system could provide the necessary amount of torque, so a major modification to the existing Pegasus FAS had to be made.

The original design used an electronic control unit (ECU), which received steering commands from the HXLV's flight computer. The ECU's signal conversion boards took the steering commands and generated analog signals that were then sent to the three electromechanical actuators (EMAs). These devices had several elements—a permanent magnet motor, reduction gear train, and position potentiometer—contained within a metal housing. The three EMAs were mounted in the booster's aft skirt, and each controlled one of the three fins. Electrical power was supplied by a battery, which was activated at the T–2 minute mark.

Fin movement was measured by the position potentiometers, which sent signals to the ECUs, which in turn forwarded the measurements to the flight computer. Simultaneously, the inertial measurement unit (IMU) was determining accelerations, rates, inputs, velocities, and position data, using both the inertial system and GPS receiver. The IMU took the different readings and generated new commands for the flight computer. This provided both new steering directions and feedback to correct errors. Inherent in the system were delays due to the time required by each component to generate the measurements and the time it took for the mechanical components to move [1, p. 12].

Engineers believed a relatively simple modification to the FAS would meet the increased-torque requirement. A brake on each EMA would be removed and replaced with a second actuator motor, electrically parallel with the first, in a torque-summing arrangement. This design would increase the hinge torque from 1850 ft · lb to over 3000 ft · lb. Even in a worst-case situation, this would mean a 25 to 30 percent torque margin. Although engineers had wanted a 50 percent torque margin, the lesser value was judged adequate.

Adding the second motor, however, required modifications to the FAS electronic control unit. The original ECU design consisted of a signal conversion board, control board, predriver board, powerboard, and motherboard. Adding the second motor doubled the current. The new design used the Pegasus heritage motherboard, signal converter board, and control board. But the higher current required redesign of the predriver board and powerboard.

The original ECU powerboard's analog systems were replaced with a digital system, which used complex programmable logic devices (CPLDs). When a command was issued by the flight computer to move the fins, the CPLDs activated opposite pairs of field effect transistors (FETs), in an "H-bridge" configuration. The direction the fin moved in depended on which pair of FETs was activated, while the amount of movement was determined by the current strength. To return the fin to a neutral position, the initial FET pair was closed and the opposite pair was activated to reverse the direction of movement.

The new ECU configuration presented several new requirements. CPLDs were low-voltage logic devices that controlled high-voltage currents. The high-current switching could create electrical effects that could cause CPLD failure. The high- and low-voltage components both were mounted on the powerboard, complicating the situation.

The final modifications to the FAS were material rather than electronic. The original casing was made of aluminum. The new version was made of stainless steel for greater strength. The FAS gears were also made of stronger metal, to better withstand the higher torque [4].‡‡‡

The MIB completed its work in early February 2002, closing the last 21 fault boxes, out of the original 613. Several teams composed of both government and contractor personnel were formed to develop fixes to specific problems uncovered during the investigation. The public version of the MIB final report was issued in May 2003.§§§ Engineers initially thought that the FAS modifications would be relatively simple. Certainly, they appeared simple in comparison to the propellant offloading. Initial impressions are often incorrect.

FIXING THE FAS

By March 2002, design work began for FAS modifications that would improve HXLV fin torque. A month later, Dryden representatives met with Parker Aerospace, the FAS contractor, to discuss the potential modifications. By the beginning of May, final specifications for the FAS actuator upgrades were distributed for review by engineers. By the end of May, the FAS design was finalized, and a review was scheduled for June 18. The second Hyper-X flight was scheduled to take place a year later.¶¶¶

By mid-August 2002, Parker had released all mechanical and electrical drawings for the redesigned FAS and the company was manufacturing controller test boards for development work, with delivery scheduled for early September. By then, drawings of the gears had also been released by Parker, and Orbital Science's subcontractor was buying the materials needed to fabricate them. By October, FAS powerboard testing was under way, and the cutting of the gears had begun. The work was on schedule for integrated testing in mid-October.****

With the advent of a new year, the old problem of schedule slippage reappeared. In early February 2003, there were delays in release of the HXLV

---

‡‡‡Yohan Lin history interview, NASA Dryden Flight Research Center History Office, p. 62; Corpening history interview, Tape 1, pp. 51, 60.

§§§Aerospace Projects Highlights, January 18, 2002, p. 1, February 1, 2002, pp. 1, 2.

¶¶¶Aerospace Projects Highlights, March 8, 2002, p. 2; April 19, 2002, p. 1; May 3, 2002, p. 1; May 31, 2002, p. 1.

****Aerospace Projects Highlights, August 16, 2002; October 4, 2002.

aero model, in final adjustments to the sensors needed for the propellant offloading test with the inert stage, and in development of the Pegasus adapter proof-test support structure and instrumentation troubleshooting. The HXLV actuator modification qualification test unit was also part of this problem. It was being fabricated at Parker's Dublin, Georgia, facility. Delivery to Orbital Sciences for environmental testing was scheduled for February 17, 2003. This was about 10 working days behind schedule, placing it on the hardware-critical path, but Orbital Sciences engineers believed the delay could be made up during final qualification testing to avoid delaying the launch date.[††††]

By mid-February 2003, burn-in testing of the qualification actuator had been completed at Parker. The actuator was then shipped on March 7, to begin environmental thermal and vibration testing at Orbital Science's Chandler facility. Despite the success of propellant offloading and other work, a 2-week slip in "the most optimistic flight date" had occurred, pushing the first possible flight from June 2003 to somewhere between August 16 and 20, 2003. The initial qualification testing was done at NTS in Santa Clarita, California. The electronic control unit successfully passed the testing, which was a major milestone; however, when actuator testing began, the vibration table malfunctioned. The NTS tests were halted, and the hardware was sent to Orbital Sciences; although the company's vibration test equipment had earlier suffered a malfunction, it was now repaired.[‡‡‡‡]

Testing at Orbital Sciences was completed in early April, with the qualification actuator successfully loaded in hinge torque to the new design goal of a 50-percent margin over maximum expected loading. Development of the new FAS design had taken 9 months. At the same time, the flight 2 actuators and controller had been completed at Parker, and were entering burn-in testing. During testing, a problem with the gear train appeared, requiring that it be replaced. Parker and Orbital Sciences engineers, along with Dryden quality assurance personnel, worked at Parker's Dublin facility to reassemble the actuator.

The equipment problems required a rearrangement of the test sequence. The actuator was sent to Chandler to begin environmental testing, preventing a schedule delay. The flight schedule was being updated each week, and assessment of the Parker vibration test equipment problems was being made [2, p. 10].[§§§§]

Checkout of the flight 2 electronic control unit went smoothly; a 100-hour burn-in test was completed during mid-June. Soon after, thermal and vibration acceptance testing of the flight 2 actuators was completed at Orbital

[††††]Aerospace Projects Highlights, February 7, 2003, pp. 4, 5.
[‡‡‡‡]Aerospace Projects Highlights, February 14, 2003, pp. 4, 5; February 21, 2003, pp. 4, 5; March 14, 2003, p. 5.
[§§§§]Aerospace Projects Highlights, April 4, 2003, p. 5; April 11, 2003, p. 4; May 3, 2003, p. 3; May 9, 2003, p. 4.

Sciences. The components were then returned to Parker for a 1-day loaded function test, before being returned to Orbital Sciences for installation in the HXLV's aft skirt. This was the last good news about the FAS engineers were to get for a long time.¶¶¶¶

The first indication of trouble came at the beginning of July 2003. While undergoing testing, the ECU failed. This was traced to a component failure that occurred as a result of a manufacturing defect. All similar components used on the Hyper-X by engineers at Orbital Sciences' Dulles, Virginia, facility were checked, and one was discovered to have a similar defect. Once the suspect components were replaced, the ECU was delivered on July 15 to Orbital Sciences.*****

New ECU problems appeared in the flight actuator controller hardware boards during the latter part of August, and several Hyper-X team members were sent to Chandler to review solutions and discuss workarounds that would keep the planned launch on schedule. NASA personnel inspected the malfunctioning boards and ordered modifications. Final modifications were being made by the end of August 2003, with retest of the boards scheduled immediately after.†††††

The schedule for the second flight had slipped considerably as a result of these issues. The original flight date had been sometime in June 2003. By early September, the tentative flight date was November 15, 2003, pending range availability. The Point Mugu range was also used for U.S. Navy tests and for Air Force F-22 test missions flown from Edwards.‡‡‡‡‡

But by the end of October, the FAS controller hardware had exhibited another problem, failing to correctly command the actuator at low temperatures. Langley, Dryden, and Parker engineers were sent to Orbital Sciences to analyze the problem. As part of this effort, the channel, which had failed at low temperatures, was instrumented in an attempt to capture the failure. Several cold temperature cycles were made, and each time the actuator worked properly. Whatever was causing the failure was subtle, because merely connecting the diagnostic equipment appeared to be enough to fix it.

By this time the FAS investigation had grown substantially. Langley, Dryden, the Marshall Space Flight Center, Glenn Research Center, Orbital Sciences, Parker, DCI, and International Rectifier were now involved. Investigation activities included looking at the test setup, actuator malfunction, and ECU malfunction. The focus of the investigation was the circuits that drove the actuator motors. This involved examining the powerboard logic used for power shutdown, and the timing circuits that controlled current

---

¶¶¶¶Aerospace Projects Highlights, June 13, 2003, p. 5; June 30, 2003, p. 3.
*****Aerospace Projects Highlights, July 3, 2003, p. 2; July 18, 2003, p. 3.
†††††Aerospace Projects Highlights, August 22, 2003, p. 3; August 29, 2003, p. 3.
‡‡‡‡‡Aerospace Projects Highlights, September 5, 2003, p. 3.

switching. Orbital Sciences-Dulles component-reliability laboratory personnel looked at the powerboard components, while Parker engineers looked at integrated circuits and programmable device firmware for any that exhibited anomalies at low temperatures. Other organizations provided input as the investigation continued. A fault tree was generated to provide a structure for the investigation process.

A separate activity was assembly of a new ECU for use on the second flight. Engineers initially estimated that the new unit would be ready by January 2004. (The primary issue was successful completion of ECU acceptance testing.) Qualification testing at ambient temperatures was finished by mid-November 2003, but the plan to test further, at cold temperatures, was delayed. That decision was based on concerns that the system had a design flaw, and the ECU would be damaged due to cold temperatures. A decision on undertaking environmental testing would be made based on results of investigations into both of these concerns.§§§§§

As November drew to a close, troubleshooting on the FAS continued. The focus had been narrowed, and progress had been made. The NASA/contractor team was working well together, and was focused on an unusual characteristic of the failure. They had discovered that the ECU controller only failed with a one-motor actuator. When fitted with a two-motor actuator, no problem appeared. But just why this was the case was unclear.

By the first week in December 2003, the cause of the failure had been identified. The CPLD in the power control boards for each fin channel would "latch up" due to high levels of electronic noise. This occurred during high-slew-rate maneuvers, which required the FAS to reach its current limit due to the start/stop/reverse inertial loads on the fins. Several potential changes were proposed that would reduce the electronic noise and resolve the actuator problem. Reviews of these proposals were held in mid-December [3, p. 10].¶¶¶¶¶

Modifications were made to the ECU powerboards at the beginning of January 2004. Transistors were added to CPLD outputs to limit, but not eliminate, electronic noise, keeping the noise below the threshold that triggered the latch-up. The CPLD firmware was also changed to eliminate programming that did not function properly across the full temperature range. Ironically, although the investigation took a significant amount of time and effort, the actual fix was relatively simple.******

The fixes then underwent a series of thermal tests. These involved 200 command-step inputs of 5 deg each, which produced the current limit states

§§§§§Aerospace Projects Highlights, October 31, 2003, pp. 2, 3; November 14, 2003, p. 2.
¶¶¶¶¶Aerospace Projects Highlights, December 8, 2003, p. 1; December 12, 2003, p. 2.
******Information provided by Dryden X-43A project manager Joel Sitz, and project engineers Yohan Lin and Paul Reukauf.

that had triggered latch-ups in earlier testing. The sequence was repeated three times, at cold, ambient, and hot temperatures, for a total of 600 cycles without a failure. The flight powerboards were modified with the fix, and the flight ECU passed functional testing and then was shipped to Dryden for the captive-carry test aboard the B-52B launch aircraft. All of the Langley, Dryden, Marshall, Parker, Dynamic Controls, Inc., XILINIX, and Orbital Sciences experts who had worked on the FAS issue agreed that the system was ready to fly. The FAS issue had delayed the second Hyper-X launch by some 6 months.[††††††]

OTHER ACTIVITIES

Though the FAS malfunction was the critical issue in the Hyper-X project's return to flight, it wasn't the only problem to appear on the first launch. Others were also dealt with as part of the return-to-flight effort. The first of these occurred within the first few seconds of the brief flight and involved the adapter's nitrogen system, which was used to circulate a water and glycol mixture to cool the engine during the ascent. The water/glycol mixture went through cooling passages in the leading edges and was dumped overboard. In videos of tests done in the 8-Ft. HTT, the water and glycol can be seen as spray coming from the engine's leading edges.

The malfunction began when a pyrotechnic valve fired. Debris caused a regulator to fail, which vented the nitrogen gas. Although the nitrogen system failure was unrelated to the loss of the booster, the MIB analyzed the issue. Results indicated that despite the regulator problem, enough nitrogen gas would still have been available for engine cooling. The failure was corrected by adding a filter to catch valve debris.

The other two problems occurred when the X-43A tore away from the adapter as the stack tumbled toward the Pacific. The adapter camera showed the right wing had turned 180 deg and was rotating freely on its shaft. The aluminum control horn that connected the shaft to the actuator had been broken by the forces on the vehicle and it began to spin. Although the design met the normal flight requirements, a review indicated the horn's strength was marginal, and there was more slippage in the design than engineers were comfortable with.

Project managers had issued orders to reduce risk wherever possible. As a result, engineers were directed to fabricate the horn out of steel rather than aluminum and to redesign the attachment mechanism so it had nearly zero slippage. What sounded like a simple change, however, proved very difficult. Both the fabrication and installation of the new horn ran into difficulties and delays, requiring several extra months to be completed.

[††††††]Aerospace Projects Highlights, January 9, 2004, p. 3.

The other X-43A-related issue was a sheet of metal that tore free of the adapter's outer skin during the stack's gyrations. As with the control horn failure, this was a direct result of the HXLV going out of control. Though the panel would not have been lost under normal flight conditions, the adapter structure was reinforced as a hedge against any possibilities of a recurrence [4, pp. 6, 8].######

Return-to-flight preparations lasted nearly 3 years, and involved a wide range of activity. This included a complete review of the Hyper-X flight profile—B-52 launch, boost, X-43A separation, and engine test—to identify risks needing analysis and correction. There were also reviews of the software and hardware; wind-tunnel tests; refinement of the mathematical models of the hardware; and tests using different computer programs to check critical areas and uncover any errors that may have been missed. The new tests involved specific mission segments as well as full "drop-to-splash" simulations, and used both nominal trajectories and new Monte Carlo simulations, using revised, more conservative, models of the different flight parameters.

The review process itself was also revised as part of the return-to-flight effort. The integrated product teams (IPTs) developed action plans listing all activity in their specific disciplines. Each of the actions was assigned to an individual whose work was tracked until the issue was closed. The IPT members also determined what areas needed additional checks, and how these were to be undertaken within the IPT.

The next step in the process was a review by the Engineering Review Board (ERB) of all IPT actions and decisions before these were implemented. The work of the IPTs and others was also reviewed by several independent groups. The first level was peer assessors and expert teams/consultants—experts in specific fields who were not part of the Hyper-X project. These individuals looked at very specific areas in great detail.

Other review bodies and organizations included the Flight Readiness Review, the Integrated Mission Assurance Review, and the Airworthiness and Flight Safety Review Board. As reviews were moved to upper levels, their focus shifted from the finer details to the broader picture. Though the upper-level groups spent less time on minutia, their level of independence was greater. Finally, at each level, the activities and decisions of each group were documented. The goal of the reviews was for "different sets of eyes" to look at every aspect of the projects at multiple, overlapping levels [4, pp. 3, 4].

TWO FUTURES

As 2004 began, activities shifted to preparing for the captive-carry flight. Once this was accomplished, the next step would be to make the second

---

######Corpening history interview, Tape no. 1, pp. 53, 54.

Mach 7 flight. Because the first launch failed within seconds of ignition, many unknowns remained with the second flight. Transonic, supersonic, and hypersonic flight had to be achieved with the HXLV. The booster then had to fly a depressed trajectory followed by a negative angle of attack. Separation had to be successful. The X-43A had to stabilize within the 2.5-s time limit. The scramjet had to ignite, establish stable combustion, and demonstrate positive thrust. Project personnel felt confident leading up to the second flight, but they also understood the risks and what was at stake. They believed that if the second flight were also lost, the third flight would be cancelled and a half-century of work would have been in vain.

But if the second flight were successful, then the design tools would have been validated, the scramjet engine would have been proven in flight, and the third flight would be attempted.

That would mean a half-century of scramjet development had finally resulted in success, and follow-on projects could be undertaken.

REFERENCES

[1] X-43A Mishap Investigation Board, "Report of Findings X-43A Mishap," Volume 1, May 8, 2003, p. 1.
[2] Bermudez, L. M., Gladden, R. D., Jeffries, M. S., McMillian, D. L., Porter, J. R., Allen, V. C., and Engelund, W. C., "Aerodynamic Characterization of the Hyper-X Launch Vehicle," AIAA 2003-7074, pp. 5, 6.
[3] Joyce, P. J., Pomroy, J., and Grindal, L., "The Hyper-X Launch Vehicle: Challenges and Design Considerations for Hypersonic Flight Testing," AIAA 2005-3333, p. 9.
[4] Reubush, D. E., Nguyen, L. T., and Rausch, V. L., "Review of X-43A Return to Flight Activities and Current Status," AIAA 2003-7085, p. 5.

Chapter 12

TOWARD THE UNKNOWN

> It's only the beginning but the implications are terrific.
>
> *Gerald Sayer, pilot of the first flight of the Gloster-Whittle E28/39 jet aircraft, May 15, 1941*

With the start of 2004, final meetings were held to close out remaining requests for action, hazard and software reviews, and trajectory analysis. The captive-carry tech brief was held on January 12, with a few minor issues remaining to be dealt with. The HXLV, adapter, and X-43A were being prepared. Control room training exercises were also under way. The control room simulations were described as "more real" than those of the first flight by some engineers. The captive-carry mission to check out the launch aircraft and stack systems under flight conditions was scheduled for Saturday, January 24, 2004. The B-52B was moved to the Dryden ramp on January 19. The Hyper-X stack was pulled from its hangar and checked out before being attached to the launch aircraft's pylon. System checks were then begun. Delays occurred, and the captive-carry flight slipped to Monday, January 26. The B-52B took off from Edwards at 3:21 p.m., flew out to the test range for the simulated launch, and returned for a landing at 5:28 p.m.* Joel Sitz, Dryden's X-43A project manager, wrote after the flight:

> I want to express my gratitude to ALL the organizations at DFRC (none can be left out) for what was a very successful Captive Carry flight for the X-43A Monday (1/26). It is quite humbling to experience the level of personal investment that people are willing to put forth to get the job done! I am proud to be a part of "Team DFRC."
>
> During Monday's flight, all flight objectives were met. Preliminary results of the flight data were reviewed this week. A more detailed review including resulting actions will be completed over the next couple of weeks. No major technical issues have been identified to date that could delay the upcoming drop-flight. The project is currently working towards a February 21 launch date.†

---

*Aerospace Projects Highlights, January 16, 2004, pp. 2, 3; January 23, 2004, pp. 2, 3; "Dress Rehearsal Flight of X-43A; Hyper-X Free Flight Soon, January 27, 2004," Release 04-03.
†Aerospace Projects Highlights, January 29, 2004, pp. 2, 3.

Following the captive-carry flight, final preparations began for the second launch. The planned HXLV trajectory was revised by Orbital Sciences to reflect February launch date weather conditions, the final aero/FAS models, and a slight reduction in the B-52B's launch speed. The HXLV ECU, rudder actuator, and rudder fin had been delivered to Dryden, and Orbital Sciences technicians were on hand for the installation. The X-43A's carbon-carbon leading edges had also been delivered and plans had been made for their installation. A control room emergency-procedures training exercise was also held.[‡]

Last-minute changes were still being made, however, that pushed the launch date to late March. One of these involved a software update to the HXLV's autopilot gains. The change was made to reflect the newly predicted in-flight fin deformations, a contributing factor in the loss of the first Hyper-X launch.

Another issue was equally serious. The data-acquisition system for a fin actuator compliance test was configured improperly, causing an error signal in the actuator control loop. The actuator motors produced torque on the gears beyond their design limit, rendering them no longer flightworthy. An investigation of the incident followed, and a report was prepared. At the same time, acceptance testing of the replacement actuator was under way at Parker's facility in Dublin, Georgia.

Once acceptance testing of the replacement actuator was completed at Parker, it was shipped to Orbital Sciences' Chandler, Arizona, facility for vibration, thermal, and altitude testing. With successful completion of tests in Chandler, the actuator was returned to Parker for final postacceptance tests and loaded functional checks. These were completed and the actuators shipped back to Orbital Sciences for final assembly, which was scheduled to begin on March 5, 2004.

Being conducted simultaneously with the actuator work was an HXLV autopilot trajectory 5.0 software update incorporating the new HXLV fin-compliance model. The autopilot gains had also been adjusted, and nonlinear analysis indicated large sideslips were reduced. Linear analysis of the gain settings was under way as February drew to a close. All that remained was a formal decision on implementing the changes for flight 2. The Flight Readiness Review (FRR) briefing on open trajectory 5.0 items was scheduled for March 10, 2004.

A separate element of the return-to-flight effort was testing to provide data that would be used to improve Hyper-X computer software. The twin goals of this risk-reduction activity were improving the vehicle's ability to respond to disturbances in flight, and better capability for predicting flight results. Among those conducted were an ejector piston test, cowl door calibration

---

[‡]Aerospace Projects Highlights, January 29, 2004, pp. 2, 3.

and free-play testing, compliance and hysteresis testing of control linkages, and system timing tests.

The scramjet's propulsion system controller (PSC) software, which managed fuel flow, prevented unstarts and flameouts, and commanded the cowl opening and closing, also underwent testing. This focused on the unstart protection algorithms, a program called Durascram. To prevent an unstart, the program continually adjusted fuel flow to compensate for changing flight conditions. It had been developed and tested as part of the original Hyper-X scramjet wind-tunnel testing. Return-to-flight software changes improved the program's capabilities. Another change made as a result of additional wind-tunnel tests was one to the X-43A's angle of attack. The data showed that engine performance increased and the risk of a flameout decreased while at greater angles of attack than those at which engineers initially expected the increase/decrease to occur. And so, in order to improve the chances of successful scramjet operation, the nominal angle of attack was increased for flight 2.

A final preparation was the use of independent simulations as cross-checks on the boost and separation maneuvers. Despite all the efforts made before the first flight to assess the boost and separation risks, these remained a concern. The new simulations incorporated the improved models, resulting in identification of areas of improvement for both the simulations and models [1].

The return to flight work was now finished. As Griff Corpening later noted, "it was once again very much time to go fly."[§]

FLIGHT 2

The launch was scheduled for Saturday, March 27, 2004. The night before the flight was very cold and windy. The ground crew had to preflight and fuel the Hyper-X in 25 to 30 kt winds. Tumbleweeds were blowing in across the desert. Ed Teets had forecasted the winds would die down by takeoff time. Corpening was sure the forecast was wrong. Early in the morning, Corpening went out to the flightline. The winds were strong, and they "were one beat up crew, but still smiling." The ground crew had done their job under extraordinary conditions. By the next morning the winds had died down, and the flight was on.

The B-52B launch aircraft took off from the Edwards runway at 12:40 p.m. and was quickly joined by a pair of NASA F-18 two-seat chase planes already in the air. The formation headed west, toward the Pacific Ocean and the Naval Air Warfare Weapons Sea Range off Point Mugu, California. As before, the launch was a complicated choreography of aircraft, involving NASA's B-52

[§]Griff Corpening history interview, December 20, 2004, NASA Dryden Flight Research Center History Office, p. 57.

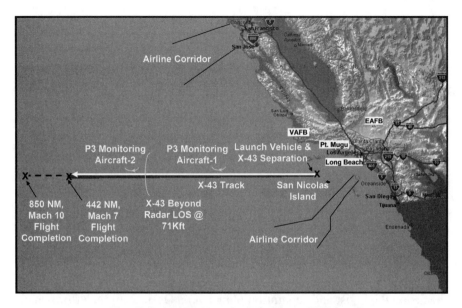

Fig. 12.1   The flight path of the Hyper-X flights.

and the two F-18 chase planes, the Navy's test range, a pair of P-3 Orion telemetry aircraft, and commercial airliners flying across the Pacific to Los Angeles. The Federal Aviation Administration had authorized a specific launch window because the stack and the X-43A would be launched and fly directly toward inbound transpacific flights. If the launch could not be made within the time period specified, it would not be made at all because the FAA could not keep the corridor open indefinitely. Springtime weather patterns were also an issue, because these affected all aspects of scheduling.

The map in Fig. 12.1 shows the complexity involved with Hyper-X flights. The launch box is near San Nicolas Island, off the California coast and north of the airline corridor into Los Angeles International airport. The X-43A flight path is so long that it extends over the horizon from the Point Mugu tracking station. A flight path of that length required the use of two P-3 tracking aircraft to maintain coverage of the test vehicle in flight. This task became even more difficult with the Mach 10 flight, which entailed a flight path more than 400 n miles longer than the Mach 7 flights.

Although the takeoff was uneventful, a major problem appeared during the flight out to the launch point. The 37,537-lb Hyper-X stack was mounted on a pylon under the B-52's right wing, between the fuselage and the inboard engine pylon. Fitting the pylon to the B-52 required that the right inboard main fuel tank be removed. As a result, the launch aircraft's left wing had two main fuel tanks but the right wing had only one.

The weight of the stack and the nonsymmetrical fuel load required that during takeoff most of the B-52's fuel would be in the two left-wing tanks.

Once the aircraft was heading toward the launch point, fuel amounting to about half the stack's weight was transferred to the single right-wing tank. Without this transfer, the B-52 would be significantly out of balance when the stack was dropped. To calculate both the fore and aft and the lateral centers of gravity, fuel quantities for each tank were entered into an Excel spreadsheet program on a laptop that the flight crew used to guide the fuel transfer.

Gordon Fullerton, serving as co-pilot in the B-52B on this flight, noticed that the fuel quantity in the right outboard tank was dropping, rather than increasing, indicating that they were burning fuel faster than they were transferring it. But the problem was worse than that: Fullerton soon realized that one of the transfer valves had not opened, and no fuel at all was moving from the left to the right wing. Soon the yellow low-level fuel-warning light came on, indicating that only 4000 lb of JP-4 fuel remained in the tank. Unless the problem was solved, the aircraft would have to turn back and land. A call went out to Gary Beard, a B-52 technician at Dryden, who suggested a different method of transferring the fuel. This was radioed up to the aircraft, where Fullerton had already begun the same procedure. To his relief, he soon saw the yellow light go out, and the flight proceeded.[¶]

Nine minutes before the planned launch time, the countdown entered its terminal phase. The Gold Room door was locked. The B-52 and the F-18 chase planes were flying at 40,000 ft and Mach 0.8. The Point Mugu tracking stations, the two Navy P-3 tracking planes, and a modified U.S. Army Grumman G-III business jet fitted with both visible-light and infrared cameras were ready. CNN was carrying the flight live, showing both the chase plane video and shots from the B-52's onboard cameras. These sent back images of the X-43A's fins and wings moving back and forth in preflight tests while controllers continued running through checklist items as the minutes slipped away. Figures 12.2 and 12.3 show the Dryden personnel and the Control Room team watching live transmissions from the B-52 and F-18s as the second Hyper-X launch nears. In Fig. 12.3, from right, front row, are Vince Rausch (NASA Hyper-X program manager), Dave Reubush (NASA Hyper-X deputy program manager), Craig Christy (Boeing), and Randy Voland (NASA Langley propulsion). In the back row, from right, are Dave Bose (AMA), John Martin (NASA Langley guidance and control), Pat Stoliker (acting NASA Director of Research and Engineering), and Bill Talley (DCI).

At the 2-min mark, the FAS batteries were turned on. To this point, had a problem occurred, the launch could have been delayed to correct it. Once the batteries were activated, however, their short operating life meant that the launch would have to be made on time, or it would be canceled. The three

---

[¶]Gordon Fullerton history interview, August 3, 2005, NASA Dryden Flight Research Center History Office, pp. 18, 19. Fullerton accrued more than a thousand hours in bombers in his stint as a test pilot at Wright Patterson Air Force Base, before he moved to the astronaut corps, and this experience paid off now.

Fig. 12.2   Dryden personnel watch video from the B-52 and F-18 chase planes as the flight 2 launch grows near.

HXLV fins were swept back and forth to check FAS operation. Because airflow was great enough at Mach 0.8, the fins' movements caused the stack to shake on the B-52B's pylon, but this was expected. The countdown now reached the 1-min mark. For some, time seemed to slow as the countdown neared zero; others reported that they felt as though they were being swept

Fig. 12.3   In the control room, senior NASA and contractor personnel also await the launch.

along in an unstoppable current. In the Gold Room, the wait was over, and mission controller Brad Neal called, "Launch, launch, launch."**

A brief, hard shock went through the B-52 as the stack was released (see Fig. 12.4). The time was 2:00 p.m. The stack fell clear of the pylon, rolling about 30 deg right wing low in the first second after the drop. This was a larger angle than expected, but once the control system was enabled, the vehicle stabilized. Before the HXLV motor ignited, the stack's free-fall seemed to drag on; but it accelerated quickly and reached supersonic speed without difficulty within about 5 s. The only indication it had become supersonic was wing rocking similar to that occurring on earlier Pegasus flights. There was no hint of the instability seen on the first launch, indicating the gain selection and the revised aerodynamic and control system models had been correct. The first barrier had been passed.††

Between L + 8 and L + 13 s, the steep pull-up maneuver began, with the stack pulling 1.7 Gs. From the B-52 crew's viewpoint, it seemed as if the HXLV was flying nearly straight up. At L + 40 the stack experienced maximum dynamic pressure of 1650 psf. The stack was by then at an altitude of about 47,000 ft and a speed of Mach 3.5. This was followed, at L + 48, by activation of the adapter's fluid control system to cool the scramjet engine

Fig. 12.4 The second Hyper-X launched from the B-52B on March 27, 2004.

**Personal observations of the CNN launch coverage.

††Dave McAllister history interview, March 2, 2005, NASA Dryden Flight Research Center History Office, p. 14. McAllister was the X-43A panel operator aboard the B-52B for the second flight.

and purge the X-43A's interior of oxygen. (On the first flight, the adapter nitrogen regulator had failed seconds after launch [1, p. 11, and 2].)‡‡

The next maneuver by the stack was the "push over," when the flight path began to level out. During the maneuver, the stack pulled –0.3 Gs. It now flew at a negative angle of attack for about 10 s, putting the heaviest loads of the boost phase on the X-43A's wings and requiring the actuator brake to absorb the wing hinge moments. Throughout the boost phase, the roll rates of the stack had remained within minimum/maximum Monte Carlo limits. HXLV burnout occurred at about L + 84, with the stack at about 89,000 ft and a speed of just over Mach 7. The stack then coasted for about 10 s before beginning the separation sequence.

The "ready to separate" signal was sent from the HXLV to the X-43A. The planned separation altitude was 93,932 ft, but the actual value was 94,069 ft. Planned separation velocity was Mach 7.075. The actual value was slightly slower, at Mach 6.946. As a result of the higher altitude, and, more important, the lower speed, the dynamic pressure was slightly lower—1037 psf—than the planned 1066 psf. The flight path elevation was 2.4 deg versus the planned 2.0-deg elevation. The booster's planned angle of attack and sideslip angles were 0 deg. The stack's sideslip angle was 0 deg, but angle of attack was 0.09 deg. All of these were well within the minimum error.

At L + 93.466, the X-43A's flight management unit sent the command to begin the separation sequence. The HXLV systems fired the explosive bolts and ejector pistons. The behavior of the pistons as they extended was close to that in nominal Monte Carlo high-fidelity separation simulations, and the two adapter cameras transmitted grainy black-and-white images as the X-43A separated (Fig. 12.5). As CFD and other tests had predicted, no "kick" occurred as the gap between the X-43A's aft fuselage and the adapter opened. But other predictions were not fulfilled.

The X-43A had 2.5 s after separation to stabilize itself at a 2.5-deg angle of attack before the cowl opened. If the research vehicle did not stabilize itself, the mission would fail. Monte Carlo data had indicated that the X-43A could nose down, then pitch up and still stabilize itself at the correct angle of attack. Some of the computer runs had shown extreme nose-down angles of attack immediately after separation, and smaller but still significant nose-up profiles. Despite wild extremes shown in a handful of runs, the nominal prediction was for a pitch down, followed by a slower increase in the angle of attack, with the correct value achieved in 1 s or less.

The X-43A actually followed the nominal prediction for the first quarter of a second, but as the predicted nominal track rose, the research vehicle continued to pitch down. As the half-second mark approached, the angle of attack

---

‡‡Gordon Fullerton history interview, August 3, 2005, NASA Dryden Flight Research Center History Office, p. 22.

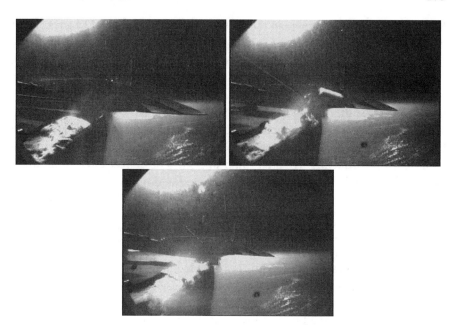

Fig. 12.5   Views from an adapter camera as the second X-43A separates from the HXLV booster.

increased to a positive alpha. Then this slowed, and the vehicle's angle of attack was outside the worst-case Monte Carlo projections. Not until 2 s after separation was the angle of attack within Monte Carlo projections. When the 2.5 s were up, the nominal prediction was met, and the X-43A was at the desired test conditions [1, p. 14, and 3].

The cowl door opened 2.5 s after separation, and airflow through the engine began. This was referred to as the "pre-experiment tare," and lasted about 5 s. The test did show one surprising result: preflight Monte Carlo data had predicted that a negative angle of attack transition would occur as the cowl opened, but the data showed that the X-43A had a positive alpha as the cowl opened, and this was stabilized by the X-43A control system within about a second and a half.

With the tare test completed, at L + 100 the fuel and silane were injected into the engine and ignited. As they were injected, the vehicle's angle of attack became negative, and then reversed to a positive value. These, too, were within predicted Monte Carlo simulations. With the engine ignition, onboard instrumentation showed an immediate acceleration of the vehicle as thrust increased. Within about a second, the thrust had increased so the scramjet was producing more thrust than the X-43A's drag. This acceleration continued for over half of the 10-s burn. The amount of thrust declined during the second half of the burn, but positive acceleration was maintained until the hydrogen fuel was shut off. Peak speed reached during the engine burn was

Mach 6.83, slightly faster than the X-15A-2's record speed of Mach 6.7, set in 1967.

With the engine burn completed, the postexperiment tare was performed. As with the earlier test, this involved measuring the pressure inside the engine without combustion so it could be compared with ground-test data. This lasted 4 s. Both the tare measurements had been performed in stable flight. The final part of the scramjet engine test was a 17-s parameter identification (PID) maneuver made with the cowl open. During the PID, the X-43A showed variations in inertial angle of attack and sideslip. These were close to preflight Monte Carlo predictions, and the vehicle remained under control. With this, the cowl closed, and the scramjet engine test was over, 39 s after separation [1, pp. 14, 15, and 4].

The X-43A began the secondary phase of the research flight. During the engine test, dynamic pressure and heat on the vehicle rose; to arrest this, the X-43A's control system increased the vehicle's angle of attack. During this "recovery" maneuver, the research vehicle underwent a series of low-amplitude angle-of-attack oscillations at a frequency of about 0.65 Hz, and the wing-spindle torque readings and actuator currents showed large spikes during these. After executing the recovery maneuver, the X-43A altered its flight path to "descent mode." During this part of the flight profile the X-43A performed four sets of research maneuvers. Each set consisted of a PID, frequency sweeps of the wings and rudders, and a push-over, pull-up (POPU) maneuver. These control inputs were done quickly, because the vehicle had a very limited ability to maintain a specific flight condition, necessary for accurate data, during descent. By making the separate inputs at different frequencies on the different control loops, it was possible to simultaneously determine the frequency response for both axes. During descent, the X-43A was under control, and was able to perform all preprogrammed test maneuvers. (As with the ascent and scramjet burn, research maneuvers were controlled by onboard systems.) This was in spite of the research vehicle's rapid deceleration. Because of the deceleration, immediately after the second set of test maneuvers was complete, the third set began.

The X-43A followed the reference trajectory throughout the research maneuvers. However, a problem occurred during the descent phase, at approximately 350 s after separation. The control system was unable to follow the reference trajectory, and, as a result, the dynamic pressure was at the lower limit at the end of the descent. Additionally, the vehicle's angle of attack reached its upper limit near the end of the flight. Another anomalous event late in the descent was a 30-deg, left-wing-down bank by the X-43A. The vehicle's control system corrected the upset without incident.

At about 492 s after separation, the final mission phase began. This was the "approach mode," just prior to ocean impact. The vehicle's control system

Fig. 12.6   The B-52B and F-18 chase planes made the traditional flyover of the Dryden main building.

Fig. 12.7   Soon after the completed flight, Vincent L. Rausch, Griffin Corpening, and Joel Sitz (left to right) talk with the press.

reduced the descent rate from 715 to 440 fps. The final data retrieved from the X-43A showed it at an altitude of 41.86 ft, with a velocity of Mach 0.92. The research vehicle had a 6.03-deg angle of attack and a 1.29-deg bank angle. The flight path angle was –25.67 deg and the descent rate was 440.53 fps. The impact point was about 400 n miles downrange from the launch point. This was short of the aim point, due to the deviation from the reference trajectory. Despite the shortfall, the impact location was within Monte Carlo prediction. No debris was recovered [1, pp. 11–19, and 4, pp. 9–19].

POSTMISSION ANALYSIS

With completion of the X-43A's successful mission, the most important result was immediately apparent. The scramjet engine had successfully operated in actual flight, and demonstrated for the first time that it could produce sufficient thrust to overcome the vehicle's drag and accelerate the X-43A (see Fig. 12.8). It was no longer an unproven laboratory curiosity. When the scramjet performance data for the pre-experiment tare, the engine burn, and the postexperiment tare were compared to the wind-tunnel data, preflight predictions, nominal predictions, and Monte Carlo data, they showed, in the words of a later report, "excellent agreement between the flight data and the predictions." The report continued: "These results substantiate hypersonic engine tools and flight scaling methodologies, satisfying the primary objective of the Hyper-X program" [1, p. 14].

The validity of these statements can be seen in graphs plotting actual flight data versus the SRGULL analysis and HXFE test data, and the flight data

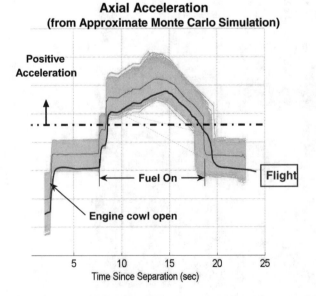

Fig. 12.8   A chart showing the scramjet burn data from the second flight.

versus predictions of the centerline wall pressures. Similar results were shown in graphs of the body and cowl side-pressure measurements of the flight data versus pressure measurements from the HXEM engine in the Arc-Heated Scramjet Test Facility (AHSTF) wind tunnel, the flight data versus HXEM in the 8-Ft. High Temperature Tunnel, and the flight data versus the HYPULSE Scramjet Model. In each case, there was little scatter between the flight pressure data and the ground data. The comparison between the flight data and the AHSTF tests indicated that the flight combustor pressure was generally higher than that of the ground tests, suggesting slightly better flight performance [1, p. 16, and 3, pp. 4, 6–8].

As it had been since the beginning of hypersonic flight, the thermal environment of the X-43A was of major concern. Thermal data, including airframe and wing-structure measurements, behavior of thermal protection system materials, and interior temperatures, showed that the overall environment had, by and large, been successfully predicted. The only unexpected results came from two thermocouples on the vertical tail. The thermocouple nearest the leading edge recorded temperatures at the high end of the predicted uncertainty. A second thermocouple in the same area also showed higher-than-expected temperatures, confirming the reading. These did not, however, exceed the uncertainty value [1, p. 14].

Another major issue was the X-43A's guidance, navigation, and control system's ability to maintain the required trajectory. Problems appeared at several points in the flight. The first, and potentially most serious, was at separation. The X-43A was very slow to reach the required engine test angle of attack. Not until the final half-second or so of the allotted time was the angle of attack within Monte Carlo data predictions, and the angle of attack did not match the nominal prediction until the end of the 2.5-s period. The research vehicle did achieve the proper angle, however, and the issue had no detrimental effect on the scramjet test. Had the delay been any longer, the flight could have failed.

Postflight analysis indicated the cause was the uncertainty model in the separation aerodynamic database, which lacked the proper value for the trim pitching-moment uncertainty. As a result, preflight simulations had not predicted the delay in stabilizing the vehicle.

A second factor was the X-43A's control system design. The angle-of-attack schedule resulted in the vehicle trimming at a lower angle of attack. Additionally, the angle of attack controller design lacked a proportional error feedback at low angles of attack. As a result, the error in the trim bias was eliminated by the integrator [1, p. 14, and 2, p. 9].

A second surprise came during the cowl opening for the pre-experiment tare. Both the Monte Carlo data and the nominal simulation had predicted that the X-43A would have a nose-down transient when the cowl door opened. In fact, the vehicle's cowl-open transient was nose up, and was completely unanticipated.

None of the Monte Carlo predictions showed anything resembling a nose-up by the vehicle in this configuration. This unexpected behavior was apparently the result of inaccurate modeling of cowl-door-opening dynamics.

The control system was able to maintain test conditions within mission requirements during the engine burn. Several minor issues arose during the scramjet test, however. Between 11 and 13 s after separation, the X-43A's angle of attack fell outside Monte Carlo predictions. At the end of the engine burn, the fuel flow was shut off, causing a transient. This was close to the lower limits of Monte Carlo prediction. Both these events were probably due to the differences between predicted engine performance, based on wind-tunnel testing, and actual scramjet flight performance. The other issue was a steady-state offset in the vehicle's sideslip during scramjet operation, which was insignificant because the angle was still within test requirements.

Furthermore, the research vehicle had been unable to maintain the reference trajectory. This was because the trajectory was developed at the start of the Hyper-X project, and had not been revised using the newer vehicle flight models. The altitude of the vehicle fell below the reference trajectory as a result of limits on the onboard-estimated dynamic pressure and angle of attack. These problems aside, the actual flight trajectory was close to the nominal path, and well within Monte Carlo predictions.

Control issues also arose during the descent. Because the vehicle had not maintained the reference trajectory, the X-43A's angle of attack reached its upper limit near the end of the flight. This, in turn, caused the angle-of-attack response during the fourth POPU maneuver to be "chipped," and prevented the controller from tracking the normal acceleration command. This was similar to the performance predicted in the nominal simulation, however, and no detrimental effects resulted.

The vehicle bank angles remained within Monte Carlo predictions until late in the flight, when the X-43A banked 30 deg left-wing down. Though the control system corrected the upset, the root cause was not fully understood. Possibilities identified by engineers included winds aloft or aerodynamic effects not included in preflight models. Perhaps significantly, prediction had indicated the vehicle's stability would be reduced at higher angles of attack when flying at transonic speeds [4, p. 15].

Data analysis from the second flight took about a month. With this completed, activities shifted to preparing for the third flight. Flight 2 had shown a scramjet was feasible, and that it could produce more thrust than the vehicle's drag. Now flight 3 would push the envelope.

TOWARD MACH 10

Due to the loss of the first vehicle, and the nearly 3-year delay that followed, the Hyper-X project was behind schedule and over budget. Project managers wanted the third flight completed by the end of September 2004. (The start of the new fiscal year 2005 budget was October 1, 2004). The difficulty of this task was apparent, inasmuch as flight 2 preparations took some 2 years. Moreover, the team could not shift its full attention to the flight 3 vehicle until after flight 2 had been completed.

A different approach and philosophy would be required if the Hyper-X team were to meet so aggressive a schedule. Flight 2 had seen extensive cross-checking to ensure nothing had been missed. With flight 3, little in the way of duplicate or outside analysis would be performed. This represented a major change from the original plan. Based on engineers' review of flight 2 data, no changes were made to the flight 3 models to incorporate new data. Instead, the team's efforts focused on the boost, separation, and engine operation through cowl closing. Any work on the descent would be done on a time-available basis. Because the engine test data were critical, and the hypersonic aerodynamic data were a secondary priority, this approach made sense [5].

As events unfolded, the situation proved less demanding than originally thought. The flight 3 vehicle had been used for some of the flight 2 tests, so it was in an advanced state of readiness by the spring of 2004. Test personnel now had extensive experience with the hardware, so they were able to complete the checkout of the third X-43A faster than they had with the second vehicle. As a result, testing that had originally been cancelled, including the aircraft-in-loop and compliance testing, was completed without compromising the schedule. Additionally, although flight 2 data were not incorporated into the software and flight models, they were used to prepare the updates and stress cases [5, p. 22].

The major change in flight 3 was in separation speed and engine test conditions. Originally, projected separation speed was to be Mach 10.07 at a dynamic pressure of 1090 psf. In the course of the project the stack weight and drag had both been increased. To reduce the total stack weight, all ballast was removed from the HXLV, and the BAM and aft skirt material were made from aluminum, rather than steel as on the Mach 7 flights. This weight reduction had other consequences, however. It shifted the stack's center of gravity a significant distance aft. This reduced the stack's static stability, reducing the autopilot margins and requiring a new set of gains. The stack-stability issues also required modifications to the angle of attack during transonic flight. By late May 2004, all transonic angle of attack conditions had not been accounted for in the Mach 10 trajectory. Efforts were under way to determine the best approach to the problem.

As the Mach 10 flight trajectory was being finalized, Hyper-X engineers held several meetings in early June to discuss the need for additional wind-tunnel tests of the 6-percent stack model. These would validate several transonic data points that had been extrapolated for the flight model being used to develop the Mach 10 trajectory.

The basic problem was that the existing wind-tunnel database had been assembled using a 15-deg angle of attack at Mach 0.8. Because of the HXLV stability issues, however, the Mach 10 trajectory required an angle of attack closer to 11-deg while at transonic speed. The wind-tunnel tests would be used to confirm the design approach [5, pp. 7, 12].[§§]

The group decided in favor of the tests despite the potential for delays, and by early July, the HXLV model was in the wind tunnel at Langley. The weekly report noted: "All data collected to date match the models being used in the current trajectory development. Trajectory development is on schedule."[¶¶]

As a result of the weight and drag increases, as well as the stability issues, the separation speed was reduced to Mach 9.6. This was effectively the speed at which the scramjet test would be made, because it wasn't possible to significantly increase the speed or dynamic pressure of the X-43A in the 2.5 s between separation and the start of the engine test. Trade studies and risk assessments would now be necessary to determine the extent of any potential reduction in scramjet performance due to lower separation velocity.

The trade studies dealt with modifications to the HXLV trajectory Mach number, dynamic pressure, and flight-path angle parameters. Risk assessment was more complex for the third flight, and looked at the shock location, peak heating rate, and the research vehicle's axial acceleration as a function of Mach number, angle of attack, and dynamic pressure. The examination led to reduction of the separation speed to Mach 9.6. Dynamic pressure was reduced to 1000 psf, the flight path angle would be 1.5-deg and the angle of attack and sideslip angle remained at 0.0 deg.

Independently of these changes, modifications had to be made to the X-43A for the more demanding flight 3 profile. The additional heating meant that rather than the Haynes metal skin and internal ribs of flights 1 and 2, the vertical fins would have to be made of solid Haynes metal, and carbon-carbon leading edges would have to be added to the fins. In addition, the carbon-carbon leading edge on the nose would have a blunter radius than that of its predecessors, after wind-tunnel tests revealed erosion of the material under the more severe flight conditions. The change in the radius reduced the amount of potential erosion. For added heat protection, all carbon-carbon surfaces were treated with a hafnium carbide coating in addition to the existing silicon carbide coating used on the flight 2 vehicle. The flight 3 scramjet also required additional heat protection, consisting of a thicker layer of zirconium applied to its underside. Flight 3 instrumentation included new leading-edge thermocouples and two high-temperature strain gages fitted to the left vertical tail. The scramjet was equipped with two skin friction sensors and two heat flux gages. The skin friction sensors measured the engine

[§§]Aerospace Projects Highlights, May 21, 2004, p. 4, and June 4, 2004, pp. 5, 6.
[¶¶]Aerospace Projects Highlights, July 2, 2004, p. 3.

wall sheer stress; the heat flux gages provided data on the engine wall heat-transfer rate.

The instrumentation changes also addressed results of the X-43A weight and balance test, which indicated that the vehicle's center of gravity was 0.51 in. aft of the allowable limit. Though the measurement was small, the amount of tungsten ballast required to correct it was significant: to bring the research vehicle within the center of gravity limits, 58 lb had to be added, yet no extra space was available to accommodate this. As a result, some of the instrumentation was removed to make room for the ballast. The flush air data sensing (FADS) system used both absolute and differential pressure sensors to determine the vehicle's angle of attack and sideslip. Flight 2 data had indicated that the more sensitive differential pressure sensors were more valuable in determining sideslip angle; the absolute sensor measurements had less value regarding sideslip measurements, so their removal would have no effect on data collected.

Though identical control laws had been used on both Mach 7 flights, significant changes to this software were required for the flight 3 vehicle. These included updates to the guidance control software, control surface calibrations, test angle of attack, fueling schedule, and igniter subsystem controller. Of particular importance was correction of the delay in reaching the proper angle of attack on flight 2. The problem was traced to a miscalculation of the pitching-moment bias. To prevent a recurrence, the flight 3 system was modified to increase the angle-of-attack integrator gain.

Flight 3 represented more of an unknown than had flight 2 because of the higher separation speed. Yet because extensive wind-tunnel testing for the Mach 7 flights had been costly and time consuming, project managers decided against a similar effort for flight 3. Instead, existing wind-tunnel data and CFD work was used to assess Mach 7-to-Mach 10 scaling issues, and the relative motion between the adapter and the X-43A. The CFD results were also cross-checked with flight 2 data. The analysis showed good agreement between CFD predictions of forces and moments and the actual data, and provided confidence in flight 3 separation aerodynamics.

A bigger unknown lay in the lack of Mach 10 scramjet ground-test data. For Mach 7 ground tests, the scramjet could undergo a full duration test—cowl opening, pre-experiment tare, complete burn, postexperiment tare, and cowl closing. These tests used the actual flight engine designs and system software, raising confidence in the potential for the flight's success.

For flight 3, however, full-duration testing was not possible. The high energy required for a sustained Mach 10 airflow was simply too great for any wind tunnel. All that could be done was to use shock tubes to momentarily produce flight conditions. The NASA HYPULSE facility at ATK-GASL and the U.S. Army's LENS shock tunnel at CalSpan in Buffalo, New York, were the sites for these tests.

Though the flight-3 scramjet burn would last about 11 s, the HYPULSE shock tunnel test run lasted 2 to 3 ms. The LENS tunnel had a much longer test time, at 12 ms. Such short runs had several consequences. Each provided only a single data point. Producing a graph of engine performance required numerous runs. Because the tests were only single-point, they could not be used to examine dynamic engine operations such as fueling schedules or cowl position transients. And because of the limitations posed by shock-tube test durations, uncertainties in the propulsion database were increased to allow for the lack of ground-test data.

There was another consequence, however, stunning in its implications, regarding the amount of test data. Flight 3 would return the largest amount of data on Mach 10 scramjet operation ever collected. The data would be of higher quality, quantity, and type than that collected from all previous wind-tunnel tests—if the flight was successful [5, pp. 7–10, 22].

While these activities were under way, ground tests of the X-43A's systems were conducted. As in preparations for earlier flights, hardware problems were discovered during the tests. A leak and function test of the X-43A's fluid system ended in failure when a leak was discovered in a motorized control valve in the hydrogen system, which adjusted the hydrogen flow rate to the scramjet. Posttest inspection revealed the cause of the failure to be copper contamination, which was traced to corrosion in the heat exchanger that regulated the hydrogen gas temperature for proper engine operation. Rather than a tank heater, the heat exchanger was used to allow the maximum amount of fuel to reach the scramjet during the burn.

The heat exchanger was replaced with a newly manufactured unit that was installed in the fuel system. To prevent a reoccurrence of the failure, the heat exchanger was kept inert with helium gas at pressures higher than those of outside air. This prevented corrosion, oxidation, or contamination. The heat exchanger was delivered in early June 2004, and subsequently passed the leak check [5, p. 10].***

RETURN OF THE FAS PROBLEM

One of the reasons Hyper-X engineers and managers believed the mid-September 2004 launch date could be met lay in the solution that had been developed for the fin actuation system (FAS) problem on flight 2. The FAS for flight 3 received the same set of transistors, to limit the amount of electronic noise that had caused the latch-ups during flight 2 qualification testing. The fix was seen as reliable, even if the reasons why it worked were not fully understood.†††

***Aerospace Projects Highlights, May 7, 2004, p. 4; May 21, 2004, p. 5; June 4, 2004, p. 3.

†††Paul Reukauf History Interview, February 17, 2005, Dryden Flight Research Center History Office, pp. 57, 58.

The HXLV powerboards for the FAS electronic control unit (ECU) were integrated and ECU burn-in tests began on May 17, 2004, at Parker's Dublin, Georgia, facility. Integration of the actuators with the ECU was scheduled for May 21. Dryden quality assurance personnel were at the facility for the powerboard inspection and to monitor testing. In early June, the actuator and ECU successfully passed pre-environmental acceptance testing. The hardware was then sent to Orbital Sciences' Chandler, Arizona, facility for environmental testing. This began on June 4, 2004, and the two powerboards promptly failed the testing. The FAS problem had reappeared.[‡‡‡]

A three-step plan was developed to address the failure. The failed ECU powerboards were replaced with spare boards. A heater plate was added to the ECU to maintain a higher temperature during flight. The heater plate was designed almost entirely with in-house parts, and was similar to the one used to heat avionics batteries in the adapter. The heater system was to be completed in mid-July. The final step was to rearrange stack integration testing of the BAM and the aft skirt. The latter would not be delivered until early August. Maintaining a mid-September launch date was now problematic. There was still a chance of meeting the flight date *if* the spare powerboards arrived on time *and* the changes in the stack-integration testing were acceptable.[§§§]

Initially, efforts to adhere to the schedule were successful. The two new powerboards were delivered to Parker in mid-June 2004 and underwent modification and testing. The use of a heater plate on the ECU was independently reviewed and approved by Orbital Sciences and NASA management. The stack integration was also rearranged so there was no expected delay in the launch date.

While the FAS issue was being worked out, progress continued with the vehicle itself. The electrical connections between the X-43A and the adapter were nearly completed. Orbital Sciences personnel were at Dryden to complete booster integration and checkout of all components except the aft skirt. Ground service equipment and BAM avionics were also checked out.[¶¶¶]

But for every advance in the launch preparations, there was a setback. This seemed to define the third X-43A's "personality." Yohan Lin, a Dryden engineer who worked on the Hyper-X system checkout, noted that each X-43A was very different. The first one was a learning experience for the checkout crew, requiring the swapping out of many parts. The second X-43A's checkout went very smoothly. The third was the most difficult to test. "It just didn't want to fly," Lin noted. Problems large and small kept occurring, some just appearing temporarily and then vanishing.[****]

[‡‡‡]Aerospace Projects Highlights, June 4, 2004, p. 3.
[§§§]Aerospace Projects Highlights, June 4, 2004, pp. 3, 4.
[¶¶¶]Aerospace Projects Highlights, June 4, 2004, p. 4.
[****]Yohan Lin history interview, August 12, 2005, NASA Dryden Flight Research Center History Office, pp. 27–29.

Two examples of this occurred between mid-June and mid-August. One began with a heating analysis of the X-43A's left wing root. Data indicated that the wing root would have to be trimmed to provide a sufficient gap to allow for thermal expansion at near-Mach 10 flight speeds. This posed a problem, because it would be necessary to move the wing beyond normal operating limits to provide enough room to trim excess material. Trimming was approved and, initially, modifications went smoothly. Once trimming was completed, the left wing was rotated back to its normal position. But in the process, the left wing and left rudder made contact.

An inspection found small dents in the surfaces of both the left wing and left rudder at the points of contact. Worse, the inspection also revealed that the rudder spindle had been damaged, leaving the rudder offset from its normal position. The damage assessment team inspected the actuators and controller, and loads on them were calculated. Analysis showed the components had not been stressed beyond their existing qualification limits; though the rudder spindle was damaged, a replacement would not be needed. Sufficient margin remained to fly the mission with a high degree of confidence in success. The left rudder and left wing were recalibrated, and the flight software was updated with the new calibration, removing the offset in the rudder position. The corrective measures received approval in mid-August [5, p. 10].††††

While this was under way, Boeing and MER technicians were at Dryden on June 30 for installation of the carbon-carbon leading edges on the nose of the X-43A. During the fit check, the team noticed that the right forward leading edge showed some delamination. The piece could not be flown, nor could it be repaired. A replacement was needed, and the extensive heat treatment and machining required to fabricate an entirely new part meant that the September launch date would not be met.

Fortunately, a spare carbon-carbon billet was available that had partially been through the heat-treating process, allowing a replacement leading edge to be fabricated without creating a schedule delay. All the carbon-carbon leading-edge pieces, including the replacement edge, underwent tap tests and thermographic inspection to ensure that none had similar delamination problems. The work was completed by mid-August, and successful fit checks of the leading edges were made [5, p. 10].‡‡‡‡

But while these issues came and went, the FAS problem remained. On June 30, the same day the carbon-carbon delamination was discovered, the FAS once again failed initial tests at Parker's Georgia facility. Two of the channels successfully passed the testing, but the third failed to follow the commands. More significant were the circumstances of the failure: the problems with the flight 2 FAS occurred at low temperatures and the newly

---

††††Aerospace Projects Highlights, June 25, 2004, p. 3; August 13, 2004, p. 4.
‡‡‡‡Aerospace Projects Highlights, July 2, 2004, p. 3; August 13, 2004, p. 4.

designed heater plate had been installed to correct this. But the flight 3 FAS had failed at room temperatures.§§§§

Within a week of the failure, Parker had collected some information about the problem. The flight 3 boards were noisier than the flight 2 boards; that is, they generated more electronic noise than had the previous set. No one at the time knew why this was so, but they did know that the problem would affect the flight, now scheduled for "late September or early October." Various options were being studied to allow adherence to the existing schedule, but to stay on track, the failure's cause would have to be identified within the next few weeks.¶¶¶¶

Troubleshooting continued at Parker's Irvine, California, facility. Two weeks after the failure, some progress had been made in developing a solution, but no major breakthroughs had been reached regarding the cause of the higher noise level.

The initial FAS failure was traced to the CLPDs that latched up when exposed to unusually high amounts of electronic noise, and the fix used on the flight 2 boards had been added to those for flight 3 with the expectation that this would also work. But the flight 3 boards contained components from a different production lot than that of the components used in the flight 2 boards, and the minor component differences were enough to cause the CLPDs to fail on vehicle 3.

After 3 weeks, the cause had been identified, a fix had been developed, and the boards had been successfully tested at low temperatures over many cycles. The solution was another design cycle aimed at lessening the amount of electronic noise to which the CLPDs would be exposed. The fix was approved by the Orbital Sciences Flight Readiness Board at the end of July, and by the NASA Flight Readiness Review Board in early August.*****

Once more, it seemed that the FAS issue had been resolved. But the following week, on August 18, the cycle began anew. The ECU and actuators were undergoing acceptance testing at the Parker-Dublin facility when, during one of the tests, the ECU suffered a failure that damaged its boards. The initial assessment was that the cause was an ECU-component failure. However, the root cause was not known. Spare boards had been developed as part of an earlier plan and these were soon being inventoried. A full recovery plan was developed and an assessment made of effects on the schedule, and a massive effort was launched by both the Hyper-X team and outside experts to solve the continuing problems.†††††

---

§§§§Aerospace Projects Highlights, July 2, 2004, p. 3.
¶¶¶¶Aerospace Projects Highlights, July 9, 2004, p. 3.
*****Aerospace Projects Highlights, July 16, 2004, p. 3; July 23, 2004, p. 3; Laurie Grindle history interview, tape 3, January 30, 2005, NASA Flight Research Center History Office.
†††††Aerospace Projects Highlights, August 20, 2004, p. 3.

The chain of events that created the anomaly began with electronic noise causing errors in the timing of CPLD opening and closing of the field effect transistor (FET) pairs. To move the fins, opposite FETs were energized. The electronic noise disrupted the timing, causing the opening and closing of the pairs to overlap. When both pairs were energized, the high voltage shorted them out and burned up the boards.

To reduce the likelihood of the timing error recurring in flight, the CPLD firmware was modified to increase the time delay between switching. The amount of time required for the FETs to turn on was also reduced. In addition, engineers added a jumper cable between the ECU heat sink and the powerboard, reducing the electronic noise from the grounding system. Changes were also made to test procedures in order to avoid damage to the components while still providing necessary stress testing.‡‡‡‡‡

The damaged hardware was returned to Parker's Irvine facility for inspection and testing after it had been analyzed at the Orbital Sciences-Dulles laboratory. The initial conclusion was that a high-current switching component failure occurred on the rudder powerboard. Part of the analysis entailed a re-review of earlier design changes, test procedures, test methods, engineering powerboard checks, and component/board heritage, but these showed no indication of design problems. While the investigation was under way, a new ECU was assembled using spare components. Testing was completed on the replacement and cleaned powerboards on August 23 at Parker-Irvine. A new ECU chassis was sent to Parker's Dublin facility and the reassembly process began on August 27. Beyond this, additional powerboards were being manufactured at Parker-ESD in New York. The schedule delay was estimated at 3 weeks, with a flight in late October still considered possible.§§§§§

There was a larger issue, however. The original FAS bore the heritage of numerous Pegasus satellite launches. When the decision was made to increase FAS torque output, this heritage was lost. The FASs for the second and third flights were effectively new, one-of-a-kind units. Neither the time nor the money was available for the full development/test cycle normally conducted on a new system.

Through careful selection of components, engineers were confident that the problems from electronic noise could be eliminated. For flight 3, they were able to reproduce the failure conditions and could show both that these conditions would not occur during the boost phase and that adequate margin existed. The flight 2 and 3 FASs had also undergone nearly a thousand high-stress cycles, validating predictions. A change to the fin sweep test made

‡‡‡‡‡Aerospace Projects Highlights, August 20, 2004, p. 3; August 27, 2004, p. 3; information from Yohan Lin, engineer with the Hyper-X project.
§§§§§Aerospace Projects Highlights, August 27, 2004, p. 3.

before the launch allowed engineers to make a final check that the FAS was operating properly.

But engineers also realized that the potential for failure had increased. On flight 2, they believed that an FAS failure could occur only in the event of a 2-deg change in the fin angle. This occurred just twice, once during the HXLV burn and a second time after separation. In the latter case, a failure would be irrelevant. The FET overlap issue meant that an FAS failure could now occur anytime after the system was activated, 2 min before the drop. Engineers had analyzed the problem, made measurements on an engineering powerboard, and undertaken circuit modeling. This added credibility to what they believed to be the root cause of the failures, but they could not give the review panels the assurance that failure had been ruled out with absolute certainty.¶¶¶¶¶

With the modifications complete, the rebuilt ECU hardware began pre-burn-in testing. This was successfully completed, clearing the way to start burn-in testing, which began at 8:20 a.m. on September 17, 2004. The burn-in was a 100-hour test that was to be completed on September 20. The FAS and ECU completed initial testing at Parker's Dublin facility on September 22. All three fin channels passed without complications. This was followed in early October by compliance testing, at Orbital Sciences, of the fin/fin actuator assemblies. Once the work was completed, all the flight hardware was shipped by October 22 to Dryden. The next step was final integration and assembly.******

Overlapping the FAS issue was a reappearance of Langley concerns regarding excessive testing of the Hyper-X vehicles and hardware. Dave Reubush later noted that "as for the third flight captive carry, this was another situation where [Langley] believed that, since we had done captive carries for the first and second flights and had flown successfully, the captive carry for the third flight would add more risk than it removed."††††††

The Langley researchers' concerns were not without merit. The B-52 launch aircraft had been flying since June 1955, nearly half the era of powered flight, and was showing its age. On the flight 2 launch, a fuel-transfer valve had stuck, creating a potentially serious imbalance problem. On March 12, 2004, 2 weeks before flight 2, a planned B-52 flight ended with a ground abort due to the failure of a spoiler activator valve. On July 22, the B-52 crew aborted a flight after takeoff when the main landing gear failed to retract. On August 19, yet another flight was aborted when the fuel shut-off valve for the #6 engine stuck in the closed position.‡‡‡‡‡‡

¶¶¶¶¶Laurie Grindle history interview, tape no. 3, January 20, 2005, NASA Dryden Flight Research Center History Office, pp. 14, 15; Aerospace Projects Highlights, September 17, 2004, p. 3.

******Aerospace Projects Highlights September 10, 2004, p. 3; September 17, 2004, p. 3; September 24, 2004, p. 5; October 8, 2004, p. 4; October 15, 2004, p. 4; October 22, 2004, p. 4.

††††††Dave Reubush e-mail to Curtis Peebles, "Re; The Issue of Testing," December 11, 2008, 2:37 p.m.

‡‡‡‡‡‡Daily Flight Log 2004, Flight Crew Branch, Dryden Flight Research Center.

In light of the B-52's increasingly frequent problems, as well as Langley's repeated anxiety regarding "over-testing," Dave McAllister, who had been the launch panel operator aboard the B-52 for flight 2, put together a benefits-versus-risks briefing regarding the captive carry. The benefits he listed were:

- Increased probability of successful launch mission
- System verification at flight conditions without X-43A fuels
- B-52 performance verified
- Range verification
- Flight operations verified

The risks included:

- Damage to launch vehicle or research vehicle during preflight and flight operations
- Damage to B-52 during preflight and flight operations
- B-52 critical systems failure (landing gear/fuel)
- Hard landing damaging pylon hooks

McAllister sought out B-52 pilots and operations engineers who had worked on earlier programs for their opinions. Every one of them said they felt the captive-carry mission was necessary before attempting the flight 3 launch. It is important to remember that Dryden engineers had been making captive-carry flights in preparation for research aircraft flights since the late 1940s. These had been successful at revealing problems before committing to a free flight. In contrast, Langley did not have this experience.

McAllister thought the Langley engineers' opposition to the captive-carry flight was due to a culture clash and schedule pressure. He likened it to "disagreements between airplane people and rocket people." His suggestion for dealing with this was to simply step back and say, "Just a culture clash." The issue was settled with approval being given for the captive-carry flight. The issue of the B-52's reliability, however, remained.§§§§§§

On August 20, the day after the B-52 abort caused by the fuel shut-off valve, the aircraft made a check flight over the Point Mugu test range and came home with no significant problems. There was a sense of relief at Dryden. The flight cleared the way for the captive-carry mission. At this point, the Hyper-X stack was still undergoing final checkout, with the captive-carry flight scheduled for September 9, 2004.

The Hyper-X stack was attached to the B-52 on August 23 in preparation for the captive-carry flight. As a first step, on August 27, the B-52 with the stack attached to the pylon taxied from the Dryden ramp to the Edwards' taxiway, then back to the Dryden ramp. Now everything was ready for the actual captive-carry flight.

---

§§§§§§Notes on phone interview with Dave McAllister December 12, 2008.

With the pressure on to complete the Hyper-X project by the end of 2004, project engineers and related personnel had to skip weekends and holidays to stay on schedule. But they preferred this to the alternative, taking time off but flight 3 not getting made. Leading up to the captive-carry flight the Hyper-X ground crew worked rotating 12-hour shifts over the 3-day Labor Day holiday weekend. On Tuesday, September 7, 2004, the B-52 and Hyper-X stack were ready to go. Just before takeoff, the #2 hydraulic pack on the aging airplane was discovered to be leaking. The packs used high-pressure engine bleed air to spin a turbine that pressurized the hydraulic fluid, and there were 10 hydraulic packs operating various systems on the B-52.

With the mothership and stack back on the side taxiway, where they were kept while fully loaded and ready for flight, the mission was delayed one day to allow repairs. The second attempt on the following day initially went well, but as the B-52 took the active runway, the #4 hydraulic pack was discovered to be leaking. Once again, the flight ended in a ground abort. The Hyper-X stack was removed from the pylon this time, and work began on checking the B-52's hydraulic system. The captive-carry flight would not be rescheduled until the situation was resolved.¶¶¶¶¶¶

By mid-September, the #4 hydraulic pack had been repaired and reinstalled, while the #2 hydraulic pack was still under repair. It subsequently passed bench checks and it, too, was reinstalled on the B-52. The captive-carry flight was then rescheduled for Monday, September 27. Once more, the Hyper-X ground crew had to work 12-hour weekend shifts to prepare both vehicles. But the B-52's hydraulic problems were still not fully resolved. During the preparations, the #1 hydraulic pack was discovered to be leaking, and the ground crew had to find a spare pack to replace it.

After all the delays, difficulties, and doubts, the B-52 finally lifted off for the captive-carry flight on September 27, over the Point Mugu range. The only significant issue was the failure of the Hyper-X GPS to realign after the B-52 power transfer. But this was not seen as a major setback, because either new procedures or the addition of a battery could solve the problem.*******

Ironically, after all the pressure to accomplish flight 3 as soon as possible, the launch was delayed by two Pegasus satellite launches, using Orbital Sciences' L1011 aircraft, on the Vandenberg Air Force Base range. What was more ironic was that both launches, originally scheduled for October 26 and October 28, were postponed due to satellite problems. The B-52 conducted a successful training and proficiency flight on October 18. The ground crew then began preparing the aircraft for flight 3, now tentatively scheduled for November 9.

¶¶¶¶¶¶Fullerton history interview, pp. 22–24; Daily Flight Log, 2004; Aerospace Projects Highlights, August 27, 2004, p. 3; September 10, 2004, p. 3; September 24, 2004, p. 4.
*******Aerospace Projects Highlights, October 1, 2004, p. 3; October 8, 2004, p. 4.

As October 2004 came to a close, there was a positive change in the mood of project personnel. Acceptance testing of the FAS was complete, and the flight hardware had been delivered to Dryden. The series of reviews leading up to the launch were also under way. The first of these was by the Airworthiness and Flight Safety Review Board, which finished work on October 18. Only minor items still required action. The board's review was followed by the Integrated Mission Assurance Review, on October 22, and then, on October 28, by the Tech Brief.†††††††

The Tech Brief was the final step before flight 3. It began at 8:30 a.m. and ran for 3 hours. Marta Bohn-Meyer, Dryden's chief engineer, presided over the meeting. Each of the engineers made presentations on outstanding issues. Bohn-Meyer then asked detailed questions; engineers had to know the situation, describe how the problem had been corrected, and show that they had settled on a solution. The Tech Brief dealt with such issues as positioning the P-3 telemetry aircraft, a display lock-up in the control room, display of incorrect data in the control room, inconsistencies in flight documentation, and remaining captive-carry open items.

Not surprisingly, the primary issue at the Tech Brief was the FAS. The actuators had been checked and passed the tests; the hardware had been delivered, checked out, and installed without incident; and flight simulations and step functions had been run, also with no indication of problems. The FAS heaters operated correctly, but the recurring failures left engineers uneasy. Nevertheless, the meeting ended with Bohn-Meyer satisfied that everything possible had been done to ensure a successful flight. Approval for the actual flight followed.

But after the formal Tech Brief was over, informal conversations continued regarding the FAS. A major topic was the difficulty of giving a meaningful estimate of the chance of a failure. One engineer remarked that it all boiled down to the basic question of calculated probability versus gut feeling. Another engineer added that had shown him that unknowns existed, but it was difficult to give a qualitative measurement of probability that actually meant something. This was most worrisome when it came to giving senior managers an accurate risk assessment, which the team had to do as one of the steps toward actual flight.‡‡‡‡‡‡‡

Flight 3 was then scheduled for Monday, November 15, 2004.

ELEVEN SECONDS INTO THE UNKNOWN

Unlike the two previous launches, flight 3 was scheduled on a weekday. This meant the Federal Aviation Administration limitations on flying through

---

††††††Aerospace Projects Highlights, October 22, 2004, p. 4; October 29, 2004, p. 3.
‡‡‡‡‡‡Curtis Peebles X-43A notes, NASA Dryden Flight Research Center History Office, p. 17.

controlled airspace over the Pacific were tighter than they had been for the weekend launches. The FAA insisted that the X-43A and HXLV be in the water by 4:00 p.m. Given the longer duration of the Mach 10 flight, this required that the launch be made no later than 3:45 p.m. That, in turn, determined how late the B-52 and its pair of F-18 chase planes would be able to leave Edwards.

With ground controllers in position in the locked Gold Room, spectators in the Blue Room, and reporters watching the activities in the Dryden cafeteria and one of the large hangars, final ground preparations moved smoothly. In fact, the crew and engineers were ahead of the timeline. Running true to form for the Hyper-X project, however, at 10:27 a.m., the first sign of trouble appeared. Telemetry indicated that the electronically scanned pressure #4 sensor (ESP#4), aboard the X-43A, was reading too high. Dave McAllister, mission manager for the flight, called for a hold; controllers in the Gold Room looked at the problem and estimated the issue would create a 40-min delay. Controllers began working to correct the problem but weren't successful. ESP#4 was connected to several onboard sensors. Engineers from both Langley and Dryden began discussing whether the flight could proceed: could the necessary data be collected even though not all the sensors would supply measurements?

The engineers decided the launch preparations could continue, because both the B-52 and the stack were ready to fly. The ground crew began the X-43A closeout at 12:25 p.m., but by now some 2 hours had passed. The time left for completing necessary preparations before the launch window closed was quickly slipping away.

Now, to the dismay if not really the surprise of many project members, a new problem appeared—the aft S-band transmitter on the X-43A apparently turned itself on. This confused the engineers because they thought it physically impossible for such a thing to happen. One possible cause was that a lanyard had prematurely separated. If this were the case, the stack would self-destruct upon being launched. To correct the problems, controllers decided to power down the X-43A for 5 min, and then reactivate its systems.

When power was restored, the X-43A's main transmitter was working, but the aft S-band transmitter was off, confirming that the lanyard had not separated. Engineers in the control room were also satisfied that the ESP #4 issue had not been triggered by the aft S-band transmitter. McAllister polled controllers for a go/no go recommendation on starting the B-52's engines, and their approval was passed on to the flight crew. Initially, the signs looked good for a flight. The launch aircraft's radar beacon was working, the GPS data were good, and control room displays were operating properly. But when checks of the X-43A systems began, a controller announced that the aft S-band transmitter had again turned on. This report was soon corrected—it was the primary transmitter that was on, not the S-band unit. However, the

P-3 aircraft that would be the primary data receiver for the flight had to depart the range at 3:15 p.m., and the control room was having trouble contacting a modified KC-135 aircraft that also was to act as a data-reception aircraft.

As a result of all the complications, only a 20-min margin in the takeoff time now remained. Controllers requested an extension from the FAA but were refused. And so the launch attempt was cancelled, because there simply was not enough time to take off, fly out to the range, and make the drop before the FAA's mandated 4:00 p.m. splashdown. The third X-43A lived up to its personality—it still didn't want to fly.

Following the flight cancellation, the B-52's engines were shut down and it was refueled with JP-4 fuel. This was another of the B-52's peculiarities. Other aircraft at Edwards and Dryden used JP-8 fuel, but this fuel damaged the rubber tanks inside the B-52's wings. The B-52B was the only aircraft at the base that used JP-4. As a result, the JP-4 had to be specially ordered. Because this was the mothership's last flight, the fuel supply on hand was low. There was enough, however, to refill its tanks once more.

The late abort made the turnaround effort difficult. All the external-access panels on the X-43A had been attached and sealed, which made it difficult to monitor the systems and control internal pressures and temperatures overnight. If the X-43A were to be powered up, all its internal systems would turn on, including telemetry transmitters, and when operational these posed a risk to ground personnel. A work-around was developed in which every few hours during the night, the X-43A was powered up using the monitor station on the B-52. The monitor station showed the vehicle's internal temperatures and pressures, and the internal heaters were turned on as needed. (That night, the wind blew fiercely and overnight temperatures dropped to 45°F.) Orbital Sciences engineers also wrapped the stack in a bright yellow thermal blanket (nicknamed the banana) and then blew hot air underneath it, keeping the stack's internal temperature at about 75°F throughout the night and early morning.

November 16, 2004, dawned with perfect weather conditions. The launch would have to be made before 3:45 p.m. in order to meet the FAA's 4:00 p.m. splashdown deadline. During the night engineers solved the aft S-band transmitter problem: it had been connected to the battery earlier in the countdown than on previous ground tests and this caused it to turn on without being commanded.

Preventing a recurrence was simple—the battery would not be connected until later in the countdown. Regarding the ESP #4 issue, engineers decided that if the malfunction recurred, the flight would proceed anyway. The engineers' best estimate was a 50/50 chance that it would operate correctly. A last change to the previous day's plans was that the modified KC-135 was not available, leaving only the P-3 for telemetry collection and relay.

At the point in the countdown when the ESP #4 problem had occurred the day before, a controller reported that the pressure checks looked good. The aft S-band transmitter also behaved, remaining off. Once final closeout activities on the X-43A were completed, at 12:34 p.m., the B-52 crew was given approval to taxi out. A minute later, the plane began rolling. The F-18s were first off the ground, at 1:03 p.m. The two chase planes made a wide turn, and joined up on the B-52 as it lifted off at 1:08 p.m. (see Figs. 12.9 and 12.10). The formation headed west, toward the Pacific and the Point Mugu range. Unlike flight 2, which had been hampered by the fuel-transfer problem, the only issue to arise on the flight out was an inoperative autopilot on the B-52, and this was corrected by putting a switch in the proper position. The ground stations and the P-3 tracking aircraft checked in as the clock moved toward zero.

At L–9 min, the Gold Room door was locked. As before, the seconds seemed to accelerate, rushing toward zero. At L–2 min, the FAS batteries were activated. The fin pins were retracted and the three control surfaces were cycled back and forth, making the stack quiver on the pylon as the sweeps were made. There were no indications of problems, but participants knew an FAS failure could now occur at any time, and everyone moved closer to the edge of their seats. At L–1 min, the "arm" light came on, and the HXLV guidance system was activated. The final seconds passed, and at 2:34 p.m. PST McAllister called out, "Launch, launch, launch!" [6].§§§§§§§

Fig. 12.9   The B-52B taking off.

§§§§§§§Curtis Peebles X-43A notes, NASA Dryden Flight Research Center History Office, p. 17.

Fig. 12.10   The B-52B with the Hyper-X stack under its wing banks away from an F-18 chase plane.

The hooks released and the stack fell away from the pylon. As on flight 2, the airflow caused the stack to roll to the right between L + 0.5 and L + 0.75 s. This had been expected, and to prevent the possibility of an FAS failure, the 1-s delay had been eliminated in the actuator system so it activated immediately upon release [2, p. 14].¶¶¶¶¶¶¶

The roll was stabilized, and the stack dropped for 5 s before the HXLV ignited. Sending out a long yellow-white flame behind it, the stack accelerated away from the B-52 (see Fig. 12.11), which was making a slow turn to the left to clear the HXLV. Between L + 8 and L + 13 s, the pull-up into climb began. During this maneuver the stack was flying at a relatively high angle of attack of 12 to 13 deg and being subjected to 2.5 Gs. As the stack reached transonic speeds the telemetry data showed a "bump," similar to what had occurred during flight 2 and on earlier Pegasus launches. As the stack continued to climb, it maintained an attitude close to the nominal ascent trajectory. The dynamic pressure on the stack peaked at 1600 psf at L + 35 s and Mach 3.2 at 54,000 ft. Immediately after the maximum dynamic pressure occurred, the stack began the pull-down maneuver, which caused a –0.5 G loading. Rather than continue in a steep climb, as with a Pegasus satellite launch, it nosed into a negative angle of attack, putting the highest loads of the flight on the X-43A's wings.

As the ascent continued, aerodynamic heating rapidly increased. At L + 43 s the adapter fluid system activated, supplying coolant to the scramjet engine

---

¶¶¶¶¶¶Reukauf history interview, pp. 61, 62.

and nitrogen gas to purge the X-43A's internal cavities. Shock impingement heating on the HXLV wing, the fin leading edges, the silicon joint between the wing fillet sidewall and the motor casing, and fin interference regions along the aft skirt and fin root added to the increasing temperatures heating the stack. In addition, the flow field around the fin and fin shaft created a complex interaction, also heating the underside of the fin, the shaft housing, and the aft skirt. To cope with this the graphite epoxy wing, fillet, and fins had been covered with a layer of insulation to prevent the graphite epoxy from exceeding 90°F during the flight. Nevertheless, due to the Mach 10 flight's higher stagnation temperatures, some areas would exceed this limit. But these exceptions had been approved based on structural loading at these points.

At L + 64 s, data indicated that local heating was causing damage to the HXLV. The stack was flying at Mach 7.4 and an altitude of about 92,000 ft when the flight dynamics data and other telemetry indicated that an "anomalous shock event" had occurred on the booster. Wing-fillet strain and temperature data showed "a potential yield condition or other localized event"; presumably, the fillet shifted position due to excess heating, although the exact cause was never identified. Although the rudder trim was affected by the "event," the stack continued to fly normally [2, pp. 13–15, and 6, pp. 3, 11].

In the Blue Room, VIPs and other guests watched large-screen displays showing computer images generated by telemetry data. They sat in awe as the Mach number increased rapidly, quickly surpassing the Mach 6.7 velocity reached by the X-15A-2 in October 1967, and continuing through Mach 7, 8,

Fig. 12.11   The flight 3 Hyper-X ignites its rocket motor and begins to accelerate away from the launch aircraft.

9, and 10. Burnout occurred about 75 s after launch. The HXLV had followed the planned flight path throughout the rocket burn.********

After burnout the stack began to coast, losing speed due to aerodynamic drag. Onboard systems monitored drag while the FAS held the stack on the proper flight path. Nominal separation conditions were planned to be: L + 89.8 s; 109,580 ft ASL; Mach 9.60; flight path angle, 1.5 deg; dynamic pressure, 1000 psf; and angle of attack and sideslip, 0.0 deg [2, p. 14].

The "ready to separate" signal was sent from the HXLV to the X-43A. Three seconds later, the X-43A's FMU responded, sending the separation command to the HXLV. In the midst of the sequence, at about 2 s before separation, telemetry data indicated five of the adapter's pressure sensors simultaneously showed a loss of pressure. Two of the sensors measured the left and right piston pressures. Yet, when the HXLV received the X-43A's separation command, the explosive bolts and pistons fired on cue, pushing the research vehicle away from the adapter and successfully separating the X-43A. The pressure readings were later determined to have been instrumentation errors.

The separation occurred at L + 88.16 s (0.64 s early). Separation conditions for flight 3 were an altitude of 109,440 ft (140 ft lower than planned), a speed of Mach 9.736 (Mach 0.136 faster than the nominal value), and a dynamic pressure of 959 psf (41.0 psf less than the nominal value). The booster angles of attack and sideslip were +0.08 deg and –0.13 deg, compared to the planned 0.0 deg for both. The two onboard cameras transmitted images of the X-43A being pushed away from the adapter. In the Blue Room, spectators began to cheer. The FAS issue had been overcome.

The 2.5-s separation event now began. To prevent a repeat of the long delay (and potential near-failure) on flight 2, the controller had been modified with an increased angle-of-attack integrator setting. Upon separation, the X-43A's angle of attack initially matched the nominal prediction. Between 0.25 and 0.75 s after separation, however, the research vehicle's angle of attack was below the nominal value. At about 1 s after separation, the X-43A's angle of attack matched the nominal prediction. Although these were slight deviations from the prediction, they were too small to have an effect on the flight.

With the X-43A in the proper angle of attack, the cowl door was commanded to open 2.5 s after separation for the pre-experiment tare. The tare test lasted 3 s, after which the fueling sequence began. The silane/hydrogen mix and hydrogen gas were injected into the engine; the silane ignited the fuel/air mixture and, once running, the engine began to produce thrust. Only a second or so after ignition, the engine was at full power.

Later analysis showed that the fueling schedule produced fuel compositions and fuel levels similar to those used in wind-tunnel tests. The silane flow was shut off after about 5 s, and the hydrogen gas combustion became

---

********Curtis Peebles X-43A notes, p. 34.

self-sustaining. The hydrogen flow rate was then decreased to match the level used for most of the wind-tunnel scramjet testing, because one of the primary reasons for the flight was validation of wind-tunnel data. The flow rate decreased as the hydrogen tank emptied, and then the fuel valves were closed. The total burn time was approximately 11 s. The anticipated disturbance caused by the engine shutdown was quickly corrected, and with the burn complete, the 6-s postexperiment tare began. The cowl then closed, and the engine test ended [5, pp. 11–13, and 7].

The difficulties inherent in successful operation of a scramjet were great. Hydrogen gas was injected into an airflow traveling at multiple Mach numbers, after which the fuel and air mixed and ignited. The combustion then had to be stable and capable of holding a flame front without causing an unstart or blowout. The exhaust had to expand out of the nozzle/afterbody to generate thrust. And all of this had to occur within 0.001 s and in a space just over a 1 ft wide and less than 3 ft long, while maintaining supersonic airflow in the combustion chamber.

The burn successfully produced sufficient thrust to cancel out the vehicle drag. During the burn, the axial acceleration was lower than the nominal prediction, but was still sufficient for sustaining cruise flight. As on flight 2, the engine data were a close match for those captured in fuel-off and fuel-on wind-tunnel tests, qualifying the concept of an airframe-integrated scramjet design, the validity of engine testing in pulse facilities, and the accuracy of the CFD techniques used to analyze wind-tunnel data. The database used to design the engine was also shown to be accurate, confirming that it could be used to make predictions with confidence.

And there was a final accomplishment. A report noted: "The cowl open duration [of flight 3] was approximately *50 times* longer than the sum total of all the final engine wind-tunnel testing" [5, pp. 13, 14, and 7, pp. 5, 6].[††††††††]

When the cowl closed, the internal pressure dropped as expected. With the engine test complete, the X-43A began its final test activities. These started when the vehicle initiated the recovery maneuver. The X-43A increased its angle of attack to 6 deg to slow its descent. On flight 2, the vehicle had experienced an oscillation in its angle of attack during the recovery maneuver. With flight 3, the vehicle experienced a nose-down attitude. In attempting to correct this, the wings triggered an oscillation in bank attitude. At the same time, engine-pressure data showed a higher reading than the nominal baseline for cowl-closed conditions, indicating that air was flowing through the scramjet, which was believed to have caused the nose-down attitude. However, cowl door actuator data indicated that the door had closed properly, and had not opened again during the descent, leaving engineers unsure of the cause.

[††††††††]TAMU AIAA Presentation, final PowerPoint slide.

Ninety seconds after separation, a disturbance occurred that triggered a large spike in wing-position data. Simultaneously with the disturbance, the remaining hydrogen/silane ignition mixture was vented overboard, and post-flight analysis concluded that the silane gas had ignited, causing the anomalous event. The timing of the hydrogen and silane vent and purge, and the engine internal-pressure data were compared. This showed that engine pressure dropped when the venting and purge ended. The event, nicknamed the "Mach 8 unpleasantness," did not recur during descent. The later report concluded, "... there is no known cause for either the initiation or cessation of this phenomenon" [5, p. 20].

Despite this unknown, the vehicle remained under control during descent, and successfully accomplished all seven parameter identification (PID) and rudder sweep maneuvers. The rudder and wing actuators responded properly to the commands issued by the onboard systems. This sequence included proper responses by the left rudder and wing, which gave no indication of adverse effects resulting from damage incurred during contact in preflight testing. The vehicle decelerated rapidly during descent, requiring that the seventh and final PID maneuver follow immediately after completion of the sixth.

These maneuvers provided data on aerodynamic performance and open-loop frequency response from hypersonic down to supersonic speeds. These maneuvers also put the X-43A on the proper descent trajectory to the impact zone. The vehicle then entered the terminal glide to impact. The P-3 aircraft was able to receive data until 721 s after separation, when the X-43A passed out of range. The last data received showed the X-43A was at an altitude of 918 ft and flying at a speed of Mach 0.72. It was about 4.5 n miles from the aim point. The total distance from launch to impact was 850 n miles, and the X-43A had covered about 600 n miles following separation. Impact occurred at about 2:46 p.m., 12 min after launch and well within the FAA requirement [5, pp. 16, 19, 20].########

THE BEGINNING OF THE FUTURE

Hyper-X flight 3 marked the culmination of more than a half-century of hypersonic research going back to postwar efforts and the origins of the scramjet. After all the false starts, dead ends, high hopes, and broken dreams, scramjet-powered vehicles had successfully flown. Not for very long, and not for very far, to be sure. But without question, they had flown. Those brief seconds marked the potential beginning of a new dimension in flight.

The heritage of the past, and the beginning of the future, came together in the sky above Dryden on the afternoon of Tuesday, November 16, 2004. The

---

########X-43A Flight #3 Fact Sheet.

Fig. 12.12   The end of the Hyper-X program was marked by a traditional Dryden flyover by the B-52B and the chase planes.

returning B-52 launch aircraft and its two F-18 chase planes made a fly-over of the center (see Fig. 12.12). This was B-52-008's last flight, but its career ended as it had begun, 45 years before. That had come on Saturday, January 23, 1960, while the Cold War raged, and while a divided world was suspended between oblivion and a gleaming future. It was a time when flying to the moon was the definition of the impossible, and the scramjet was an immature technology with great potential and many doubters.§§§§§§§§

On that day, B-52-008 was new. Tucked under its wing was another hypersonic vehicle, the X-15 #1. Piloting the X-15 was North American Aviation's A. Scott Crossfield. The launch was successful and Crossfield reached a top speed of Mach 2.53 and an altitude of 66,844 ft before gliding back to a landing on the lakebed. That flight was but one step in the speed and altitude build-up for the X-15 program. And it marked the beginning of the future.¶¶¶¶¶¶¶¶

§§§§§§§§Watching the B-52B takeoff on January 23, 1960, was an Army officer named Roy Bryant, who was assigned to what was then the NASA Flight Research Center. On November 16, 2004, Bryant watched the B-52B's last flight as its project manager. He subsequently retired after 48 years at the center. He died on May 30, 2005.

¶¶¶¶¶¶¶Aerospace Projects Highlights, November 19, 2004, p. 4; *X-press*, June 17, 2005, pp. 4, 15.

REFERENCES

[1] Marshall, L., Corpening, G., and Sherrill, R., "A Chief Engineer's View of the NASA X-43A Scramjet Flight Test," AIAA-2005-3332, pp. 10, 11.

[2] Joyce, P. J., Pomroy, J. B., and Grindle, L., "The Hyper-X Launch Vehicle: Challenges and Design Considerations for Hypersonic Flight Testing," AIAA 2005-3333, pp. 11, 12.

[3] Ferlemann, S. M., McClinton, C. R., Rock, K. E., and Voland, R. T., "Hyper-X Mach 7 Scramjet Design, Ground Test and Flight Results," AIAA 2005-3322, p. 5.

[4] Bahm, C., Baumann, E., Martin, J., Bose, D., Beck, R. E., and Strovers, B. "The X-43A Hyper-X Mach 7 Flight 2 Guidance, Navigation, and Control Overview and Flight Test Results," AIAA 2005-3275, pp. 9–14.

[5] Marshall, L. A., Bahm, C., Corpening, G. P., and Sherrill, R., "Overview with Results and Lessons Learned of the X-43A Mach 10 Flight," AIAA 2005-3336, p. 7.

[6] Vachon, M. J., Grindle, T. J., St. John, C. W., and Dowdell, D. B., "X-43A Fluid and Environmental Systems: Ground and Flight Operation and Lessons Learned," AIAA 2005-3337, pp. 8, 9.

[7] Ferlemann, P. G., "Comparison of Hyper-X Mach 10 Scramjet Preflight Predictions and Flight Data," AIAA 2005-3352, p. 4.

Chapter 13

Looking Back—Looking Forward

> It is not really necessary to look too far into the future; we see enough already
> to be certain it will be magnificent. Only let us hurry and open the roads.
>
> *Wilbur Wright*

The Hyper-X project is now part of aerospace history. The Langley, Dryden, and contractor personnel who worked on it have gone on to other projects or have retired. As the first decade of the 21st century ends, what is there to learn from the project, and the long quest for the scramjet engine? As a backdrop to this question it is important to remember that engineering projects and technological advancements do not exist in isolation.

Rather, they can only be understood when they are seen as part of larger social, cultural, political, scientific, and technical events. In this context, a new technology may be impractical simply because the knowledge, tools, need, and political and social factors that would make it a reality do not yet exist. A project's fate may turn on whether a sufficient number of congressmen decide that voting "no" rather than "yes" would appeal to a greater number of their constituents. Histories of technology have long shown that invisible influences can play extraordinarily powerful roles in the development, adoption, or rejection of a given technology. In some cases, it is the invention that shapes the social, political, scientific, and technical context, not the other way around. It was sociologist and economist Thorstein Veblen who coined the expression "invention is the mother of necessity," as a counter to the more familiar version of this phrase. His point, and that of many historians of technology since, is that there is no necessary logic to technological development: B does not necessarily follow A; we can go from A to D without seeing B and C along the way. Only by paying attention to the actors, the often unseen forces, can we understand how technology is developed—or not, as the case may be.[*]

---

[*]Consider, for example, *Inventing Accuracy: A Historical Sociology of Nuclear Missile Guidance* (Cambridge: MIT Press, 1993), in which Donald McKenzie's argues that the need for missile guidance in the post-World War II era was unclear, and that no *need* existed for the precision the engineers offered, so they "created" one. "A 'need' would have remained latent and without influence—like the 'need' of an anti-gravity device, for example—if the sense of inertial navigation's impossibility had not been undermined. The ultimate task of the inventors of inertial navigation, in their heterogeneous engineering, was to harness their schemes for black boxes to the interests of state power" (p. 29).

The Hyper-X project endured a protracted gestation of nearly half a century, during which time many factors affected its development. Early technological enthusiasm was matched with unmet expectations, leaving many unconvinced by future hype. To be sure, unlike the challenge of Apollo, technology did not yet exist to see the promises through. But cultural disenchantment with enormous technological systems also helped undercut possible funding that could have advanced the state of the art in the 1970s. Political will to support pure research did not transfer from the Apollo program to hypersonic research any more than it did to further human planetary exploration, or even commercial supersonic transportation, in that decade.

Despite the apparent rule of ever-increasing speed evidenced in the progressively faster airliners since the end of World War I, Americans did not vest hypersonics with the same importance as they had the race to the Moon. Economic, political, and cultural forces weighed in against such an expensive investment—one that had no obvious need—until the Reagan presidency, which saw large scale spending on the promise of Mach 7 commercial flight service.

Once again, however, promises exceeded technical capabilities, and $1.5 billion spent on a paper airplane led many who controlled funding to tighten the purse strings when hypersonics was mentioned. Although for several decades many in NASA have sought cheaper access to space, there is an undeniable element of an "invention looking for a need" when it comes to hypersonics. These factors, and more, have played a role in the history of Hyper-X.

HYPERSONIC ORIGINS

The Hyper-X represented the merging of two different technological threads. The first of these was the development of air-breathing reaction motors. The idea of the ramjet originated during the early years of powered flight, at a time when wood and fabric airframes were state of the art. Obviously, a propulsion system that operated at supersonic speeds was not practical at this technological level, and not merely from a materials perspective. The quest for faster speeds and higher altitudes during the first four decades of powered flight was evolutionary. Piston engines became more powerful, biplanes gave way to monoplanes, and all-metal construction replaced wood and fabric. Despite these changes, fundamental aircraft design features were retained.

Not until the late 1930s did aircraft performance finally reach the point of diminishing return. As aircraft reached transonic speeds, the amount of drag they experienced increased rapidly and their propellers lost efficiency. To counter this, piston engines had to be made considerably more powerful.

These larger engines weighed more and burned more fuel, but, in the end, provided only marginal performance increases [1, 2].

What was required was not an evolutionary improvement in existing technologies. Rather, aviation needed to undergo the revolutionary change represented by air-breathing reaction motors—turbojet, turbo ramjet, and ramjet engines. With this revolution came the discovery of new aerodynamic phenomena such as roll coupling and pitch up, not previously experienced. The old rules were no longer valid, and the cost of defining the new rules would be high.

The second thread in the Hyper-X story was the development of large rockets. The rocket engine is capable of achieving very high speeds with a relatively small vehicle, although its efficiency is much less than that of a jet engine. The V-2 was militarily ineffective, but it showed the potential of rockets as weapons of war, and of spaceflight as a near-term possibility. As with jet aircraft and later, supersonic flight, the V-2 also represented a revolutionary change in technology. The hypersonic speeds rockets were soon achieving represented a new operating environment, and few research tools were available in the 1950s with which to produce design data.

Hypersonic speeds were far more demanding in terms of ground-test facilities. Wind tunnels were built that could operate for sustained periods at low-hypersonic velocities. But even today, speeds of Mach 10 or higher can be produced only for durations measured in milliseconds. An added complication was the chemical and thermal phenomena encountered at hypersonic speeds. These issues led to disagreements over the most appropriate research tools for hypersonic flight—ground-based tests, or flying subscale reentry vehicles launched by rockets?

The idea of merging the two disparate threads of airplanes and rockets had been implicit from the beginning of manned heavier-than-air flight. The goals were always to fly faster, higher, and farther. By the end of World War II, piston engine aircraft were flying at altitudes of 45,000 ft. At such altitudes the pilots saw a curved horizon. Looking up, pilots saw a blue-black sky. As yet still out of reach, space now beckoned.

By the mid-1950s, the "Round 1, Round 2, and Round 3" concept had developed within the NACA, charting a path to spaceflight as an extension of aviation. It was an approach that also fit the NACA research culture. This was a focus on incremental advances, and reflected an engineering conservatism within the NACA.

Round 1 encompassed research with the X-1 series, the D-558-II, and the X-2 rocket planes. Round 2 was the X-15, development of which began in the mid-1950s. This aircraft could reach the edge of space, exposing the pilot to weightlessness for 2 min or more. Round 3, still in the study phase in the mid- and late-1950s, would see winged vehicles fly into space on a suborbital trajectory. These would glide down to a landing like earlier research aircraft.

Among Round 3 concepts to be developed was the Air Force Dyna-Soar, a small delta-wing vehicle carrying a lone pilot and launched atop a modified ICBM [3].

Although not actually part of the Round 1, 2, and 3 concepts, the Aerospaceplane studies of the late 1950s and early 1960s most closely fit Round 3 criteria. Aerospaceplanes would be aircraft, powered by scramjets, capable of taking off from a runway and being flown into orbit. Given the advances in aircraft performance, the pace at which these had been achieved, and the technological optimism of the 1950s, this seemed to be the route humans would take into space. But the future would take a different path.

FROM SPUTNIK TO APOLLO

The path the future would take was shaped by the launch of the Soviet satellite Sputnik 1, the shock it created in Western countries, and the inability of President Dwight D. Eisenhower to successfully deal with the political consequences. Although he approved the management structure for U.S. space activity that still exists today, Eisenhower never intended the United States to go as far or as fast as it did over the next half century. A week after the Sputnik 1 launch, in an October 9, 1957, press conference, Eisenhower said, "The value of that satellite going around the Earth is still problematic." He concluded: "So far as the satellite itself is concerned, that does not raise my apprehension. Not one iota" [4].

The president had attempted to ease public anxieties over Sputnik, but his assurances that the U.S. satellite effort was "well-designed and properly scheduled," and that the U.S. ballistic missile program was unaffected by Sputnik made him seen unsure, lacking in leadership, and detached from the situation facing the United States. This applied to his actions as well as his words.

During a meeting with Republican congressional leaders on February 4, 1958, the idea of an unmanned U.S. Moon probe was raised. The president responded by calling such missions "costly ventures where there was nothing of value to the Nation's security." On March 24, Eisenhower approved the launch of U.S. Air Force and Army Moon probes. The formal announcement was made 2 days later, and referred to "launching a number of small unmanned space vehicles" as part of "efforts to determine our capability of exploring space in the vicinity of the Moon." This was described as "an orderly program for space exploration." The way the issue was handled by Eisenhower—refusing to consider the attempt, delayed approval, and, finally, minimizing its importance—negated any domestic political or international benefit from the launches [4].

Eisenhower's difficulties in countering the effects of Sputnik opened the way for a powerful but somewhat amorphous groundswell in the United

States in support of an extensive U.S. space effort. This arose not out of scientific interest, because many leading academic physicists and chemists, including those advising Eisenhower, were hostile toward space activities. During President's Science Advisory Committee (PSAC) meetings on February 7 and March 12, 1958, George Kistiakowsky said, "we are in the middle of a great tragedy," and complained about spending large sums of money on space, rather than what he saw as more serious and deserving scientific research. I. I. Rabi agreed, going so far as to call this a "distortion of moral values." Hans Bethe felt that public enthusiasm toward space was "misguided," but also believed it would be "a real mistake" to resist it. PSAC chairman James Killian added that public interest could "work to benefit the rest of science." Rather, Soviet exploitation of space as a symbol of its growing military power introduced space into the Cold War [4, pp. 102, 105, and 5]. Domestically, this translated into both public interest in space and public frustration at Eisenhower's policies. In political terms, it also meant space would be an issue in the next presidential election.

John F. Kennedy stressed the importance of space exploration in the 1960 campaign, but on becoming president, he initially did little to change Eisenhower's policies. The scientists who had advised Eisenhower were now advising Kennedy, and they had not changed their views.

What forced President Kennedy into action was the launch of Vostok 1 on April 12, 1961, carrying Maj. Yuri Gagarin into orbit. Kennedy understood the political necessity of directly challenging the Soviets in space. He was fortunate in that the stars aligned when the Apollo decision was made. The public was supportive of an expanded space program, due in part to Eisenhower's passive approach to the repeated Soviet successes since Sputnik 1. Kennedy also had the support of Vice President Lyndon B. Johnson, who had been the senate majority leader up to the 1960 election, and a strong backer of space activity. Johnson was able to assure Kennedy of bipartisan political support for Apollo. Just as important, the technology was now available to build the huge rocket and complex spacecraft needed to go to the Moon. Finally, the successful suborbital flight of Alan Shepherd had raised the public's spirits following Gagarin's orbital mission, and that enthusiasm was manifested in general public support of a more ambitious U.S. space program.

Kennedy addressed the Congress on May 25, 1961, announcing that the United States would put a man on the Moon by the end of the decade. It was a bold challenge, with a goal that was easy to understand.

The Apollo program formed the core of the U.S. civil space effort in the 1960s. Reaching the Moon was now part of the struggle between East and West. Science would play a role in Apollo, and knowledge gained through the missions would revolutionize human understanding of the solar system. But the "why" of Apollo was the Cold War. Early work in the program played out against major Cold War events, including the Bay of Pigs invasion of Cuba, construction of the Berlin Wall, the Cuban missile crisis, and growing U.S. involvement in Vietnam.

Kennedy's assassination on November 22, 1963, transformed the president into a mythic figure, and Apollo became both Kennedy's legacy and his memorial. The new president, Lyndon B. Johnson, had been an early advocate of major U.S. efforts in space, had led Senate hearings on U.S. space and missile activities following the launch of Sputnik 1, and had recommended that Kennedy approve Apollo. He would see the goal through.

AERONAUTICS AND THE SCRAMJET IN THE AGE OF APOLLO

The Apollo program coincided with fundamental changes in U.S. aeronautical research. The turbojet revolution began in the late 1930s, as aircraft reached transonic speeds in dives. Less than a decade later, in 1946, Army Air Forces P-80 pilots were routinely flying at transonic speeds. A mere 3 years after that, in 1949, production F-86A fighters flown by squadron pilots were exceeding Mach 1 in dives. In the fall of 1954, the initial production F-100As entered operational service. Now Air Force pilots could exceed Mach 1 in level flight. As the 1950s drew to a close and the 1960s began, Mach 2 flights were made in operational F-104s, F-105s, F-4s, and even B-58 bombers [6].

Future aircraft were to fly at even higher speeds. The B-70 bomber was designed to cruise at Mach 3, and the U.S. Supersonic Transport (SST) then in the planning stages would cross the Atlantic at nearly the same speed. Jet airliners had entered service at the end of the 1950s. They cut travel time in half compared to propeller airliners. Additionally, a jet airliner's ride was smoother and more comfortable due to the higher altitudes at which it flew. A jet operated above the turbulence and bad weather that propeller airliners' passengers had to endure. Jet airliners made it possible to travel the world with a level of speed, comfort, and luxury not available before.

The SST promised to do to jets what jets had done to propeller planes. Speed was seen as the critical factor. Flying from New York to London would take about an hour and a half or two hours. In the technological enthusiasm of the early 1960s, the operating costs of the SST, the complexity of the vehicle itself, and the consequences of its operation were not an issue.

Also on the drawing boards was the X-20 Dyna-Soar, which would go into orbit and glide back to Earth for a conventional landing. Operational versions of the Dyna-Soar were expected to undertake strike and reconnaissance missions, satellite inspections, and space station resupply and crew transfers. And looking farther into the future, the Aerospaceplane studies offered the vision of an orbital airplane. There seemed no end to the possibilities.

In reality, the turbojet revolution ended on December 22, 1964, the day the SR-71 made its first flight. There were several reasons for this. Perhaps most important was that aviation technology had reached a plateau. Mach 2 was the maximum practical speed for an operational combat aircraft. And although operational aircraft could reach Mach 2, they could maintain such speeds for only a brief period, for a final dash to the target or to escape an attacker.

The SR-71 showed what was required for the next quantum leap in aircraft performance. To withstand the aerodynamic heating of prolonged flight at Mach 3, the fuselage would be made of titanium. This metal was expensive and difficult to fabricate. Heating also required that special fuel, oil, and hydraulic fluid be developed. The aircraft was expensive to build, operate, and maintain. What justified this was the SR-71's mission: peacetime overflights of "denied areas" defended with surface to air missiles (SAMs). Between its speed of over a half-mile per second and its early stealth technology, the SR-71 was invulnerable to attack. The efforts and expense required to build and operate a special mission aircraft like the SR-71 were not justified for conventional military aircraft; the SR-71's complexities were virtually impossible to contemplate in the world of commercial passenger service.

Shifts in military strategy and defensive technology also played a role in the end of the turbojet revolution. One of these was SAMs. To counter their threat, bombers had to fly at low altitude, which put the aircraft below the minimum effective altitude for a SAM. More important, ballistic missiles had by then become the primary means of delivering nuclear weapons, diminishing the importance of manned strategic bombers. As a result, the XB-70 program was limited to two prototypes for SST research, and the B-58 was retired after a short service life. The quest for higher speeds and altitudes, which had driven aircraft development for over half a century, had effectively ended.

This technological sea change also had social, military, and political aspects. Mach 3 flight was technologically and materially possible, but the value of the extra Mach number was outweighed by the costs and difficulties that gaining it entailed. By the first flight of the SR-71, military strategy was shifting away from nuclear tactics. Existing long-range ballistic missiles and B-52 bombers filled this role. Attention was focused toward what were called "brushfire" wars, such as the one in South Vietnam. Ironically, propeller-powered B-26s and armed T-28 trainers, not supersonic fighters, flew the early air-support missions there.

Another factor was that of the service life of aircraft. Between the end of World War II and 1964, the Army Air Forces/Air Force had deployed four generations of jet fighters—the F-80, F-86, F-100, and the Mach 2 generation (F-104, F-105, and F-4 series). As a new generation of airplanes appeared, the previous one was retired, transferred to Air National Guard units, or supplied to Allied countries. This changed with the fourth generation of fighters. Rather than replacing these with an entirely new aircraft, incremental improvements and upgrades were made to existing designs. The F-4 series included the Navy F-4A, F-4B, F-4J, F-4N, and F-4S, while the Air Force operated F-4C, F-4D, F-4E, and F-4G versions. First flown in the late 1950s, the last F-4s left U.S. service in the mid-1990s. As of 2009, Greece and Japan were both still flying F-4s. (The year 2009 was also the 50th anniversary of the F-4's entry into service. This is equivalent to the Sopwith Camel still being in Royal Air Force service in 1967.)

With the end of the technological revolution seen between the late 1940s and early 1960s, the focus was no longer on new realms of flight. Rather, it was on better radar and missiles, computerized fly-by-wire controls, increased maneuverability, and more efficient wing designs that reduced fuel consumption. These changes meant that advanced, cutting-edge high-performance technology was of lower priority than improved capabilities and efficiency.

The Apollo program had a similar effect on space exploration. The most significant change driven by Apollo was abandonment of the Round 1, 2, and 3 concepts. A winged spacecraft was too complex for reentry from the Moon, but a capsule was ideal. The X-15 continued to fly until nearly the end of the 1960s, but the Dyna-Soar was cancelled after Kennedy's death and the Aerospaceplane studies also ended in the early 1960s.

This was an example of how social forces were driving technological decisions. Round 1, 2, and 3 was abandoned in favor of a Moon trip because this was politically necessary/expedient. Gone, as a result, was the initial constriction of an orbital space station that was meant to serve as a stepping-stone to lunar and planetary exploration.[†]

Apollo also affected work on lifting bodies. There was neither time nor money in the age of Apollo for a full-scale lifting-body effort. The lifting bodies were technology demonstrators, testing different vehicle shapes to determine the advantages and shortcomings of each. Like earlier research aircraft, the effort looked to possible future use. By keeping the cost down, limiting the speed to supersonic, and not trying to build an orbital spacecraft from the start, the lifting body program was politically and economically feasible. The effort also avoided the fate of Dyna-Soar, which was bogged down in issues of experimental versus operational roles, cost effectiveness, and military roles and missions.

The scramjet overlapped both the aviation and space fields, and was affected by the shifting landscape in both. The engine offered great possibilities, but many researchers doubted its feasibility. But though lifting-body concepts had been proven in wind tunnels, the scramjet was still largely theoretical. The Aerospaceplane effort showed much work still remained to be done before a flyable vehicle could be built. During the turbojet revolution, the impetus was on developing more advanced, higher-performance

---

[†]For examples of early concepts of how space travel would develop, see the Disney Tomorrowland space documentaries from the mid- and late 1950s. These looked at the development of space vehicles, the biomedical unknowns of spaceflight, what missions would be undertaken in space, and flights to the Moon and Mars. The documentaries were the first to introduce these ideas to a mass audience.

technology. This had driven the rapid advances during the 1950s. With the plateau in aircraft performance reached with the SR-71, however, this factor evaporated.

Also problematic was the fact that scramjet technology still lacked the tools, concepts, and experience levels necessary to gain the data needed to build a workable design. But an effort focused solely on research into these areas was unlikely to gain political support, because it would not fill a near-term need. A flight test program was seen as necessary to meet this political need, yet the existing scramjet engine designs were not close to being operational. (The pod-design HRE was suitable for research purposes, but investigators knew from the start it would never constitute a workable engine design.) The Scramjet Incremental Flight Test Program was beset by cost increases, higher-than-expected drag, and reductions in research funding. As a result, it was cancelled. With the heat damage to the X-15A-2, which was to carry the HRE, and the loss of X-15 #3, the HRE was reduced to a ground-test effort.

Despite the disappointing results of early work on scramjets, some advances had been made in their technology as well as with the tools needed to develop it. More important, the conceptual breakthrough of the airframe-integrated scramjet engine had been made by the late 1960s. Although this represented a starting point for a workable scramjet design, it still remained an immature technology. The technology, research tools, and basic data were still lacking. Without the need for ever higher aircraft performance, and with space activities focused on Apollo, there was no support for anything but a token research effort.

This was overshadowed by a growing unease with the sweeping promises of technology. The publication of Rachel Carson's *Silent Spring*, in 1962, revealed an unexpected and dangerous side to our technological accomplishments. Her book also helped launch the environmental movement, which had as a major theme the interconnectedness of all things, and the possibly negative consequences of even the best-intended actions. Meanwhile, an increasing number of U.S. troops committed to the Vietnam War, a conflict that dragged on year after year without resolution, suggested to many that there was something hollow to promises of U.S. technological improvements, which seemed better at destruction than construction. By the first Earth Day in April 1970 some even argued that technology would destroy life on the Earth.

Fears of a nuclear apocalypse are a prominent example of this. The transition from the hope for this new nuclear world—exemplified by Project Plowshare (a name coined in 1961), which was part of a larger initiative, dubbed Peaceful Nuclear Explosions, which involved using nuclear explosives to construct canals and artificial harbors—to anxiety about our self-destruction, came quickly. The shift began as early as 1951 with the release of *The Day the Earth Stood Still*, followed by *Them!* (1954), and *On the*

Beach (1959). This angst is carried forward in films such as Stanley Kubrick's dark comedy, *Dr. Strangelove, or How I Learned to Stop Worrying and Love the Bomb* (1964), *Fail Safe* (1964), *The Bedford Incident* (1965), *The Planet of the Apes* (1968), *Soylent Green* (1973), and *Logan's Run* (1976).

E. F. Schumacher's *Small Is Beautiful: Economics as If People Mattered* became, in many respects, a touchstone work of the period: "Ever bigger machines," he argued, "entailing ever bigger concentrations of economic power and exerting ever greater violence against the environment, do not represent progress: they are a denial of wisdom. Wisdom demands a new orientation of science and technology towards the organic, the gentle, the non-violent, the elegant and beautiful" [7, p. 20]. The Green Revolution, which depended on technology to increase crop yield in the Third World, had several decades to go before its costs would become apparent.

The technological optimism of the 1950s and early 1960s was gone. In such an environment, science, engineering, and technology, and their most visible symbol, the Apollo program, became a target. Though there had always been politicians and scientists who opposed Apollo, their voices now became more strident and more successful in waging attacks against it. Throughout Apollo's development, Congress had made periodic cuts but overall, the program continued to receive adequate funding. Then, in 1967, following the Apollo 1 launch pad fire, Congress made deeper cuts. Despite this, Apollo would go to the Moon. But congressional actions foretold the program's future decline [8].

AFTER APOLLO

Symbolic of the change in mood regarding technology's promises was the cancellation of the government funding for the SST program in 1971, due in part to environmental concerns over its effects on the ozone layer. Without this support, Boeing and the other contractors pulled out.

The American SST's demise reflected several factors shaping technological activities, including social forces in shaping technology's development, evidence of political power in deciding the fate of something new and expensive, how the need for government funding for technological development both expands and limits what activities can be undertaken, the importance of the environmental movement in adding a significant voice to the cancellation movement, and the fact that technological development does not necessarily proceed in a logical manner.

The following year, Apollo 17 made the last manned landing on the Moon. President Nixon had decided not to undertake a new major effort in space despite the remaining Saturn Vs, spacecraft, and existing infrastructure, all of which had been paid for. Instead, the U.S. manned space program would continue (in low Earth orbit only), with the Skylab space station (1973–74) and the joint U.S./USSR Apollo Soyuz Test Project (1975). Despite congressional efforts to end the space shuttle program, work began in 1972, but the program never enjoyed the political or funding support that Apollo did. The result was that project funding was marginal for meeting the demands of the program's development. This was exacerbated by the economic difficulties of the 1970s, with both inflation and a stagnant economy. Not until 1981 was the first shuttle orbital mission flown. In contrast, the United States went from its first manned space flight to landing on the Moon in just 8 years.

The rocket-plane program ended with cancellation of the proposed X-24C/National Hypersonic Flight Research Facility aircraft in 1977. This vehicle would have continued the work in hypersonics begun with the X-15, and could have been used in research with scramjets and other high-speed air-breathing propulsion systems. But both NASA and the Air Force were facing budget difficulties. The NHFRF had become too costly as a result of trying to meet a range of disparate research goals. And, as was often the case, the vehicle was perceived as lacking a near-term goal. Again, social factors, such as competing funding demands during a time of economic difficulties, continued hostility toward technology, and the shift from the pioneering spirit of Apollo to the shuttle's goal of making low-earth orbit more accessible, reducing the cost of spaceflight, and a focus on useful applications of space technology, were aligned against a major scramjet effort.

For all these reasons, the mid-1970s to the mid-1980s saw only limited activities in scramjet research. By the mid-1980s, however, these small efforts had advanced the state of the art considerably. Just as before, this sparked a shift toward an optimistic assessment of the technological possibilities of the scramjet, similar to that of the Aerospaceplane in the late 1950s and early 1960s. The advances in scramjet technology and research tools looked promising for development of a scramjet-powered vehicle that would achieve the goal of a single stage-to-orbit vehicle. Such an approach offered the possibility of low-cost access to space for both commercial and military activities. This accomplishment took place at a time of mixed portents. There was a more positive attitude toward technology, but the deep divisions within the United States still remained, and, indeed, had grown sharper.

An example of this was the effort to build a space station. What became the International Space Station (ISS) was approved by President Ronald Reagan, continued during the presidency of George H. W. Bush, and was modified during Bill Clinton's two terms. Construction did not begin until late in Clinton's second administration, and continued through President George W. Bush's two terms. It was not until Barack Obama's administration that the ISS was nearly complete. The difference was that the stars never aligned for U.S. space activity in the decades following Apollo. The level of bipartisan support was never present for the ISS in the way it was for Apollo, and there were many who automatically opposed its development.

Although the ISS survived, the NASP was a program doomed to failure by flawed initial premises. The early design lacked many features such as landing gear and thrusters that a real airplane would need. More problematic, the NASP's highly optimistic performance and weight estimates did not hold up to scrutiny. The vehicle's size, drag, and weight went up as the scramjet's specific impulse dropped. Even after extensive work, and with a more limited equatorial launch profile, the vehicle's performance was still insufficient for going into orbit.

The technological enthusiasm on the part of the NASP advocates echoed the claims of the scramjet advocates in the early 1960s with the Aerospace-plane and the subscale tests. In part, this is necessary for any new, unproven technology. In the end, however, enthusiasm goes only so far. The failure to deliver on the promises meant scramjet research was again reduced to a low level of activity.

Beyond the optimistic assessments of the technology, and how social factors in the larger society determine the success or failure of a project, NASP had a basic structural flaw. Specifically, a project's goals must be clearly articulated, and the participants must understand these goals and know what to do to meet them. With NASP, these goals were not met. The project was different things to each group working on it—a technology research program, an experimental aircraft, a trans-Pacific airliner, and an orbital vehicle. The Apollo program, in contrast, had had a simple mission statement: "To land a man on the Moon and return him safely to the Earth." These 14 words defined every aspect of the spacecraft, booster, and mission profile, leaving no question about what Apollo was and what it was not.

Although the NASP program did not produce an actual vehicle, it did provide the tools and the knowledge needed to design, build, and fly a scramjet-powered vehicle. The NASP effort fortunately coincided with the development of small but very fast and powerful computers. These had greater capabilities than the mainframe computers of the 1960s (which often filled entire rooms), and made it possible to write CFD programs that could analyze and refine complex aerodynamic and combustion data. For all the expense and disappointments of the NASP program, it had finally made a workable scramjet engine a possibility by bringing the technology of age.

There remained another, possibly more difficult, task: developing a plan that would result in a successful vehicle, while avoiding the pitfalls that had doomed all the previous efforts. This involved a project whose price tag wasn't too high, that wouldn't take too long, and that could demonstrate to the U.S. political leadership that progress was being made.

The solution was NASA Administrator Daniel S. Goldin's "faster, better, cheaper" approach. Best known for its application to robotic planetary missions, it also reflected trends in wider society. A number of management concepts became popular in the 1980s and 1990s, including "reinventing government" as a means of reducing bureaucracy; "industrial policy," which involved direct government support for emerging technologies; and "privatization" that focused on using contractors rather than civil servants for federal work. Although applied broadly within the U.S. government, within NASA these techniques were intended to make the agency more efficient and to allow missions to be developed more quickly and flown at lower cost. Boosting their appeal, these techniques were quickly embraced because of difficulties NASA had experienced during previous decades in gaining political support.

Part of the "faster, better, cheaper" concept was the DF-9 study. This created a realistic hypersonic waverider design. Rather than attempting to build this large and complex vehicle, the DF-9 served a different purpose: to serve as the basis for a cheap subscale research vehicle to test a scramjet engine in flight.

The overall results of "faster, better, cheaper" were mixed. Before the advent of Goldin's philosophy, the U.S. unmanned planetary program was effectively dead. Goldin's mantra revived U.S. planetary missions, which continue to the present day. However, the failure rate of these missions was higher than that of planetary probes in the 1960s. A popular comment that circulated in this period about "faster, better, cheaper" stated that you could accomplish two of the goals, but not all three.

For the history of the Hyper-X project, the results of "faster, better, cheaper" were a milestone in flight technology. The DF-9 conceptual study, combined with "faster, better, cheaper," accomplished what for nearly a half century had been a story of disappointment and failure. These two efforts made air-breathing hypersonic flight possible. Compared to the Aerospaceplane, NASP, and even the X-24C/NHFRF, the Hyper-X was cheap enough to avoid attempts to kill it, even after the first failure, but was still capable enough to provide the data needed to prove the concept.

"Faster, better, cheaper" was introduced to address a problem that surfaced within NASA in the years after Apollo: mission creep. What would start out as a simple mission to send a satellite to Mars, for instance, would grow into an enormous project as scientists added more and more experiments to the initial satellite. Costs soared, schedules slipped, and in the end everything rode on a single launch; failure meant failure of everything on the mission. "Faster, better, cheaper" had its own risks, of course. But with more launches of smaller, more narrowly defined missions, a failed launch cost less in money and in tangential experiments because the eggs weren't all riding in one basket.

A simple but telling example of preventing mission creep involved the question of the X-43A's length. Once this was set at 12 ft, no changes were considered by project management. This prevented endless design iterations, each of which would require resources and time but would not move the project any closer to flight. But with the vehicle length fixed, engineers had a difficult time squeezing the components into such limited space. A decision to avoid a potential critical problem that could endanger the project resulted in difficulties nevertheless. These problems could be, and were, successfully dealt with. Delays occurred, but they were not of a scale that could endanger the Hyper-X project. Had the vehicle been allowed to grow, they easily could have.

CREATING THE HYPER-X

The most basic requirement for success was keeping a project focused on core objectives. This was accomplished with the Hyper-X by keeping it a subscale test of a scramjet engine made to prove the validity of design tools and ground-test data. Adding "nice to have" items was avoided, in particular, the suggestion that the vehicle should be recovered after flight. Recovery was admittedly appealing, because it would allow examination of the effects of high temperatures and engine operation. But although a physical inspection might have indicated the cause of the "Mach 8 unpleasantness," it would not have justified the difficulties, expense, and increased risk of failure.

The secondary mission, collection of hypersonic flight data, did not interfere with the scramjet test. Once the X-43A completed the engine burn, the mission's primary goal had been met. Should a failure then occur, nothing would have been lost. The research vehicle had the capability to perform the maneuvers. Although adding this activity did require resources to develop the systems and software, do the testing required, and carry out the reviews, it made sense to take advantage of the opportunity to acquire the data.

The separation issue held the potential for endless debate, proposals, and bright ideas, none offering a resolution. The procedure followed was to create the baseline drop jaw design and undertake wind-tunnel tests to identify its flaws. As a result, the drop jaw was modified into a fixed design. To eliminate concerns regarding a possible kick as the X-43A separated from the adapter, CFD testing was used to analyze forces generated by the sudden, high-speed airflow through the expanding gap. These tests indicated pressure would equalize too quickly for instability to occur.

Project managers and engineers had to accept an ambiguous situation. All the data indicated that separation would be uneventful. The wind-tunnel, CFD, and hardware tests indicated it would work. But no one really knew until the separation was attempted.

The Hyper-X project relied on the use of existing systems that had an extensive heritage of success. This was seen as a way of both improving the odds of success and keeping costs low. If existing systems could be used to successfully execute the mission, there was no point in spending added time and money to develop new ones.

The HXLV used the heritage Pegasus systems. Normally, a research vehicle, be it manned or remotely piloted, would have backup systems. The Pegasus systems, and therefore the HXLV as well, used "single string" systems, meaning there would be no backups should a failure occur on one or more systems. Developing and testing new systems that would contain backups was deemed too difficult and costly, especially in light of the limited Hyper-X budget. Project managers bet that existing systems had sufficient reliability to operate properly even when used outside their normal design envelope.

How this affected project decisions was shown in the choice made between off-loading propellant or using a lower/slower launch profile for the Mach 7 flights. Off-loading was considered too risky, but simulations and modeling of the lower/slower profile showed it could be done successfully. This option was not selected casually—several different analyses were done. What those involved didn't realize was that the computer models and wind-tunnel tests could not identify the nonlinearity at transonic speeds. As a result, flight 1 was doomed as soon as it left the hooks.

The FAS modification illustrated the reverse, where playing it safe led to problems. After the loss of flight 1, analysis showed a possibility that the existing FAS would not provide sufficient torque to counter excessive dynamic pressure. In this case, the heritage of the FAS was not sufficient to allay concerns about a possible mishap. This was because if flight 2 failed, it was the belief among Dryden personnel that the project would be cancelled.

The solution to increasing the torque was seen as easy, but created seemingly endless difficulties. It is worth noting that each of the three flights was made with a different FAS design. Ironically, neither flight 2 nor flight 3 ever came close to needing the added torque capability. The launches could have been successfully completed using the stock Pegasus FAS design. Instead, months were spent trying to fix related problems and failures. This delayed the launches, increasing overall project costs. It also meant long days and sleepless nights for engineers and managers.

Though in retrospect the lower/slower launch profile should have raised doubts, the FAS issue was more complex. Project personnel had been confident the lower/slower launch on flight 1 would be successful. Issues such as the separation and scramjet operation were seen as encompassing more significant unknowns. Yet the booster failed before it even reached supersonic speed. The MIB determined this was due in part to the fins lacking enough torque to stabilize the HXLV under the higher than expected dynamic pressures.

When faced with the possibility of this failure's recurrence, and with a solution available, it was inevitable that the option of a higher-torque FAS design would be approved. Developing the new FAS was a logical response to a potential failure that could have doomed the effort. Project personnel could not have known in advance that the flights would not experience worst-case conditions. And so a seemingly simple change led to a great deal of difficulty.

These examples point out the limitations of using heritage systems. Though they offer a low-cost and proven means of fulfilling a mission, when used in a role they were never designed for there is always the potential for complications. At the least, engineers must be sure their assumptions are valid.

That use of heritage data can mean being taken down the wrong path is not a new development. Otto Lilienthal was the leading aeronautical researcher at the end of the 19th century. In building and flying a series of successful gliders, Lilienthal had assembled tables of lift data derived from different airfoil shapes. Due to his gliders' successes, other researchers, including the Wright brothers, assumed the tables were accurate. The Wright brothers used the heritage airfoil-lift data to design their early gliders. When they began flying their gliders, they soon realized the wings were not producing the calculated amount of lift.

The Wrights built a small wind tunnel to measure lift produced by different model airfoils. These tests showed that Lilienthal's airfoil tables overestimated the amount of lift being produced. The Wright brothers had to start over, and assemble their own lift data. This led to improved performance in their new gliders, and eventually to successful powered flight.

THE HUMAN SIDE OF ENGINEERING

Beyond the technical lessons learned in the course of the Hyper-X project, there is also the human element. The lone inventors who pioneered both early flight and rocket development have been replaced by large government research centers and private contractors. But though the scale of the effort has changed, engineering remains a very human activity.

The success of the Hyper-X was due to the efforts of those who worked on the project. This reflected the creation of an integrated team in early stages of the project, working toward a common goal, who merged their individual abilities to allow team members to become more than the sum of their parts. This task is intrinsically difficult, and extends beyond issues inherent in dual agency cultures. NASA government personnel and private contractors work side by side, but report through separate chains of command. A NASA employee cannot order a contractor to perform any task, but can only make a request. Some tasks, such as sending a destruct command, can only be done by a government employee.

Yet when engineers gathered for the various group meetings held each week, it was impossible to tell a NASA employee from a contractor. The same was true in the control room. No one considered who had a NASA badge or a contractor badge. It was a single team, and it had to be: a single error or oversight by any team member could mean failure.

Aerospace projects are usually on the cutting edge of the possible. Engineers are attempting things that have never been done before. Eventually, they run out of data and enter the unknown. In modern risk-averse society, the idea that failure teaches more than success is heretical. Failure is to be avoided at all costs, and those who are blamed for the failure are criticized and punished.

The Hyper-X team experienced the full spectrum of failure's bitterness. After years of work, they witnessed the results of their efforts vanish in 13.5 s. This was followed by a period during which the project was no longer under engineers' control; rather, they were answering to the MIB. The team had to correct the problems and deal with new issues such as the FAS. These activities meant long hours, lost weekends and holidays, and missed family events. For the team to remain cohesive under such circumstances required not only leadership from above, but also courage from within. Each team member had to maintain a positive attitude. They had to believe the flaws could be fixed, and that they could succeed.[‡]

The qualities these attitudes entail were summed up in a September 12, 1962, speech by President Kennedy at Rice University in Houston, Texas. Explaining the Apollo decision, the president said: "We choose to go to the Moon in this decade and do the other things, not because they are easy, but because they are hard ... because that challenge is one that we are willing to accept, one we are unwilling to postpone, and one which we intend to win" [9].

SHADOWS OF THE FUTURE

The two successful flights of the X-43A brought a resurgence of hypersonic and scramjet research, both in the United States and in other countries. The most significant activity currently under way is the Air Force X-51A project (see Fig. 13.1). Its goal is to extend the scramjet burn time from the 10 to 11 s of the X-43A flights to 300 s or more—a full 5 minutes. It's also hoped that the X-51A will demonstrate a practical scramjet engine that could burn conventional JP-7 jet fuel and accelerate the vehicle through several Mach numbers. This program is an outgrowth of the proposed X-43C, which was to be powered by three hydrocarbon-fueled scramjets. The advantage is that such fuel is far easier to handle than hydrogen. The X-43C project was cancelled before progressing very far. But the engine had been designed, and this became the basis for the SJX61 scramjet used on the X-51A.

The prototype scramjet was tested in the NASA 8-Ft. High Temperature Tunnel. The welded Inconel engine experienced temperatures in excess of

---

[‡]Yohan Lin interview, November 16, 2006.

Fig. 13.1   The follow-on to the Hyper-X/X43A effort is the U.S. Air Force X-51A vehicle.

1832°F. Unlike the water/heat sink–cooled Hyper-X scramjet, the SJX61 scramjet used JP-7 fuel in a regenerative cooled system. The fuel flowed through the engine walls, carrying off heat and cracking the JP-7 fuel so it burned more easily in the scramjet's supersonic airflow. The amount of heat carried off matched the amount the burning fuel added to the engine. This is referred to as achieving *thermal balance*, which is one of the keys to a practical scramjet. Except for the fuel pump component, the test engine was the flight configuration engine.

The 8-Ft. HTT tests included 34 cycles at speeds of Mach 4.6 and 5. These speeds were selected because Mach 4.6 is the scramjet start speed after separation from the booster rocket, and Mach 5 is the speed at which fuel is "staged," or injected at different points within the engine. The wind tunnel tests concluded in June 2007 after 40 cycles. The final series was made at Mach 6.5, which was close to the maximum top speed the X-51A could reach. The total burn time for the test engine was 17 min [10].

The X-51A's fuselage is a waverider design, referred to as a "cruiser," and weighs 4000 lb. Its shape is different from that of the earlier X-43A vehicle in several ways: the nose is angular and designed to ride on the shock wave to reduce drag. Its fuselage is boxier than the sleek surfboard design of the X-43A. At the same time, there are similarities. Weighing 150 lb, the X-51A's nose cap is made of tungsten and experiences a 1500°F surface temperature; its fins and engine are made of Inconel. The X-51A fuselage structure is made of aluminum. Due to aluminum's low melting point, this is covered with special Boeing materials developed for use on the shuttle. These include

BRI16 tiles around the engine and inlet, lightweight sprayable ablative materials on the upper surfaces, and a high-density honeycomb around the nozzle. The X-51A's total length is 14 ft, only a little longer than the X-43A.

A total of four X-51A flights are planned using an Air Force B-52H launch aircraft. The four flights will be made over the Point Mugu test range, with both ground stations and P-3 aircraft monitoring the flight. The launch will take place at an altitude of 49,500 ft above the Pacific Ocean. The stack will fall for some 4 to 5 s, after which the booster rocket will fire. The test booster is a modified Army Tactical Missile System booster fitted with steel external skin, a titanium rocket nozzle, an interstage, and additional stabilizing fins. The rocket motor weighs 2275 lb and will burn for 30 s, reaching Mach 4.6 to 4.8. During the boost phase, air will flow through the engine and out a duct in the interstage. This will warm the engine and the fuel circulating through it, cracking the JP-7.

In a throwback to one of the X-43A separation concepts, the X-51A will roll inverted so the inlet is on the top of the vehicle and the vehicle is flying at a positive angle of attack. After burnout, four bolts will fire, separating the X-51A, which will coast for a half-second. Ethylene will then be injected into the scramjet and ignited. The burning ethylene will heat the JP-7 fuel flowing through the engine sidewalls. The heated JP-7, the same fuel used by the SR-71, will then be injected and burned. Thermal balance should be established within a few seconds. The ethylene will be shut off and the vehicle will accelerate using only the JP-7.

The X-51A will carry about 270 lb of JP-7, sufficient for about 300 s at full power, giving it a top speed in excess of Mach 6. As the vehicle accelerates, the engine will shift to full supersonic combustion. The JP-7 supply will be exhausted after about 350 miles. After this point the X-51A will glide down, performing test maneuvers until it hits the Pacific Ocean [10–13]. (The X-51A's riding of its own shockwave to reduce vehicle drag is similar to what the XB-70 did in the 1960s.)

As part of the X-51A design process, project manager Charles Brink sought to eliminate unnecessary risks by prohibiting use of all new and untested components outside the critical path of the scramjet engine. The four flights are being made to test the fuel-cooled scramjet engine. Brink stated that if a failure occurred, it had better be due to the engine, and nothing else. The digital engine control system on the X-51A is from an F-35 fighter. The scramjet's igniter comes from an even more unlikely source—a TF33 turbojet engine from an old C-141 cargo plane.

The same philosophy is applied to the structural design of the vehicle. Instead of exotic materials, contractors were directed to use conventional materials, with thermal ablative coating added for heat protection. An aluminum frame surrounds the engine bay. In higher-temperature areas, such as near the nose, titanium and Inconel are used. And of course, tungsten forms the nose itself. Brink noted, "I didn't want to have a scramjet experiment and spend all the money fixing [an unrelated] structural problem."[§]

Risk was also reduced by adding margin above that indicated as necessary by the design analysis. Estimates indicated about 2 to 4 lb of ethylene would be needed for the ignition and heating of the JP-7 fuel. To add margin, the "scuba tank," as Brink calls it, holds a full 6 lb of ethylene. The risks are clear—missile development programs typically experience failures on half of early launch attempts, and like the X-43A, the X-51A cruiser uses only single-string systems. The lone back-up system on the vehicle is a redundant destruct system.[§]

The X-51A represents another step in the development of scramjet-powered hypersonic vehicles. The X-43A was designed specifically to test design tools developed in the ground-based tests and CFD analysis. In contrast, the SJX61 scramjet is fuel-cooled and uses the thermal-balance technique. These features represent a design more akin to that of an operational engine.

The 300-s burn time is several orders of magnitude longer than that of the X-43A. This duration, at a speed of over a mile per second, approaches the requirements for an operational air-to-ground missile, as well as the burn time of the first stage of an orbital launch vehicle.

Use of JP-7 hydrocarbon fuel on the X-51A also represents a major breakthrough, because it avoids the difficulties of handling and using liquid hydrogen. This comes at the cost of a potential loss in performance, however. The successful development of a thermal-balance cooling system would represent a major breakthrough, because this is critical to an operational scramjet.

But although the X-51A represents another step toward the potential offered by the scramjet engine, there remains a long journey to the future. Burn times have to be extended from several minutes to tens of minutes, and, finally, to an hour or more. There is a need to develop a propulsion system that dispenses with rocket boosters altogether. This would involve a multi-mode propulsion system capable of taking off from a standing start and reaching hypersonic speeds. Another major development effort, as well as innovative solutions, would be needed to meet the challenges.

Other requirements include development of structures capable of withstanding prolonged heating and loads while still being light enough to allow the vehicle to fly its mission. Vehicle systems are also needed to control, monitor, and operate the vehicle. Building the Hyper-X required that design and test tools first needed to be developed. Without shock tunnels, CFD programs, the airframe-integrated scramjet concept, and other concepts and tools, turning the Hyper-X into reality would not have been possible. Future vehicles will need more advanced versions of these tools before becoming a

---

[§]Trimble, S., "X-51A: Jury-Rigged for Hypersonic Success," http://www.flightglobal.com/blogs/the-dewline/2009/03/as-the-first-mach-6.0.html Printout.

reality. And without the Hyper-X, subsequent programs would have to start from scratch, rather than capitalize on what an extraordinary team of engineers already has accomplished.

Finally, this future vehicle will not exist in isolation. It will be part of a complex mix of social, cultural, and political factors. These, more than the available technology, will decide the fate of the future vehicle, just as they decided the fates of those projects that preceded the Hyper-X. The future is an unknown place. We cannot predict its nature. We can, however, draw from the past to grasp its possibilities.

REFERENCES

[1] Hallion, R. P., *Supersonic Flight Breaking the Sound Barrier and Beyond*, Brassey's, Washington, DC, 1997, pp. 15–18.

[2] Hanson, J. R., *Engineer in Charge: A History of the Langley Aeronautical Laboratory 1917–1958*, NASA SP-4305, 1987, pp. 249–259.

[3] Hallion, R. P., and Gorn, M. C., *On the Frontier: Experimental Flight at NASA Dryden*, Smithsonian Books, Washington, DC, 2003, pp. 76–80, 101–106, 130.

[4] Divine, R. A., *The Sputnik Challenge: Eisenhower's Response to the Soviet Satellite*, Oxford University Press, New York, 1993, pp. 7, 8, 100, 101, 109.

[5] President's Scientific Advisory Committee, "Introduction to Outer Space," March 26, 1958, http://history.nasa.gov/sputnik/16.html [retrieved 25 September 2009].

[6] Knaack M. S., *Post-World War II Fighters, 1945–1973*, Office of Air Force History, Washington, DC, 1986, pp. 4, 54, 115, 116, 178, 189, 193, 205, 266.

[7] Schumacher, E. F., *Small Is Beautiful: Economics as If People Mattered*, Hartley & Marks, Point Roberts, WA, 1990, p. 20.

[8] Miller, R. R., "Exploring the Impact of Reinventing Government on NASA's Space Program," PhD dissertation submitted to North Central University School of Business and Technology Management, Prescott Valley, Arizona, October 2008, p. 62.

[9] "Address at Rice University on the Nation's Space Effort," John F. Kennedy Presidential Library & Museum, Historical Resources, http://www.jfklibrary.org/Historical+Resources/Archives/Reference+Desk/Speeches/JFK/003POF03SpaceEffort09121962.htm [retrieved 18 August 2009].

[10] Warwick, G., "X-51A to Demonstrate First Practical Scramjet," *Flight International*, July 20, 2007, http://www.flightglobal.com/articles/2007/07/20/215592/x-51a-to-demonstrate-first-practical-scramjet.html [retrieved 12 February 2008].

[11] Coppinger, R., "AFRL Starts Building First X-51A Ramjet Demonstrator," *Flight International*, June 3–9, 2008, p. 32.

[12] Coppinger, R., "Waverider faces long odds," *Flight International*, April 7–13, 2009, p. 23.

[13] Norris, G., "Scramjet Reality," *Aviation Week & Space Technology*, March 30, 2009, pp. 32, 33.

SUGGESTED READING

For scholarship on the history of technology, see

Cowan, R. S., "How the Refrigerator Got Its Hum," *Social Shaping of Technology: How the Refrigerator Got Its Hum*, edited by D. McKenzie and J. Wajcman, Open University Press, Philadelphia, 1985, pp. 202–208.

Bijker, W. E., Hughes, T. P., and Pinch, T. J. (eds.), *The Social Construction of Techno-logical Systems: New Directions in the Sociology and History of Technology*, MIT Press, Cambridge, MA, 1987.

Giedion, S., *Mechanization Takes Command: A Contribution to Anonymous History*, Oxford University Press, Oxford, 1948.

Hughes, T. P., *Networks of Power: Electrification in Western Society, 1880–1930*, Johns Hopkins University Press, Baltimore, 1983.

Smith, M. R., *Harpers Ferry Armory and the New Technology*, Cornell University Press, Ithaca, NY, 1980.

Winner, L., *The Whale and the Reactor: A Search for Limits in an Age of High Techno-logy*, University of Chicago Press, Chicago, 1988.

For more on Americans' feelings about technology during the mid-twentieth century, see, for example,

Bush, V., "Science—The Endless Frontier 1945," Eisenhower, D. D., "Farewell Address 1961," and Fulbright, J. W., "... the adherence of the professors 1967," *Major Problems in the History of American Technology*, edited by M. R. Smith and G. Clancey, Houghton Mifflin, Boston, 1998, pp. 429–439.

Farber, D. R., and Baily, B., *The Columbia Guide to America in the 1960s*, Columbia University Press, New York, 2003.

For scholarly works on the American SST and its cancellation, see

Conway, E. M., *High-Speed Dreams: NASA and the Technopolitics of Supersonic Transportation, 1945–1999*, Johns Hopkins University Press, Baltimore, 2005.

Horwitch, M., *Clipped Wings: The American SST Conflict*, MIT Press, Cambridge, MA, 1982.

INDEX

Note: n = Footnote; *Italicized* page number = Figure

SUPPORTING MATERIALS

Many of the topics introduced in this book are discussed in more detail in other AIAA publications. For a complete listing of titles in the AIAA Library of Flight Series, as well as other AIAA publications, please visit www.aiaa.org.

AIAA is committed to devoting resources to the education of both practicing and future aerospace professionals. In 1996, the AIAA Foundation was founded. Its programs enhance scientific literacy and advance the arts and sciences of aerospace. For more information, please visit www.aiaafoundation.org.